Chemical Instabilities

Applications in Chemistry, Engineering, Geology, and Materials Science

NATO ASI Series

Advanced Science Institutes Series

A series presenting the results of activities sponsored by the NATO Science Committee, which aims at the dissemination of advanced scientific and technological knowledge, with a view to strengthening links between scientific communities.

The series is published by an international board of publishers in conjunction with the NATO Scientific Affairs Division

A	Life Sciences	Plenum Publishing Corporation
B	Physics	London and New York
C	Mathematical and Physical Sciences	D. Reidel Publishing Company Dordrecht, Boston and Lancaster
D	Behavioural and Social Sciences	Martinus Nijhoff Publishers
E	Engineering and Materials Sciences	The Hague, Boston and Lancaster
F	Computer and Systems Sciences	Springer-Verlag
G	Ecological Sciences	Berlin, Heidelberg, New York and Tokyo

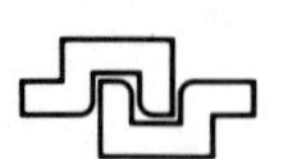

Series C: Mathematical and Physical Sciences Vol. 120

Chemical Instabilities

Applications in Chemistry, Engineering, Geology, and Materials Science

edited by

G. Nicolis

and

F. Baras

Faculté des Sciences, Université Libre de Bruxelles,
Campus Plaine, Brussels, Belgium

D. Reidel Publishing Company

Dordrecht / Boston / Lancaster

Published in cooperation with NATO Scientific Affairs Division

Proceedings of the NATO Advanced Research Workshop on
Chemical Instabilities: Applications in Chemistry,
Engineering, Geology, and Materials Science
Austin, Texas, U.S.A.
March 14-18, 1983

Library of Congress Cataloging in Publication Data
Main entry under title:

Chemical instabilities.

(NATO ASI series. Series C, Mathematical and physical sciences ; v. 120)
Proceedings of the NATO Advanced Research Workshop on Chemical Instabilities: Applications in Chemistry, Engineering, Geology, and Materials Science, held in Austin, Tex., March 14–18, 1983, sponsored by NATO and other organizations.
"Published in cooperation with NATO Scientific Affairs Division."
Includes index.
1. Chemistry, Physical and theoretical—Congresses. 2. Self-organizing systems—Congresses. I. Nicolis, G., 1939– . II. Baras, F. (Florence), 1958– . III. NATO Advanced Research Workshop on Chemical Instabilities: Applications in Chemistry, Engineering, Geology, and Materials Science (1983 : Austin, Tex.) IV. North Atlantic Treaty Organization. V. Series: NATO ASI series. Series C, Mathematical and physical sciences ; vol. 120.
QD455.2.C47 1984 541 83-24563
ISBN 90-277-1705-2

Published by D. Reidel Publishing Company
P.O. Box 17, 3300 AA Dordrecht, Holland

Sold and distributed in the U.S.A. and Canada
by Kluwer Academic Publishers,
190 Old Derby Street, Hingham, MA 02043, U.S.A.

In all other countries, sold and distributed
by Kluwer Academic Publishers Group,
P.O. Box 322, 3300 AH Dordrecht, Holland

D. Reidel Publishing Company is a member of the Kluwer Academic Publishers Group

Printed in The Netherlands.

TABLE OF CONTENTS

PART V - MATERIALS SCIENCE

PART V – [illegible]

FOREWORD

On March 14-18, 1983 a workshop on "Chemical Instabilities : Applications in Chemistry, Engineering, Geology, and Materials Science" was held in Austin, Texas, U.S.A. It was organized jointly by the University of Texas at Austin and the Université Libre de Bruxelles and sponsored by NATO, NSF, the University of Texas at Austin, the International Solvay Institutes and the Exxon Corporation.

The present Volume includes most of the material of the invited lectures delivered in the workshop as well as material from some posters, whose content was directly related to the themes of the invited lectures.

In recent years, problems related to the stability and the nonlinear dynamics of nonequilibrium systems invaded a great number of fields ranging from abstract mathematics to biology. One of the most striking aspects of this development is that subjects reputed to be "classical" and "well-established" like chemistry, turned out to give rise to a rich variety of phenomena leading to multiple steady states and hysteresis, oscillatory behavior in time, spatial patterns, or propagating wave fronts.

The primary objective of the workshop was to bring together researchers actively engaged in fields in which instabilities and nonlinear phenomena similar to those observed in chemistry are of current and primary concern : chemical engineering (especially surface catalysis), combustion (dynamics of ignition, flame stability), interfaces (emulsification, dendritic growth), geology (regularly repeated patterns of mineralization in a variety of space scales), and materials science (dynamical solidification, behavior of matter under irradiation).

We expect that the present Volume will acquaint researchers and advanced students with this important class of phenomena and with the mathematical, physical, and numerical techniques used to study them. Most of the basic problems in the area of nonequilibrium phenomena in nonlinear systems have not yet received extensive treatment. Cross-fertilization between different subjects is certainly one of the most promising avenues along which progress is likely to occur. We therefore hope that this Volume will serve as a jumping off point for new directions and new points of view.

Brussels, September 1983.

G. Nicolis and F. Baras

ORGANIZING COMMITTEE

Prigogine I., Center for Studies in Statistical Mechanics, University of Texas at Austin, U.S.A., and Faculté des Sciences, Université Libre de Bruxelles, Brussels, Belgium.

Hlavacek V., Department of Chemical Engineering, State University of New York, Buffalo, U.S.A.

Horsthemke W., Department of Physics, University of Texas at Austin.

Nicolis G., Faculté des Sciences, Université Libre de Bruxelles, Brussels, Belgium.

Ortoleva P., Department of Chemistry and Geology, Indiana University, Bloomington, Indiana, U.S.A.

Reichl L., Center for Studies in Statistical Mechanics, University of Texas at Austin, U.S.A.

Schechter R., Department of Petroleum Engineering, University of Texas at Austin, U.S.A.

Turner J.S., Center for Studies in Statistical Mechanics, University of Texas at Austin, U.S.A.

LIST OF PARTICIPANTS

Anderson D.
Geology Department
University of Illinois, Urbana
245 Natural History Bldg
1301 West Green Street
Urbana, IL 61801 U.S.A.

Balakotaiah V.
Chemical Engineering Department
University of Houston
Houston, TX 77004
U.S.A.

Baras F.
Université Libre de Bruxelles
Campus Plaine, CP n°231
Boulevard du Triomphe
1050 Bruxelles
Belgium

Bertrand G.
Université de Dijon
Faculté des Sciences Mirande
F 21004 Dijon Cedex
France

Bilgram J.H.
Labor für Festkürperphysik
ETH
CH 8093 Zürich
Switzerland

Billia B.
Laboratoire de Physique Cristalline
Faculté des Sciences
St. Jérôme
13397 Marseille Cedex 13
France

Biolsi L.
Chemistry Department
University of Misouri-Rolla
Rolla, MO 65101
U.S.A.

Borckmans P.
Université Libre de Bruxelles
Campus Plaine, CP n°231
Boulevard du Triomphe
1050 Bruxelles
Belgium

Boudreau A.E.
Dpt of Geological Sciences, AJ-20
University of Washington
Seattle, WA 98195
U.S.A.

Britten J.A.
Dept. of Chemical Engineering
Campus Box 424
University of Colorado
Boulder, CO 80309
U.S.A.

Brown J.R.
Chemical Engineering Department
University of Notre Dame
Notre Dame, IN 46556
U.S.A.

Bruno C.
5A Magie Apts
Faculty Rd
Princeton, NJ 08540
U.S.A.

Chadam J.
Department of Mathematics
Indiana University
Bloomington, IN 47401
U.S.A.

Cho S.H.
Department of Chemical Engineering
Queen's University
Kingston
Ontario K7L 3N6
Canada

Clavin P.
Laboratoire de Dynamique et de
Thermophysique des Fluides
Centre Saint Jérôme
Rue Henri Poincaré
13397 Marseille Cedex 4 France

Corbet A.B.
Department of Chemical Engineering
Stanford University
Stanford, CA 94305
U.S.A.

Dagonnier R.
Mons University
19 Avenue Maistriau
700 Mons
Belgium

de Fontaine D.
Department of Materials Science
and Mineral Engineering
Hearst Mining Building
University of California
Berkeley, CA 94720 U.S.A.

Detsch R.M.
Physics Department
University of Mississippi
University, MS 38677
U.S.A.

Dewel G.
Université Libre de Bruxelles
Campus Plaine, CP n°231
Boulevard du Triomphe
1050 Bruxelles
Belgium

D'Netto G.A.
Chemical Engineering Department
University of Notre Dame
Notre Dame, IN 46556
U.S.A.

De Luca L.
Departimento di Energetica
Politecnico di Milano
32 Piazza L. Da Vinci
20133 Milano
Italy

Dupeyrat M.
Laboratoire de Chimie Physique
Rue Pierre et Marie Curie 11
75 231 Paris Cedex 05
France

Epstein I.R.
Department of Chemistry
Brandeis University
Waltham, MA 02254
U.S.A.

Erneux T.
Department of Engineering Sciences
and Applied Mathematics
Northwestern University
The Technological Institute
Evanston, IL 60201 U.S.A.

Garcia A.L.
Department of Physics
University of Texas at Austin
Austin, TX 78712
U.S.A.

Geiseler W.
Technical University of Berlin
Institute of Technical Chemistry
Sekr. TC3
Strasse des 17 Juni 135
1000 Berlin, FRG

Graham W.R.C.
Department of Chemical Engineering
University of Alberta
Edmonton, Alberta
Canada T62G6

Gray P.
Physical Chemistry Department
University of Leeds
Leeds LS2 9JT
U.K.

Griffiths J.F.
Dept. of Physical Chemistry
The University of Leeds
Leeds LS2 9JT
U.K.

Guy B.
Ecole des Mines
158 Cours Fauriel
42023 Saint-Etienne
France

Woods Halley J.
School of Physics and Astronomy
University of Minnesota
Minneapolis
Minnesota 55455
U.S.A.

Hlavacek V.
Department of Chemical Engineering
SUNY at Buffalo
Clifford C. Furnas Hall
Buffalo, New York 14260
U.S.A.

Huerta M.A.
Department of Physics
University of Miami
Coral Gables, FL 33124
U.S.A.

Kondepudi D.K.
Center for Studies
in Statistical Mechanics
University of Texas
Austin, Texas 78712
U.S.A.

Krantz W.B.
Department of Chemical Engineering
Campus Box 424
University of Colorado
Boulder, CO 80309
U.S.A.

Laidlaw W.G.
University of Calgary
Calgary, Alberta
Canada T2N IN4

Larter R.M.
Department of Chemistry
Purdue University
P.O. Box 647
Indianapolis, IN 46223
U.S.A.

Lemarchand H.
Laboratoire de Chimie Générale
Université Pierre et Marie Curie
4 Place Jussieu
75230 Paris Cedex 05
France

Luss D.
Department of Chemical Engineering
University of Houston
Houston, Texas 77004
U.S.A.

Manley O.
Office of Basic Energy Sciences
U.S. Department of Energy
Washington, D.C. 20545
U.S.A.

Maselko J.
Department of Chemistry
Brandeis University
Waltham, MA 02254
U.S.A.

Matkowsky B.J.
Department of Engineering Sciences
and Applied Mathematics
Northwestern University
The Tecnological Institute
Evanston, IL 60201 U.S.A.

McCormick W.D.
Department of Physics
University of Texas
Austin, Texas 78712
U.S.A.

Merino E.
Department of Geology
Indiana University
Bloomington
Indiana 47405
U.S.A.

Moore C.H.
Indiana University
3900F S. Rendy Lane
Bloomington, IN 47401
U.S.A.

Müller-Krumbhaar
Kernforschung
IFF
D-5170 Julich
F.R.G.

Nandapurkar P.
Chemical Engineering Department
SUNY at Buffalo
Amherst, NY 14260
U.S.A.

Nelson G.
Department of Physics
University of Texas
Austin, Texas 78712
U.S.A.

Nicolis G.
Université Libre de Bruxelles
Campus Plaine, CP n°226
Boulevard du Triomphe
1050 Bruxelles
Belgium

Ortoleva P.J.
Department of Chemistry
Indiana University
Bloomington
Indiana 47401
U.S.A.

Owens A.D.
Department of Geology
Indiana University
Bloomington
Indiana 47405
U.S.A.

Prigogine I.
Université Libre de Bruxelles
Campus Plaine, CP n°231
Boulevard du Triomphe
1050 Bruxelles
Belgium

Ray W.H.
Chemical Engineering Department
University of Wisconsin
1415 Johnson Drive
Madison, WI 53706
U.S.A.

Razon L.F.
Chemical Engineering Department
University of Notre Dame
Notre Dame, IN 46556
U.S.A.

Reichl L.E.
Department of Physics
University of Texas
Austin, Texas 78712
U.S.A.

Rimini E.
Instituto di Fisica
57 Corso Italia
I 95129 Catania
Italy

Ringland J.
Department of Physics
University of Texas at Austin
Austin, Texas 78712
U.S.A.

Ross J.
Department of Chemistry
Stanford University
Stanford, CA 94305
U.S.A.

Rössler O.E.
Universität Tübingen
Institut für Physikalische und
Theoretische Chemie
7400 Tübingen 1
F.R.G.

Riuz-Diaz R.
Physics Department
University of Texas
Austin, Texas 78712
U.S.A.

Sandfeld A.
Chimie Physique II, CP n°231
U.L.B. Campus Plaine
Boulevard du Triomphe
1050 Bruxelles
Belgium

Schieve W.C.
Department of Physics
University of Texas
Austin, Texas 78712
U.S.A.

Schmitz R.A.
Department of Chemical Engineering
University of Notre Dame
Notre Dame
Indiana 46556
U.S.A.

Sheintuch M.
Department of Chemical Engineering
University of Houston
Houston, Texas 77009
U.S.A.

Sigmund P.M.
Department of Chemical Engineering
University of Calgary
2500 University Drive
Calgary, Alberta, T2N IN4
Canada

Simoyi R.
Department of Physics
University of Texas
Austin, Texas 78712
U.S.A.

Steinchen A.
Université Libre de Bruxelles
Campus Plaine, CP n°231
Boulevard du Triomphe
1050 Bruxelles
Belgium

Swift J.
Department of Physics
University of Texas
Austin, Texas 78712
U.S.A.

Swinney H.L.
Department of Physics
University of Texas
Austin, Texas 78712
U.S.A.

Subramaniam B.
Department of Chemical Engineering
University of Notre Dame
Notre Dame, IN 46556
U.S.A.

Sutton S.J.
Department of Geology
University of Cincinnati
Cincinnati, OH 45221
U.S.A.

Tsang T.T.H.
Department of Chemical Engineering
University of Texas
Austin, Texas 78712
U.S.A.

Turner J.S.
Department of Physics
University of Texas
Austin, Texas 78712
U.S.A.

Van Den Broeck C.
Dept. Natuurkunde
Vrije Universiteit Brussel
Pleinlaan 2
1050 Brussel
Belgium

Velarde M.G.
U.N.E.D.-Facultad Ciencias
Apartado Correos 50 487
Madrid
Spain

Vidal C.
Centre de Recherchc Paul Pascal
Domaine Universitaire
33405 Talence Cedex
France

Williams F.A.
Department of Mechanical and
Aerospace Engineering
Princeton University
Princeton
New Jersey 08544, U.S.A.

Yip S.
Department of Nuclear Engineering
24-211 Massachusetts Institute
of Technology
Cambridge
Massachusetts 02139 U.S.A.

Yutani N.
Department of Chemical Engineering
Kansas State University
Manhattan, KS 66506
U.S.A.

INTRODUCTION

It is a great pleasure to write a short Introduction for the Proceedings of the workshop on Chemical Instabilities. I would like, first, to thank the organizers, and mainly Grégoire Nicolis and Jack Turner. I would also like to express my gratitude to the organizations which have made this symposium possible, and specially the university of Texas at Austin, the Exxon corporation, the NATO division of scientific affairs, the NSF and the Instituts Internationaux de Physique et de Chimie Fondés par Ernest Solvay.

Of course, a conference like this, dealing with subjects which have been of interest for me during a long period of time, brings to my memory some reminiscences.

The first conference dealing specifically with nonequilibrium processes was probably the conference I organized in Brussels in 1946. It was in fact the first of a long series of meetings in statistical physics, which is still going on. At that time, I was just appointed as secretary of the committee of IUPAP for thermodynamics and statistical mechanics, and in this quality I organized some conferences until 1956. May I mention that in this series we had a conference in Firenze (1949), in Paris (1952), in Tokyo (1953) and in Brussels (1956). As you know, the regularity of such meetings is now established ; the last was held in Edinburgh this year.

The conference of 1946 had a rather mixed audience, which contained both classical thermodynamicists and people who were already involved in nonlinear theory. I tried to orient the major part of the conference towards nonequilibrium thermodynamics, and I remember that my lecture was dealing mainly with the property of minimum entropy production near equilibrium and with the relation between entropy production and kinetic theory.

I must say that the impression my lecture produced was, so far as I remember, rather mediocre. One outstanding thermodynamicist stood up and made a remark whose essential content was the following : "I am really astonished that this young man is interested in nonequilibrium processes. Indeed, everybody knows that nonequilibrium processes are transients : if you wait enough, you come to equilibrium. Why not study directly the final state of the system, rather than be interested in these transients ? ".

I was so upset by this remark that I was unable to answer. I should have said : "It is true that nonequilibrium processes are transients. But we all are transients, and therefore it seems

G. Nicolis and F. Baras (eds.), Chemical Instabilities, xxiii-xxvi.

only appropriate that we study this type of phenomena". In any case, this reaction expressed the distrust of physicists towards nonequilibrium processes.

I had a similar experience in 1954 when I spoke, probably for the first time, about the possibility of oscillating reactions. At that time, I had published a short paper with Radu Balescu on the possibility that far from equilibrium we could have chemical oscillations, in contrast with what happens near equilibrium. This work was connected with my involvment in the so-called "universal evolution criterion", derived with Paul Glansdorff. My lecture of 1954 had no more success than the one of 1946. The chemists were very skeptical about the possibility of chemical oscillations ; and in addition, said an outstanding chemist, even if it would be possible, what should be the interest ? The interest of chemical kinetics was at that time the discovery of well-defined mechanisms, and specially of potential energy surfaces, which one could then connect with quantum mechanical calculations. The appearance of chemical oscillations or other exotic phenomena seemed to him to be of no interest in the direction in which chemical kinetics was traditionally engaged. All this has changed, but to some extent the situation of chemistry in respect to physics remains under the shadow of this distrust of time.

Everybody knows the sentence of Dirac, stating that since the discovery of Schrödinger's equation, chemistry has become part of quantum mechanics. This statement is in my opinion quite misleading. It is similar to a statement which would assert that since the formulation of classical mechanics by Newton, Hamilton and others, there are no more problems in macroscopic physics. But over the last decades, we have seen striking and quite unexpected developments in classical dynamics, starting with Poincaré and continuing with Kolmogorov, Sinai, Smale and others. This change is reflected in the textbooks. In traditional presentations, classical mechanics was described as a closed discipline. But if you look at the beautiful book by Arnold on **Mathematical Methods of Classical Mechanics** , you find the statement that modern mathematics does not permit an exhaustive solution of problems in systems involving two degrees of freedom ! This is quite a change.

Coming back to chemistry, the sentence of Dirac neglects a basic difference between quantum mechanics and chemistry. Quantum mechanics in its traditional form decribes reversible evolutions : in short, the rotation of a state vector in the Hilbert space. But chemistry in concerned with irreversible processes. A chemical reaction is a typically irreversible process, driving a system towards equilibrium : there is no possible reduction of chemistry to quantum mechanics.

For the historians of science our century will probably present a special interest, because in a short period, our view of nature has undergone a fundamental change. If you look back at the beginning of the century, you see that physicists were unanimous to believe that the fundamental laws of nature were deterministic and time-reversible. The two great revolutions of XXth century science, relativity and quantum mechanics, did not seem to change this opinion.

Today, the situation is quite different. The most active part of quantum theory is concerned with transformations of elementary particles, and this implies a kind of new chemistry ; similarly, relativity is now involved in the history of the universe. So, from a deterministic and reversible starting point, we go to a description involving time-evolution, and certainly also a number of stochastic features, not only through the formalism of quantum mechanics, but also through the various symmetry-breaking processes which have been established in various contexts.

If you had asked a physicist fifty years ago "what can we know about the universe ? " he would generally have said that of course we do not know enough about elementary particles ; that we begin only to know the stages of cosmological evolution ; but that our knowledge of the phenomena between these two scales is quite satisfactory.

Today, a growing minority, to which I belong, does not share this optimism. There are many interesting problems which correspond to a new physics on our own scale.

This new field is in a state of explosive development, if I can say so. A few years ago, when we studied a simple model, known since in the literature as the Brusselator, we had to do the mathematics by ourselves ; when we needed bifurcations, we had to adapt the mathematical tools in our own amateurish way. Today, this field is in full blossom. I think that this is an interesting example of interaction between physics and mathematics. In fact, the new developments in nonequilibrium physics, coupled with advances in modern dynamics, had a revigorizing influence on nonlinear mathematics, which we could compare to the progress in mathematics induced earlier by other fields like relativity or quantum theory.

Whatever the forthcoming developments may be, some of our ideas about nature have been deeply modified. Let me, to conclude, take the idea of complexity. The classical view was that only biological or social systems were complex ; today, we see that the roots of complexity lie already in Physics and in Chemistry. Simple systems such as a fluid or a dense gas, have at equilibrium a simple behavior, described in terms of equations of state or simple macroscopic laws. If we place the same systems in non-equilibrium conditions, we may induce a completely different behavior, involving bifurcations, spatial or temporal organization, in short complex behavior.

This should not lead to any reductionist view of biological or social systems. If some part of the complexity which was thought to be an exclusive attribute of these systems can be found in physical and chemical systems, we may hope to get a better understanding of the specificity of social and biological systems.

I believe that these remarks may put the problems studied in this workshop in a somewhat broader perspective. About the direct application of these concepts, I think I do not have to go into details. The methods discussed in this workshop are already in use in various fields. These new developments will for sure contribute to a better understanding of our world.

Brussels, September 1983.

I. Prigogine

PART I

CHEMICAL KINETICS

THE SEARCH FOR NEW CHEMICAL OSCILLATORS

Irving R. Epstein

Department of Chemistry, Brandeis University

Before 1980, the only chemical oscillators of nonbiological origin had been discovered by chance or were variants of the two accidentally discovered systems. Since that time, experimental and theoretical advances have made it possible to systematically design new oscillating chemical reactions.

By studying appropriately perturbed autocatalytic reactions in a flow reactor (CSTR), we have produced a family of some 20 new oscillators containing chlorite ion. In addition, several new bromate oscillators (e.g., BrO_3^--Mn^{2+}-Br^- and BrO_3^--I^-), which are simpler in composition and mechanism than the classic Belousov-Zhabotinskii system, have been discovered. Mechanistic considerations have given rise to a tentative classification of oxyhalogen based oscillators which shows the linkage and differences between the various classes.

Current work focuses on the study of more complex phenomena such as birhythmicity, compound oscillation, tristability and spatial pattern formation, particularly in systems of coupled oscillators.

1. INTRODUCTION

Until about 25 years ago, chemical oscillation was the subject of considerable controversy and skepticism among chemists. While many chemical engineers were familiar with reactor oscillations (and in fact devoted considerable effort to eliminating them), the majority of chemists were convinced that sustained periodic behavior in relatively simple chemical systems must be

G. Nicolis and F. Baras (eds.), Chemical Instabilities, 3–18.

artifactual at best, fradulent at worst, a thermodynamic impossibility.

The development first of belief and then of interest on the part of chemists in oscillating reactions was spurred by two major developments, one theoretical, the other experimental. Studies in the field of nonequilibrium thermodynamics (Glansdorff and Prigogine, [1]; see Procaccia and Ross, [2] for a review) established that, sufficiently far from equilibrium, chemical oscillation was indeed consistent with the laws of thermodynamics. The accidental discovery in the Soviet Union (Belousov, [3]) of a reaction which gave easily observable oscillations at room temperature evoked the interest of several chemists, first as an amusing lecture demonstration and then as a subject of serious research. It is interesting that the first homogeneous chemical oscillator (Bray, [4]), also discovered by serendipity almost 40 years before the Belousov-Zhabotinskii (BZ) reaction, received little attention until after the BZ system had become a major focus of research.

2. THE SITUATION PRIOR TO 1980

Belousov's reaction, the metal ion (generally Ce^{3+}) catalyzed bromination of an organic substrate, most often malonic acid, by bromate, was developed experimentally by Zhabotinskii ([5]). It was, however, the publication (Field, Koros and Noyes, [6]) of a detailed mechanism for the system and of a simplified three-variable model (the Oregonator, Field and Noyes, [7]) of that mechanism that spurred interest in the BZ system as a prototype of periodic chemical behavior.

At roughly the same time that awareness of oscillatory phenomena was growing rapidly among chemists, oscillatory phenomena in biochemical systems were also becoming a subject of serious study (Chance et al., [8]). The most notable studies in this area have been those on glycolytic oscillation (Hess and Boiteux, [9]), but periodic behavior appears to be ubiquitous in the chemistry of living organisms.

As more chemists began to study the BZ and Bray reactions, several variants (Orbán and Körös, [10] ; Bowers, et al., [11]) and hybrids (Briggs and Rauscher, [12]) of these reactions were developed. However, no chemical oscillators with chemistry fundamentally different from that of the two accidentally discovered prototypes were found, and one might summarize the sources of oscillatory reactions prior to 1980 as:

1. Accident
2. Biology
3. Variation on a theme

3. THE SEARCH FOR NEW OSCILLATORS

By the mid 1970's, a number of workers had become interested in the nature of chemical oscillation and attempted to derive necessary and sufficient conditions for a system of chemical reactions to show periodic behavior. While a number of significant results and approaches (e.g., Clarke, [13]; Tyson, [14]) came out of these studies, it became increasingly clear that rigorous specification of such a set of conditions was an extremely difficult, perhaps impossible goal.

Furthermore, the existing sources of new oscillators did not provide encouragement for those wishing to test general theories of chemical oscillation. Serendipity, in many ways the most fruitful source, had furnished only two examples and was too unreliable to depend upon for more. Oscillators derived from living systems, though plentiful, appeared to be too complex for the kind of detailed mechanistic analysis needed. Variants and hybrids of known reactions were too similar to their parent systems to provide additional insights into the nature of chemical oscillation.

At this stage it was becoming clear that what was required was a more systematic and constructive, though possibly less rigorous, approach to developing new chemical oscillators. This task was undertaken by our group at Brandeis, which sought to integrate insights from earlier work with knowledge about the detailed chemistry of several promising inorganic reaction systems.

The first step in the search procedure is to put into practice the result derived from nonequilibrium thermodynamics that oscillation is a possibility only in a system sufficiently far from equilibrium. In order to maintain a nonequilibrium state we utilize a tool long familiar to chemical engineers and adapted to the study of chemical oscillation by the Bordeaux group (Pacault et al., [15]), the continuous flow stirred tank reactor (CSTR). A schematic diagram of a stirred tank reactor is shown in Figure 1.

A second requirement is that the system contain some sort of feedback, i.e., that the product of some reaction exert an influence upon its own rate of production. Probably the simplest type of feedback and the one most commonly found in chemical oscillators is autocatalysis. The simple model (Lotka, [16]) containing two coupled autocatalytic reactions shown in Table 1 was probably the first "chemical" mechanism to give sustained oscillations. While autocatalysis is far more prelavent in biological than in chemical systems, autocatalytic chemical reactions do exist; many of them are listed as "clock reactions" in collections of lecture demonstrations.

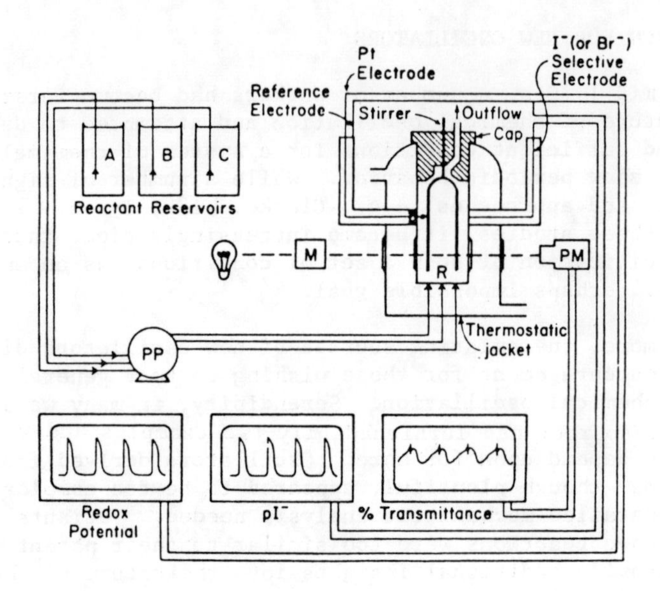

Figure 1. Schematic diagram of a CSTR. In the configuration shown, up to three different solutions can be pumped (by the peristaltic pump, PP) into the reactor. R. The detectors shown in the diagram are: light absorption (M, monochromator: PM, photomultiplier), platinum (redox), and iodide (or bromide) selective electrodes.

Table 1. The Lotka Mechanism

$$A + X \rightarrow 2X \qquad (L1)$$

$$X + Y \rightarrow 2Y \qquad (L2)$$

$$Y \rightarrow P \qquad (L3)$$

Concentration of reactant A maintained constant

The next concept which proves useful in developing new chemical oscillators is that of bistability along with the related notion of hysteresis. An open chemical system, such as a reaction in a CSTR, may have two (or more) different stable steady states under the same external constraints, i.e., values of the input flow rate, reservoir concentrations, temperature and pressure. In such a situation, transitions from one state to another show hysteresis, occurring at different points depending upon the direction in which the constraints are changed. An example of this behavior is shown in Figure 2. As the iodide concentration in the input flow ($[I^-]_o$) is gradually increased, the steady state levels of I_2 and I^- in the reactor rise smoothly until, at a critical value $[I^-]_o = 7 \times 10^{-4}$ M, the system undergoes a sharp transition to a new branch of steady states, labeled SSII. On decreasing $[I^-]_o$, we observe the reverse transition at the critical value $[I^-]_o = 3 \times 10^{-4}$ M. The region 3×10^{-4} M $\leq [I^-]_o \leq 7 \times 10^{-4}$ M is a region of bistability.

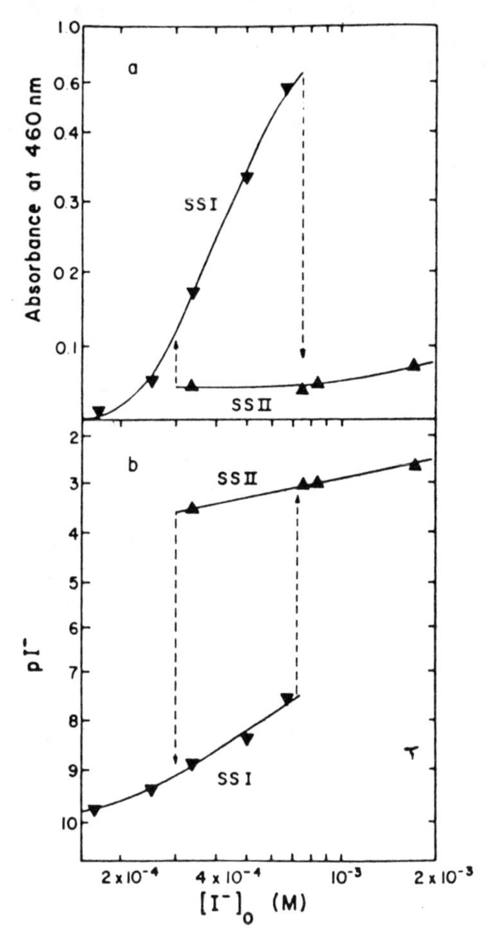

Figure 2. Steady state hysteresis phenomena in the absorbance at 460 nm and the iodide concentration in the CSTR as a function of $[I^-]_o$, with $[ClO_2^-]_o = 2.5 \times 10^{-4}$M, pH 3.35, $1/\tau = 5.4 \times 10^{-3} s^{-1}$, and T = 25°C, where τ = reactor residence time. Dashed arrows indicate spontaneous transitions between states.

In order to transform a bistable system into an oscillator, we use an approach suggested by a simple model calculation by Boissonade and De Kepper [17], and illustrated in Figure 3. We start from a bistable system (Figure 3a) in which X_o is a constraint parameter such as $[I^-]$ or $[I_2]$ in the reactor. Now suppose that there exists another species Z with the following properties: a) Z reacts with one or more input or intermediate species to produce X; thus Z effectively increases X_o. b) Z affects the two different branches of steady states by significantly different amounts; this is plausible, because as shown in Figure 2, concentrations in the reactor may differ by several orders of magnitude between the two branches. c) Z reacts slowly in the sense that the characteristic time for the production of X from Z on each branch is significantly longer than the relaxation time of the system after small perturbations from that branch. Such a species Z is dubbed a feedback species.

In Figure 3b, the effect of introducing a specific concentration of Z, say Z_o, into the input flow is shown by the horizontal dot-dash lines on each branch when the input concentration of X is C. Consider how the system will behave under such conditions. Initially (since relaxation to the steady states is the more rapid process), it will attain the pseudo-state state value of $Y = Y_c$. Then, as Z reacts, the system will "see" an apparent input of $X_o = C_1$. Y will decrease along the upper branch of steady states (solid line) as the system "tries" to reach C_1. However, when Y reaches the value corresponding to the input $X_o = A$, the upper branch disappears, and the system undergoes a spontaneous transition to the lower branch of states (dashed line). Now, the apparent value of X_o is C_2, so Y increases as the system proceeds back to the left. At $X_o = B$, however, the lower branch disappears, and we jump back to the upper branch, where the system heads back to the right, seeking again to reach $X_o = C_1$. These transitions recur indefinitely, leading to the periodic behavior shown in Figure 3c.

If we consider how the system behaves as a function of the input concentrations, we find Figure 3d, the "cross-shaped phase diagram". With $Z_o = 0$, the region of bistability is $B \leq X_o \leq A$. As Z_o increases, the upper limit of stability of the upper branch and the lower limit of stability of the lower branch both shift to lower X_o, the former by more than the latter. The net effect is to decrease the width of the bistable region. At a critical value, the bistable region disappears, and a region comes into being in which neither steady state is stable. This is the region of oscillation in the X_o - Z_o constraint plane.

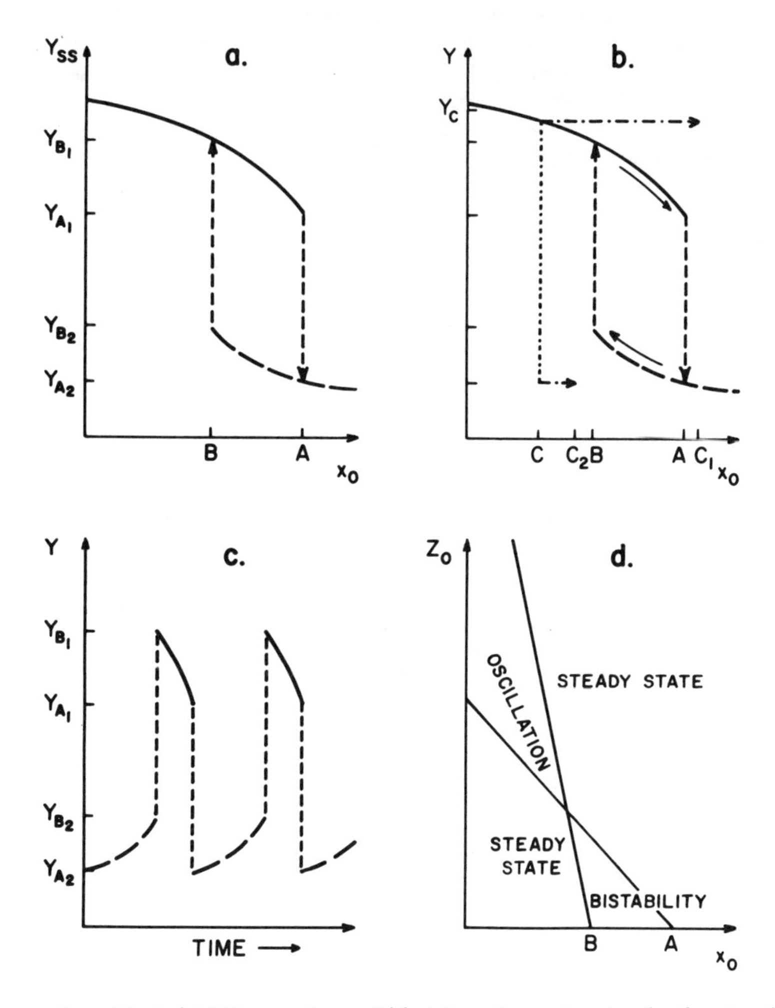

Figure 3. Bistability and oscillation in a typical chemical system (see text). a) Simple bistability; b) Effect (-·-) of adding a feedback species; c) Behavior of the system in b) as a function of time; d) Cross-shaped phase diagram.

Our systematic search procedure may thus be summarized as: a) choose an autocatalyic reaction R; b) run R in CSTR and seek conditions under which the system is bistable; c) choose a feedback species Z which perturbs the system by different amounts on the two branches of steady states; d) by increasing the input of Z into the CSTR, seek the critical point at which bistability disappears and oscillations begin.

The first system on which this approach was tried (De Kepper, Epstein and Kustin, [18]) employed two coupled autocatalytic reactions, chlorite plus iodide, and arsenite plus iodate, which have key intermediates in common. As Figure 4 shows, the chlorite-iodate-arsenite system did indeed prove to oscillate, constituting the first systematically designed chemical oscillator. More recently, by starting from the fundamental or minimal chlorite-iodide bistable system and adding different feedback species, it has been possible to generate a family of nearly 20 different chlorite-iodine species oscillators (Orbán et al., [19]). In addition, two iodine free chlorite oscillators involving thiosulfate (Orbán, De Kepper and Epstein, [20]) and bromate (Orban and Epstein, [21]) have been found.

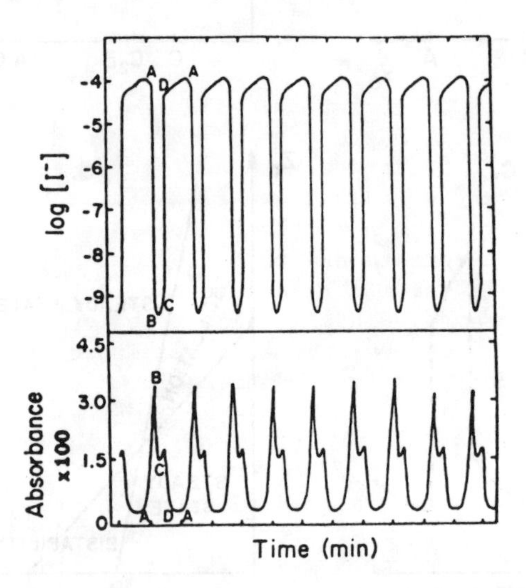

Figure 4. Oscillations of the iodide concentration and absorption per cm path length at 460 nm (proportional to $[I_2]$) for $[KIO_3]_0 = 24 \times 10^{-3}$M, $[As_2O_3]_0 = 2 \times 10^{-3}$M, $[NaClO_2] = 2 \times 10^{-3}$M with $[Na_2SO_4]_0 = 0.1$M, $[H_2SO_4]_0 = 0.01$ M, residence time = 400 s and T = 25°C. Concentrations are given in the reactor after mixing but before any reaction takes place.

While work of this type has focused primarily on chlorite systems, several new bromate oscillators, simpler in their chemistry and mechanism than the BZ system, have also been discovered. The first of these, the "minimal" bromate oscillator (Orbán, De Kepper and Epstein, [22]) consists of bromate, bromide and either cerous or manganous ions flowed into the CSTR. It is essentially the BZ reaction with malonic acid replaced by an input of bromide (which in the batch reaction is generated by a series of reactions involving malonic acid). The mechanism of this reaction is far simpler than that of the BZ reaction (Field, Körös and Noyes, [6]), since only inorganic species are involved. In fact, oscillations in this system were first predicted from numerical simulations (Bar-Eli, [23]) before they were found experimentally. From this minimal oscillator, it is possible to generate a new family of chemical oscillators consisting of bromate, the metal ion and a reducing agent capable of generating bromide from bromate at an appropriate rate (Alamgir, Orbán and Epstein, [24]). Another new bromate oscillator, possibly related more closely in a mechanistic sense to the chlorite-iodide reaction than to the BZ system, consists of bromate and iodide in the CSTR (Alamgir et al., [25]).

4. RELATED PROBLEMS

While the goal of deliberately designing new chemical oscillators has been attained, there remain many related problems of significant interest. Designing more new oscillators, particularly involving the chemistry of nonhalogen elements such as sulfur, nitrogen or carbon, remains a worthwhile challenge to the chemist. It is possible that alternate search procedures not based on the cross-shaped phase diagram might lead to oscillators with quite different phase diagrams. There is also much to be done both in terms of deepening our understanding of the known oscillators and seeking allied but more complex phenomena.

4.1 Mechanisms

Mechanisms have been developed which yield predictions in reasonable agreement with experiment for bromate (Field, Körös and Noyes, [6]; Geiseler and Bar-Eli, [26]; Orbán and Epstein, [27]) and iodate (Sharma and Noyes, [28]; Noyes and Furrow, [29]; De Kepper and Epstein, [30]) oscillators. However, no quantitative results have yet appeared for the chlorite oscillators. Recent unpublished work at Brandeis suggests that the fundamental chlorite-iodide oscillator (Dateo et al., [31]) can be understood in terms of a mechanism involving the key binuclear intermediate $IClO_2$. In contrast to the mechanism for the BZ reaction, this model would require only singlet, non-radical species, thus constituting a fundamentally different pathway to oscillation.

4.2 Classification

A major step toward understanding the bromate oscillators was the formulation by Noyes ([32]) of a mechanistically based classification scheme for the known (batch) bromate oscillators. Since that time, a variety of additional oxyhalogen (bromate, chlorite, iodide) systems have been shown to oscillate in the

CSTR, and it is an interesting question whether they can be fit together into some coherent classification scheme. In Figure 5, we show a preliminary attempt at such a scheme. Note that the systems treated by Noyes occupy only that part of the scheme involving bromate and organic reductants.

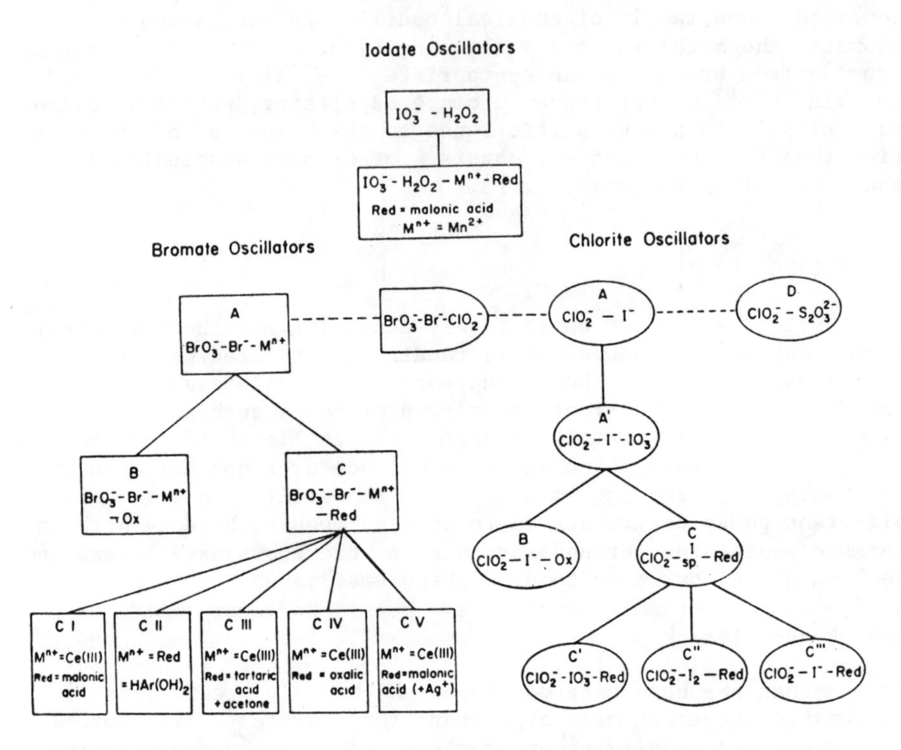

Figure 5. Schematic classification scheme for homogeneous chemical oscillators. Red = reducing agent. Ox = oxidizing agent. Dashed lines indicate possible mechanistic connections between different types of oscillators.

4.3 Related Phenomena

We have focused thus far almost exclusively on simple periodic oscillation and on the related phenomenon of bistability. There exist, however, a variety of related more complex phenomena which are increasingly becoming the subjects of investigation in this field.

Simple bistability between two stationary states is but one example of multistability. Other multistable phenomena of interest include hard excitation (Figure 6), i.e., bistability between a stationary and an oscillatory state, birhythmicity, (Figure 7), i.e., bistability between two different oscillatory states, and tristability (Figure 8).

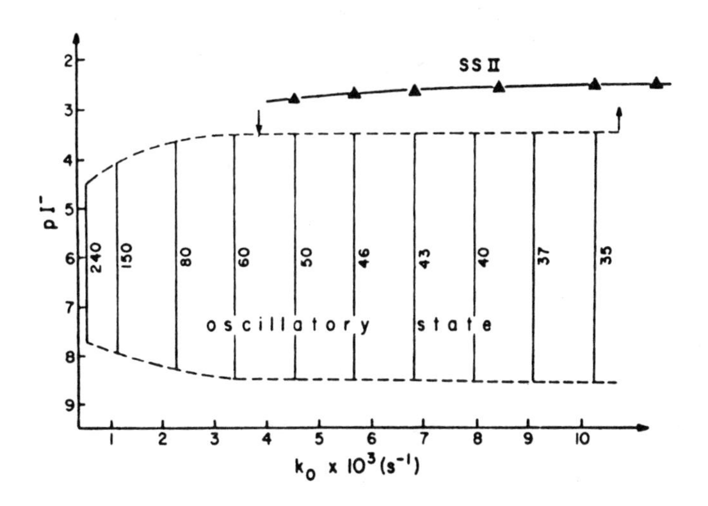

Figure 6. Hysteresis in the transition between steady state SSII and the oscillatory state as a function of flow rate k_0 with $[I^-]_0$ = 0.0065 M, $[ClO_2]$ = 0.002 M, pH 1.56, and T = 25°C. Envelopes of vertical segments show upper and lower limits of pI^- in the oscillatory state. Numbers next to these segments indicate period of oscillation in seconds. Arrows indicate spontaneous transitions between states.

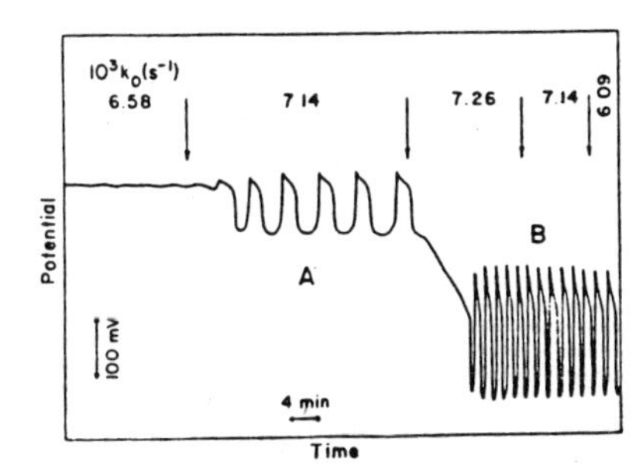

Figure 7. Birhythmicity in the chlorite-bromate-iodide system. Potential is that of Pt electrode vs Hg/Hg_2SO_4 reference electrode. At times indicated by the arrows, flow rate is changed. Flow rate is shown at top. Note that A state and B state are both stable at $k_0 = 7.14 \times 10^{-3} s^{-1}$. Fixed constraints: T = 25°C, $[I^-]_0 = 6.5 \times 10^{-4}$M, $[BrO_3^-]_0 = 2.5 \times 10^{-3}$M, $[ClO_2^-]_0 = 1.0 \times 10^{-4}$M, $[H_2SO_4]_0$ = 0.75M.

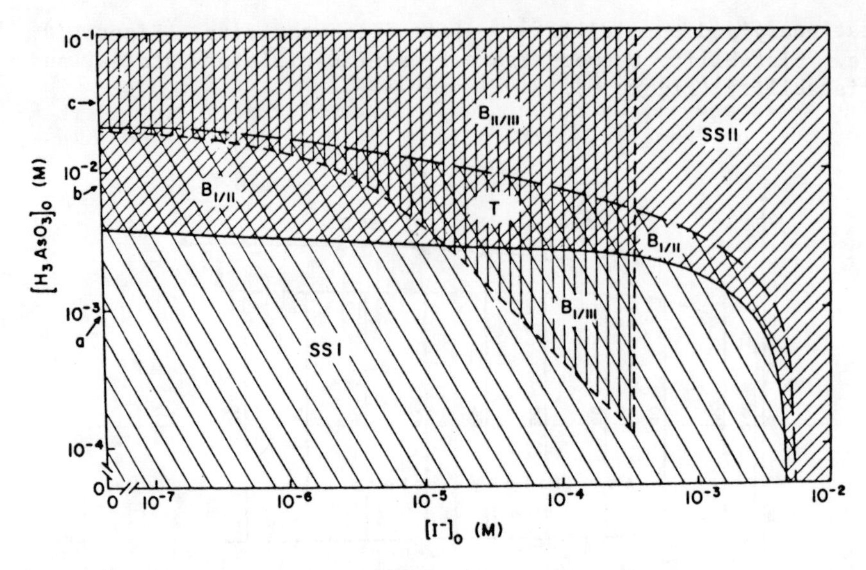

Figure 8. Section of the phase diagram of the chlorite-iodate-arsenite system showing tristability in the $[I^-]_0$-$[H_3AsO_3]_0$ plane. [hatch] SS I stable. [hatch] SS II stable. [hatch] SS III stable. B_{ij} indicates region of bistability of SS_i and SS_j. T indicates region of tristability. Fixed constraints: $[ClO_2]_0 = 2.5 \times 10^{-3}$M; $[IO_3^-]_0 = 2.5 \times 10^{-2}$M, pH = 3.35, $k_0 = 5.35 \times 10^{-3} s^{-1}$, T = 25°C, k_0 is reciprocal of reactor residence time.

In an unstirred closed configuration an oscillator or excitable solution may give rise to wave phenomena and spatial pattern formation. One example of such behavior in a chlorite oscillator is shown in Figure 9.

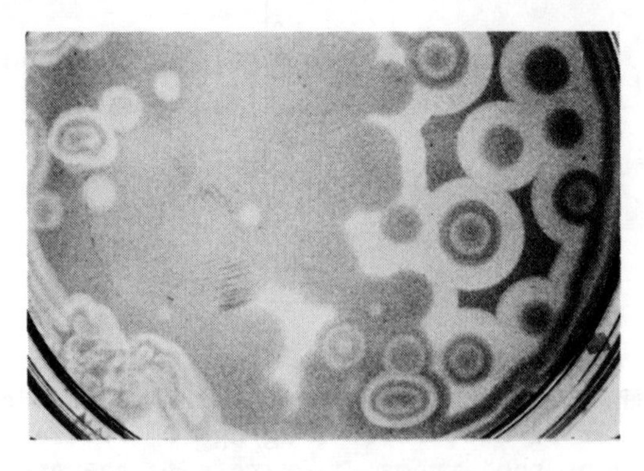

Figure 9. Spatial wave pattern observed at 5°C in a thin (2 mm) layer of reactive solution with initial composition $[CH_2(COOH)_2]$ = 0.0033 M, [NaI] = 0.09 M. $[NaClO_2]$ = 0.1 M, $[H_2SO_4]$ = 0.0056 M, and starch as indicator.

Periodic chemical oscillations may be complex, and complex oscillations need not always be periodic. These observations are illustrated for the chlorite-thiosulfate reaction in Figures 10 and 11, where we see first complex periodic and then aperiodic or "chaotic" oscillation.

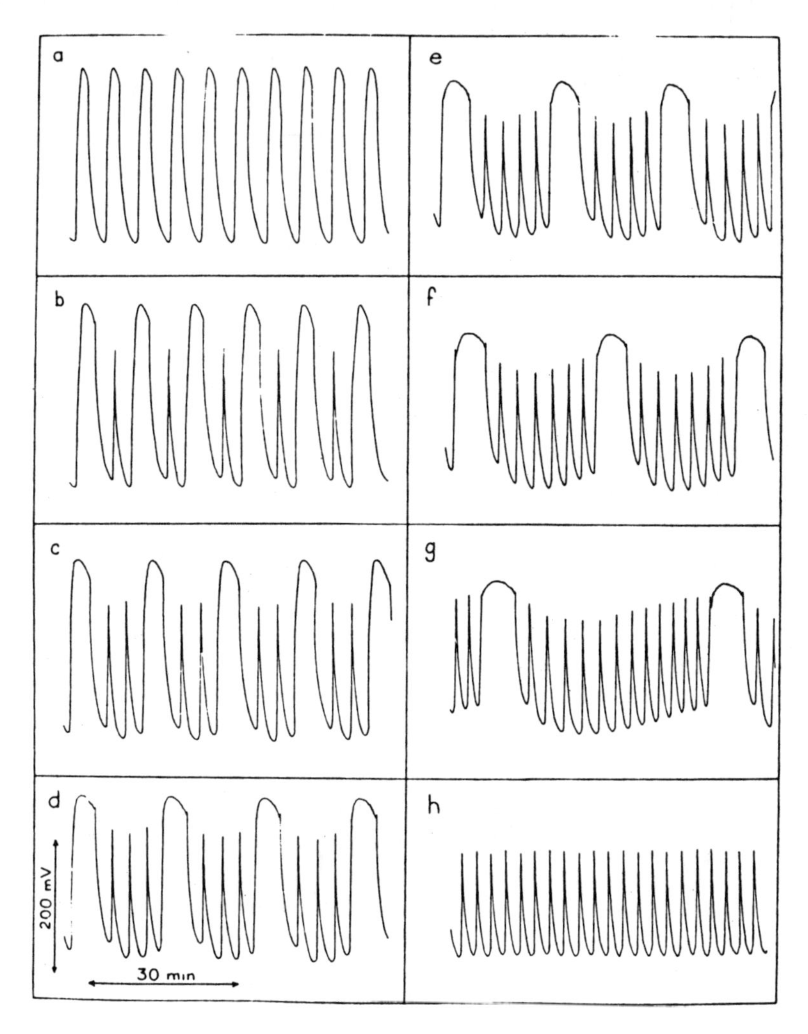

Figure 10. Complex periodic oscillations in the chlorite-thiosulfate system with $[ClO_2^-]_0 = 5 \times 10^{-4}M$, $[S_2O_3^{2-}]_0 = 3 \times 10^{-4}M$, pH 4, T = 25.0°C. Residence times: a) 5.9 min, b) 9.5 min. c) 10.8 min d) 13.5 min e) 15.8 min f) 20.6 min g) 26.3 min h) 47.2 min.

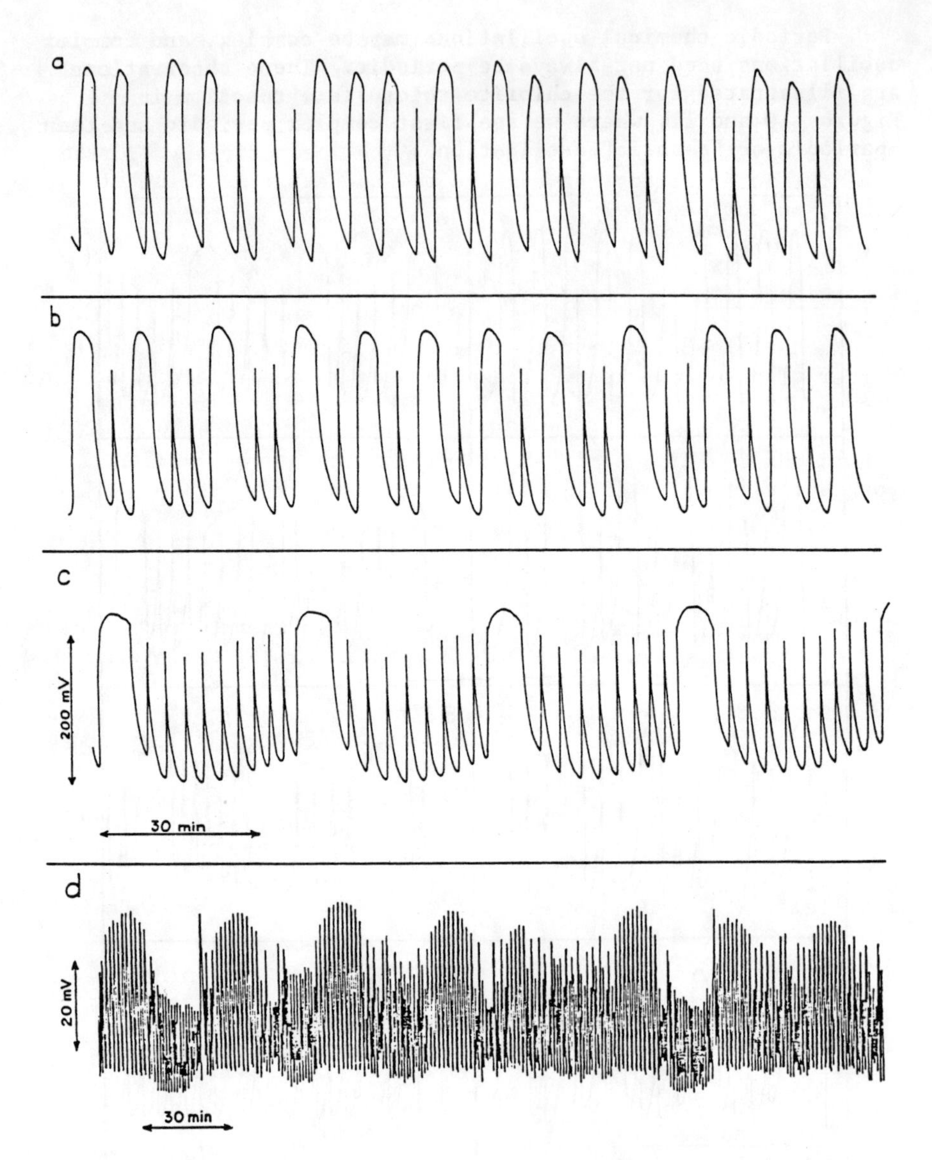

Figure 11. Aperiodic oscillations (chaos) in the chlorite-thiosulfate system. In a),b),c) conditions are as in Figure 10, except for residence times. a) (L,L+S), k_0^{-1} = 6.66 min.; b) (L+S, L+2S), k_0^{-1} = 10.4 min.; c) (L+8S,L+9S), k_0^{-1} = 23.6 min.; d) another type of chaos : $[ClO_2^-]_0$ = 2.33 x 10^{-3}, $[S_2O_3^{2-}]_0$ = 1.07 x 10^{-3} M, k_0^{-1} = 41.7 min. (Orban and Epstein, 1982 ; Maselko and Epstein, unpublished)

Many of the phenomena discussed above which has been found experimentally in chemical systems have been predicted by earlier model calculations. It appears that the wealth of dynamical behavior available to chemical systems may be almost as large as the variety that can be concocted by the theoretician. Further fruitful interactions between theory and experiment in this area are to be anticipated.

Acknowledgments

Much of the work described here was performed together with a number of able coworkers. These include Mohamed Alamgir, György Bazsa, Christopher Dateo, Patrick De Kepper, Kenneth Kustin, Jerzy Maselko and Miklós Orbán. This research was funded by grants CHE-7925375 and CHE-8204085 from the National Science Foundation.

References

1. Glansdorff, P. and Prigogine, I., 1971, "Thermodynamic Theory of Structure, Stability and Fluctuations", Wiley Interscience New York.
2. Procaccia, I. and Ross, J., 1977, Science 198, pp. 717
3. Belousov, B.P., 1959, "Ref. Radiats. Med. 1958", Medgiz, Moscow, pp. 145
4. Bray, W.C., 1921, J. Am. Chem. Soc. 43, pp. 1262-1265
5. Zhabotinskii, A.M., 1964, Biofizika 9, pp. 306-311
6. Field, R.J., Körös, E. and Noyes, R.M., 1972, J. Am. Chem. Soc. 94, pp. 8649-8664
7. Field, R.J. and Noyes, R.M., 1974, J. Chem. Phys. 60, pp. 1877-1884
8. Chance, B., Pye, E.K., Ghosh, A.K. and Hess, B., 1973, "Biological and Biochemical Oscillators", Academic Press, New York
9. Hess, B. and Boiteux, A., 1971, Ann. Rev. Biochem. 40, pp. 237-258
10. Orbán, M. and Körös, E., 1978, J. Phys. Chem. 82, pp. 1673-1675
11. Bowers, P.G., Caldwell, K.E. and Prendergast, D.F., 1972, J. Phys. Chem. 76, pp. 2185-2186
12. Briggs, T.S. and Rauscher, W., 1973, J. Chem. Ed. 7, pp. 496
13. Clarke, B.L., 1975, J. Chem. Phys. 62, pp. 773-775
14. Tyson, J.J., 1975, J. Chem. Phys. 62, pp. 1010-1015
15. Pacault, A., Hanusse, P., De Kepper, P., Vidal, C., and Boissonade, J., 1976, Acc. Chem. Res. 9, pp. 438-445
16. Lotka, A.J., 1920, J. Am. Chem. Soc. 42, pp. 1595-1598
17. Boissonade, J. and De Kepper, P., 1980, J. Phys. Chem. 84, pp. 501-506

18. De Kepper, P., Epstein, I.R. and Kustin, K., 1981, J. Am. Chem. Soc. 103, pp. 2133-2134
19. Orbán, M., Dateo, C., De Kepper, P. and Epstein, I.R., 1982, J. Am. Chem. Soc., pp. 5911-5918
20. Orbán, M., De Kepper, P. and Epstein, I.R., 1982a, J. Phys. Chem. 86, pp. 431-433
21. Orbán, M. and Epstein, I.R., 1982, J. Phys. Chem. 86, pp. 3902-3910
22. Orbán, M., De Kepper, P. and Epstein, I.R., 1982b, J. Am. Chem. Soc. 104, pp. 2657-2658
23. Bar-Eli, K., 1981, in "Nonlinear Phenomena in Chemical Dyna mics", Vidal, C. and Pacault, A., eds., Springer-Verlag, Berlin, pp. 228-239
24. Alamgir, M., Orbán, M. and Epstein, I.R., 1983, J. Phys. Chem., in press
25. Alamgir, M., De Kepper, P., Orbán, M. and Epstein, I.R., 1983, J. Am. Chem. Soc. 105, pp. 2641-2643
26. Geiseler, W. and Bar-Eli, K., 1981, J. Phys. Chem. 85, pp. 908-914
27. Orbán, M. and Epstein, I.R., 1983, J. Phys. Chem., in press
28. Sharma, K.R. and Noyes, R.M., 1976, J. Am. Chem. Soc. 98, pp. 4345-4361
29. Noyes, R.M. and Furrow, S.D., 1982, J. Am. Chem. Soc. 104, pp. 45-48
30. De Kepper, P. and Epstein, I.R., 1982, J. Am. Chem. Soc. 104, pp. 49-55
31. Dateo, C., Orbán, M., De Kepper, P. and Epstein, I.R., 1982, J. Am. Chem. Soc. 104, pp. 504-509
32. Noyes, R.M., 1980, J. Am. Chem. Soc. 102, pp. 4644-4649

CHEMICAL DISSIPATIVE SYSTEMS AND THE MEASURE OF CHAOS

C. Vidal and A. Lafon

Centre de Recherche Paul Pascal, domaine universitaire
33405 Talence Cédex (France)

This contribution deals with the problem of measuring the level of chaos of a dynamic regime from its power spectrum. A numerical investigation shows that the variations of the largest Lyapunov exponent are well-reproduced by an entropy-like function H, easily deduced from the Fourier spectrum. Therefore this function might be computed whenever λ cannot be determined, which is especially the case so far from experimental time series. An example is given for a set of dynamic regimes exhibited by the Belousov-Zhabotinsky reaction.

1. INTRODUCTION

Our understanding of turbulence was completely renewed when Ruelle and Takens [1] proved that no more than 3 degrees of freedom are needed to allow for chaotic behaviour in axiom - A dynamical systems. The relevance of the theoretical approach developed earlier by Landau for depicting aperiodic phenomena was thus questionned. Since the original paper by Ruelle and Takens, a lot of work has been devoted to the study of chaotic motions occuring in dynamical systems with few degrees of freedom. From a theoretical viewpoint, three scenarios leading to chaos have been recognized, involving different types of bifurcation: Hopf, pitchfork or saddle-node [2]. Experiments have been performed in several fields, mainly Hydrodynamics [3] and Chemistry [4]. They have largely confirmed the degree of relevance of the modern approach to turbulence, at least from a qualitative point of view [5].

As always in physical Sciences, studies must now go beyond the qualitative features and reach a more quantitative stage. One im-

G. Nicolis and F. Baras (eds.), Chemical Instabilities, 19–32.

portant problem, in this regard, is to figure or, even, to measure the level of disorder of a chaotic regime. Theory provides us with several quantities but, as we shall see below, usually their determination cannot be carried out from experimental time series.

2. HOW TO MEASURE THE POWER OF CHAOS ?

If one agrees with the conjecture that a typical chaotic attractor has a fractal structure, then it seems quite natural to look at the fractal dimension of the attractor to characterize chaos. However severe computanional difficulties are met in calculating this dimension from the time series of a single variable usually recorded in most experiments. No algorithm is known so far, which would led to reliable results. Furthermore it has been shown numerically by Greenside et al [6] that box-counting techniques cannot be helpful in analyzing experimental data, thanks to their very low convergence rates. From the physical viewpoint, one can attribute this to the strong contraction occuring in any direction perpendicular to the flow, when dissipative systems are considered. Thanks to this property, the sheet structure of the attractor remains out of reach of our experimental tools. A precision of measurements, much higher than that available nowadays, would be required to overcome this difficulty. Thus the measure of the fractal dimension of the attractor appears almost hopeless, at least in experimental cases.

Another quantity which may be used to measure chaos is the largest Lyapunov characteristic exponent. Not only a positive exponent λ tells us that nearby trajectories diverge in phase-space, but also the value of λ is related to the rate of this divergence: the greater is λ, the faster the system forgets its initial conditions. The information content of this parameter is therefore clearly linked to the main direction of the flow or, in other words, to the motion along the sheets of the attractor. It turns out that such a situation is much more favourable, because it does not matter anymore if the fractal structure of the attractor cannot be experimentally detected. In principle, one should be able to measure this exponent from any time series and several attempts have already been performed to this end. In the field of Chemistry, the Belousov-Zhabotinsky (BZ) reaction is the most widely studied chemical system. It has led to different values, namely: 0.5 ± 0.1 [7], 0.62 [8] and, lastly, 0.3 ± 0.1 [9]. This scatter points out the difficulty of getting a reliable value. Indeed the calculus does not rely on the phase-space trajectories themselves. Rather it involves the first return map (1D map) drawn from a Poincaré section of the attractor. Because the result is largely dependent on the shape of this 1D map, the analytical form of the curve chosen to fit experimental points is the key-point of the procedure. Accordingly, the exponent so computed is very sensitive to any kind of uncertainty coming from experimental noise, lack of accuracy, numerical treatment of data, etc. The main consequence is that one cannot expect that a Lyapunov exponent determined in such a way will enable us to really discriminate between different chaotic regimes. It has recently been suggested [10] to reach the largest Lyapunov exponent through the metric entropy. Some preliminary attemps indicate that such a procedure would be more powerful than the above reported one [11]. But this point has still to be confirmed by further studies.

Meanwhile one can also try to do something else in order to evaluate the level of chaos. One simple idea comes from the fact that the power density of the Fourier spectrum is frequently used to characterize dynamic regimes. In particular, the emergence of a so-called "broad-band" is commonly considered as the signature of chaos. Hence it seems reasonable to take the Fourier spectrum as a basis for the search of a relevant parameter. When introducing a new quantity, the first question to be addressed obviously deals with its theoretical background and meaning. However we will skip this fundamental aspect, whose discussion goes far beyond the scope of the present paper. Here we want to adopt a more pragmatic point of view and to simply examine whether three suggestions, made over the recent past years, might provide us with useful information about the degree of disorder.

3. THREE QUANTITIES DEDUCIBLE FROM THE FOURIER SPECTRUM

Having in mind applications to experimental data, we are essentially concerned with discrete Fourier spectra, usually coming from a FFT routine. For such spectra it has been proposed to compute:

- the number of degrees of freedom [12] , or else its logarithm DL:

$$DL = \log \frac{\left(\sum_{i=1}^{n} p_i \right)^2}{n \sum_{i=1}^{n} p_i^{\,2}}$$

- an entropy-like function H [13] :

$$H = -\frac{1}{n} \sum_{i=1}^{n} p_i \log p_i$$

- the fractal dimension d of the Fourier spectrum [14] .

In the above relationships, p_i is the power density at frequency $i.\Delta f$ (i: positive integer, Δf: frequency resolution of the spectrum) and n is the total number of discrete components over which the summations are carried out.

It is almost obvious that DL = 0 for a white noise and that DL = -log n for a pure sine wave. The H function varies also whithin exactly the same limits, provided the following normalization condition:

$$\sum_{i=1}^{n} p_i = n$$

is applied. This allows for a direct comparison between DL and H.

Because these two quantities are insensitive to any permutation of their components - which would correspond to drastically different dynamic regimes - Perdang [14] has suggested that the fractal dimension of the Fourier spectrum be determined; indeed the length of the spectrum depends on the relative positions of its components. However we must emphasize that it has never been established that the Fourier spectrum of a chaotic regime is actually a fractal set. On the contrary Wolf and Swift [15] have been able to derive an analytical expression accounting for the shape of a chaotic spectrum. Even though their calculation involves some numerical approximations, serious doubts do exist on the fractal character of such a spectrum. Nonetheless we will follow Perdang and assume that, when the Fourier spectrum is a fractal set, its length $L(\Delta f)$ depends on the frequency resolution Δf according to:

$$L(\Delta f) \propto a(\Delta f)^{1-d}$$

a being a constant and d the fractal dimension. When this relationship applies, a log-log plot of L versus Δf leads to a straightline whose slope is equal to 1-d. The spectrum of a periodic or a quasi-periodic regime being a set of isolated peaks - that is to say an ordinary curve - the corresponding fractal dimension is equal to 1. In that case the log-log plot is expected to yield a horizontal line, as it does (see fig. 1). On the contrary, a fractal embedded in the plane will give rise to a straightline with negative slope, since d is then greater than 1 and less than 2 (fig. 2).

The recipe to be applied in order to draw the log-log plot is a rather long story, explained in detail elsewhere [16,17]. Henceforth we only give below a short summary:

- i) compute an ensemble-averaged spectrum with the highest frequency resolution available Δf_o

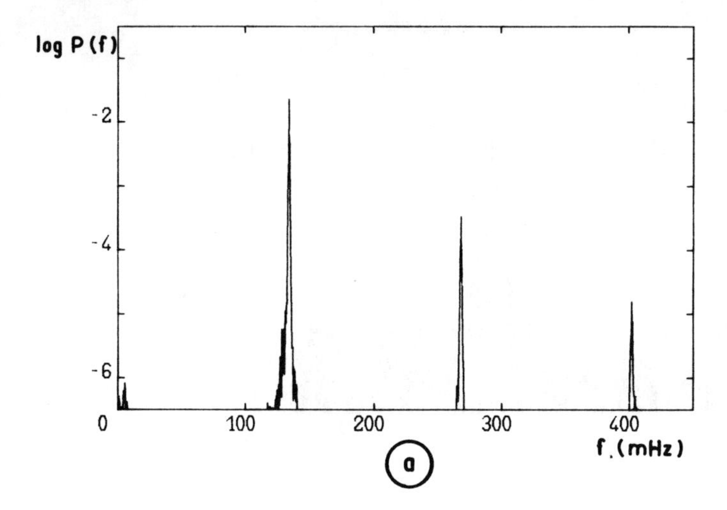

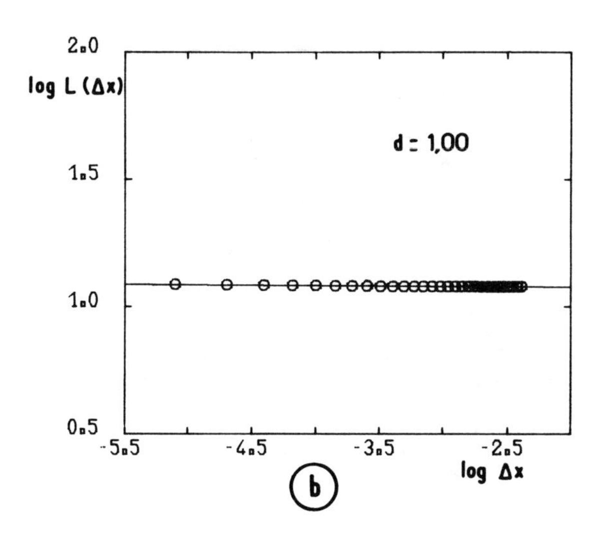

Figure 1 : Fourier spectrum (a) and log L(Δx) versus log Δx plot (b) for a periodic regime (d = 1) of the BZ reaction.

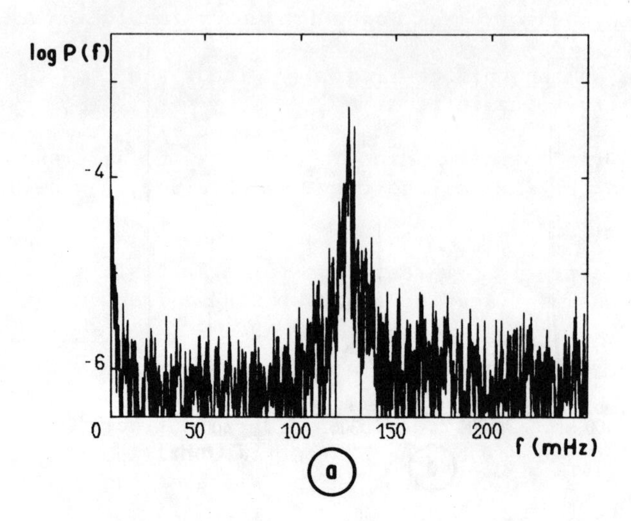

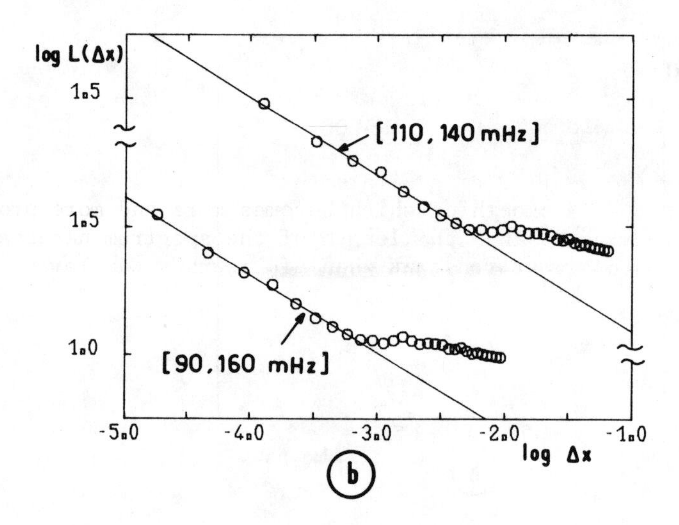

Figure 2 : Same graphs as fig. 1, but for a chaotic regime. In plot (b) it is seen that the slope is the same, regardless the width of the frequency window, provided the main peak of the Fourier spectrum is included in it.

- ii) generate a spectrum of lower frequency resolution Δf_k

- iii) choose an appropriate frequency window and find the largest power density value in it

- iv) renormalize to unity both the ordinate and abscissa axis and measure the length $L(\Delta x)$ of the curve so-obtained, (Δx being the normalized resolution)

- v) repeat the previous three steps for a suitable set of frequency resolutions Δf_k (keeping always the same frequency window)

- vi) lastly draw the log-log plot of $L(\Delta x)$ versus Δx.

Figures 1 and 2 show that such a procedure does work when applied to time series yielded by the BZ reaction. On figure 2 we note that the slope is not dependent on the width of the chosen window, provided the main peak of the Fourier spectrum is included in it. Of course a deviation is observed there, at the right hand part of the plot. But two reasons account for this fact:

- the relationship $L(\Delta x) \sim (\Delta x)^{1-d}$ is a limit law valid when Δx goes to zero. Hence, only the lowest values of Δx have to be taken into consideration

- the algorithm applied to generate the sequence of spectra (step ii) introduces a smoothing which becomes more and more pronounced as Δf_k increases. Then the length of the spectrum necessarily tends to remain almost constant when Δf_k reaches too large values.

4. A NUMERICAL TEST

Since we do not really know whether these three parameters are relevant in measuring chaos, it seems appropriate to examine how they behave in a well-known case. This is the reason why, in a first step, we have studied numerically the following set of differential equations:

$$\begin{aligned} \dot{X} &= - Y - Z \\ \dot{Y} &= X + a Z \\ \dot{Z} &= b + XZ - c Z \end{aligned} \qquad (1)$$

This model, introduced a few years ago by Rössler, is among the simplest ones exhibiting chaos. We chose it because the largest Lyapunov exponent has already been computed by the Santa-Cruz group [12] over a wide range of the control parameter c (for a=b=0.2). Because the theoretical meaning of λ is not questionable, its variations (see figure 3) can be used as a reference to evaluate those of DL, H and d.

The differential equation set (1) has been numerically integrated by means of a standard routine, scanning point by point the control parameter c over the range 4.1 to 5.2. After cancellation of the transients, the Fourier spectrum of X was taken and then analyzed in the above-described way. Figure 4 presents two spectra so-obtained, one for a periodic regime, the other for a chaotic one. The quantities DL, H and d were determined over the same frequency range (i.e., over a unique frequency window); the first two according to their definition relationships, the last through log-log plots (as shown in the right hand part of figure 4).

A thorough comparison of the variations of λ (fig. 3) with those of DL, H and d (fig. 5) leads to the following conclusions. Only the H function exhibits at the right locations the five peaks downwards displayed by λ. One peak is missing in the graphs of both DL and d. Moreover d even presents spurious peaks, so that the correlation between λ and d is quite poor. One also notes that the transition from periodicity to chaos (around c = 4.2) is not very abrupt as a function of λ; this is more or less the case for H too. Though the amplitude of the fifth peak at c = 4.88 is not large enough, this H function offers a rather good image of λ's variations.

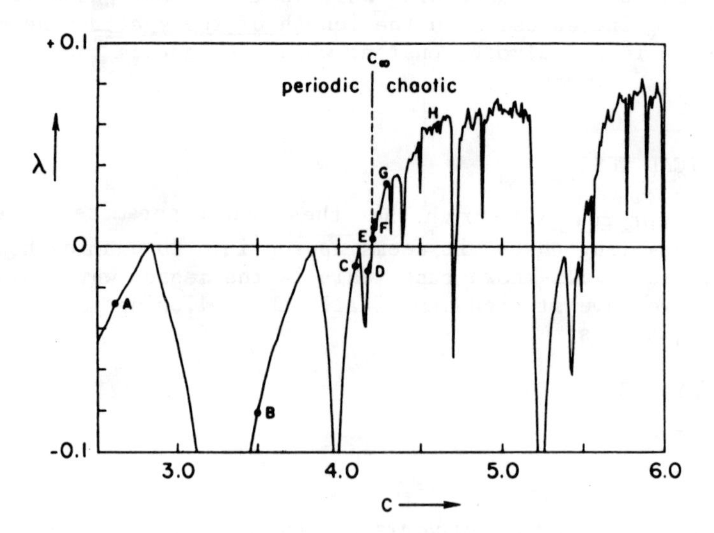

Figure 3 : Largest Lyapunov characteristic exponent λ versus parameter c for the differential equation set (1) . Taken from reference [12] .

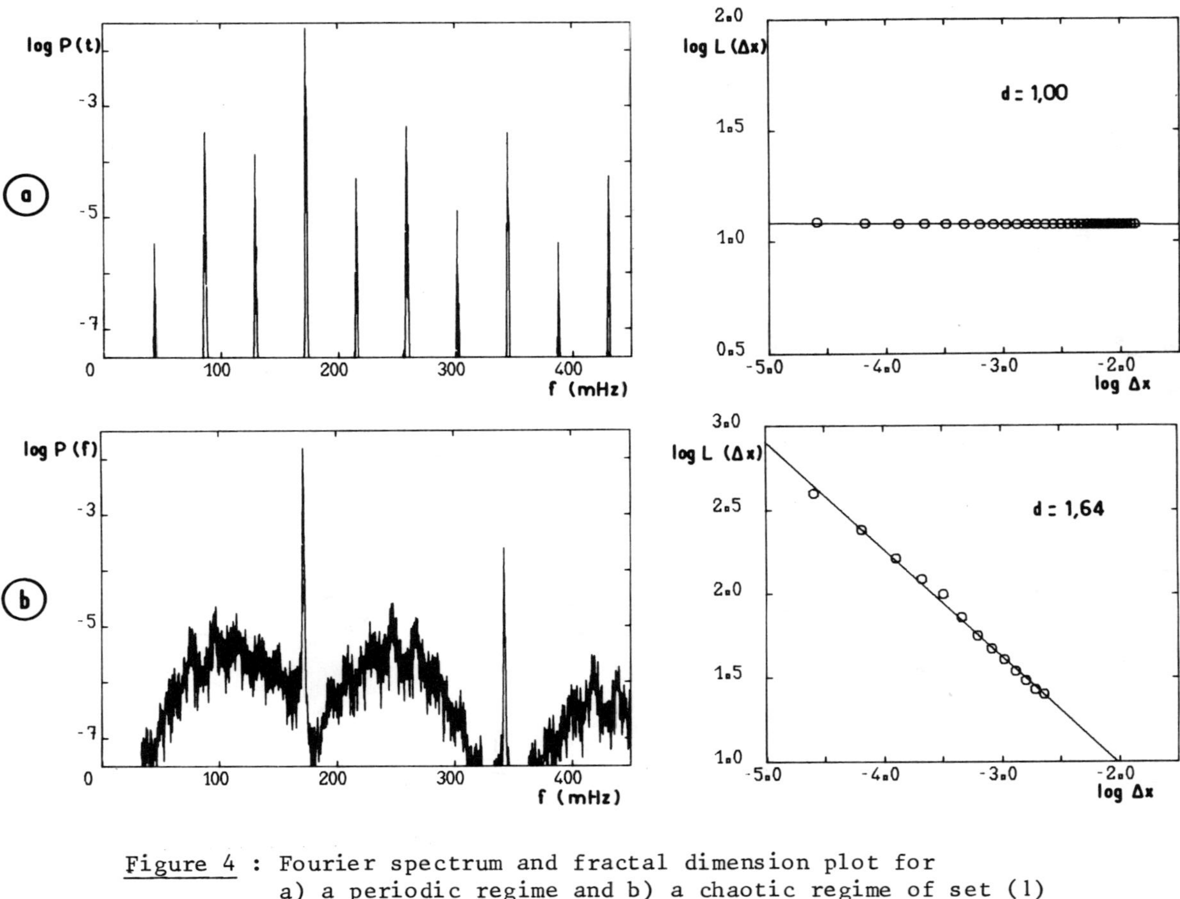

<u>Figure 4</u> : Fourier spectrum and fractal dimension plot for a) a periodic regime and b) a chaotic regime of set (1)

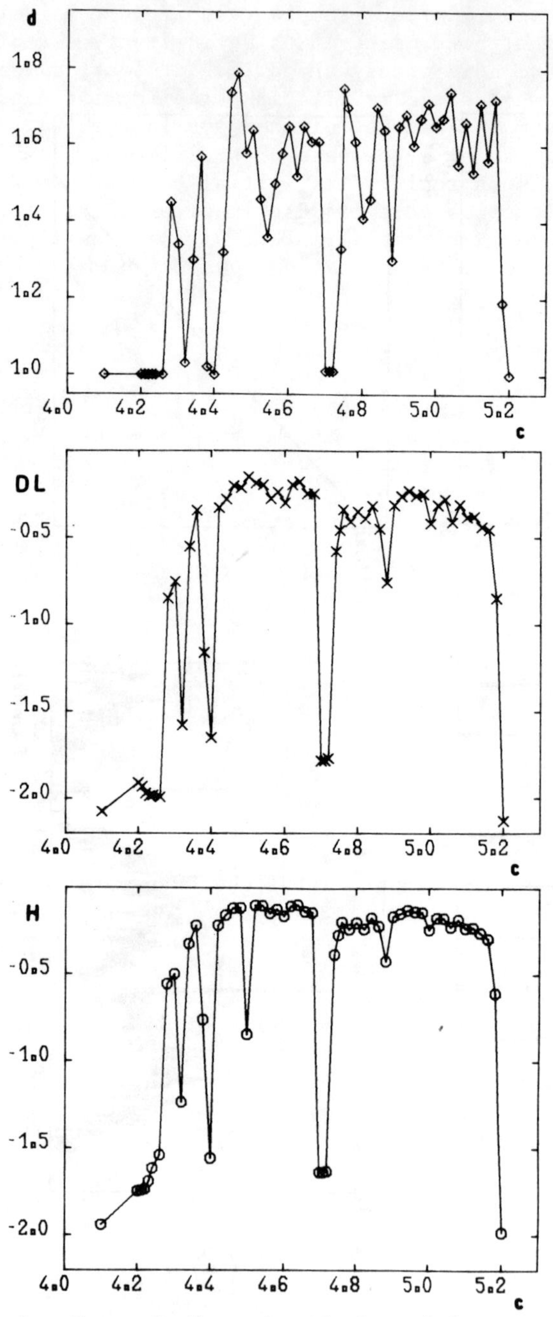

Figure 5 : Fractal dimension d, logarithm of the number of degrees of freedom DL and entropy-like function H versus parameter c. These pictures are to be compared to λ (fig.3).

Thanks to the very limited example taken into consideration, this conclusion cannot be considered as definitively established; of course, it must be more widely supported, at least by other numerical studies. Nevertheless we will make the heuristic assumption that H displays the main features of the ways into which λ varies. Accordingly an increase or a decrease of H will mean that the system forgets more or less rapidly its initial conditions or, in other words, that chaos is more or less important. In practice the great difference comes from the fact that H can be easily computed from experimental time series, whereas λ cannot as yet.

5. APPLICATION TO THE BZ REACTION

The BZ reaction, performed in a CSTR, is well-known to give rise to bifurcations between different dynamic regimes when the mean residence time of chemicals is changed [4]. Of course, temperature and inlet concentrations must be adequate. Using an experimental set-up, detailed elsewhere [18], we have studied the behaviour of the BZ reaction under the following experimental conditions:

$$[CH_2(COOH)_2]_o = 0.2 \text{ M}$$

$$[Na\,BrO_3]_o = 0.036 \text{ M}$$

$$[H_2SO_4]_o = 0.5 \text{ M}$$

$$[Ce_2(SO_4)_3]_o = 2.5\ 10^{-4} \text{ M}$$

$$T = 40\,°C$$

The control parameter μ - reciprocal of the mean residence time in the CSTR - has been scanned from 0.16 up to 0.36 min^{-1}; this corresponds to "short" residence times, roughly 3 to 6 minutes. The recorded signal (optical density at 340 nm, due to light absorption by Ce^{4+} ion) was taken each 0.1 s during experiments lasting for 2 hours or more. Then the Fourier spectrum of 64 K experimental data points was calculated by a standard FFT routine. Finally DL, H and d were determined in the above mentionned way. But this time no comparison with the Lyapunov exponent is allowed, since we are unable to compute this parameter.

Within the range of μ explored, periodic as well as chaotic regimes are encountered. Figure 6 shows that the correlation between H and DL is pretty nice in that case. Therefore, one can presume that they both give a good picture of what happens in this sequence of experiments. At first glance at figure 6, one could think that the fractal dimension d varies more or less in the same manner as DL and H. However this is not true and, on the contrary, a clear discrepancy is worthy of notice at one point. For $\mu = 0.24\ min^{-1}$,

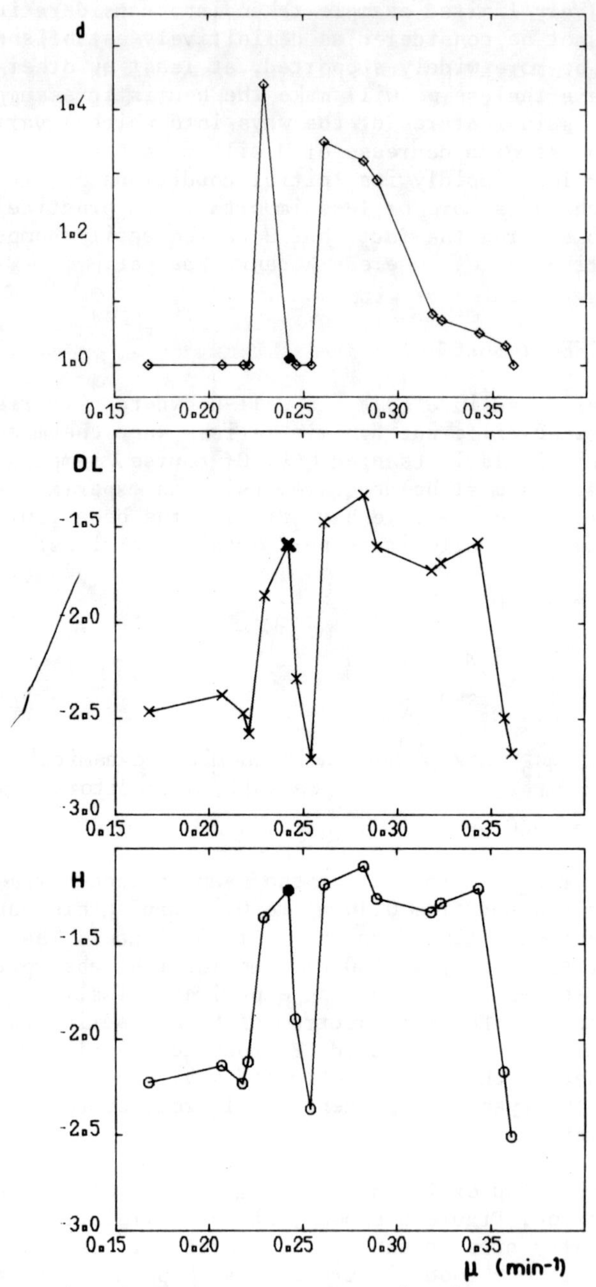

Figure 6 : d, DL and H versus μ (reciprocal of the mean residence time) for a sequence of experiments on the BZ reaction. Note the discrepancy between H and DL on one hand and d, on the other, at $\mu = 0.24\ min^{-1}$.

H and DL tell us that the regime is chaotic, whereas the fractal dimension d is equal to 1 (see fig. 6). The relevance of d is thus questionned once again, although this unique observation does not enable us to really support the opinion that the Fourier spectrum of a chaotic regime is not necessarily a fractal set.

6. CONCLUDING REMARKS

The main conclusion which comes from this work is that the entropy-like function H might provide us with an estimate of the Lyapunov exponent variations. However the proof remains far from being given and more attention should be paid to this function in the future. From a theoretical viewpoint, its basis and relevance call for clarification. On the other hand, the determination of H certainly deserves to be systematically carried out in any kind of experiment dealing with chaos. This is the price to pay for firmly establishing the usefulness of H in measuring the level of chaos.

REFERENCES

1. Ruelle, D., Takens, F., 1971, Comm. Math. Phys. 20, p. 167.
2. Eckmann, J.-P., 1981, Rev. Mod. Phys. 53, p. 643.
3. Gollub, J.P.,, Swinney, H.L., 1978, Phys. Today 31, p. 41.
4. Vidal, C., in "Nonlinear phenomena in chemical dynamics", 1981, Springer-Verlag (Vidal, C., Pacault, A., editors), p.49.
5. Swinney, H.L., Gollub, J.P. (editors) "Hydrodynamic instabilities and the transition to turbulence", 1981, Springer-Verlag.
6. Greenside, H.S., Wolf, A., Swift, J., Pignataro, T., 1982, Phys. Rev. A 25, p. 3453.
7. Roux, J.C., Turner, J.S., Mc Cormick, W.D., Swinney, H.L., in "Nonlinear problems: present and future" 1982, North-Holland (Bishop, A.R., Campbell, D.K., Nicolaenko, B., editors), p.409.
8. Hudson, J.L., Mankin, J.C., 1981, J. Chem. Phys. 74, p. 6171.
9. Roux, J.C., Simonyi, R.H., Swinney, H.L., 1983, Physica D, in press.
10. Farmer, J.D., 1982, Physica D 4, p. 366.
11. Packard, N.H., private communication.
12. Crutchfield, J.P., Farmer, J.D., Packard, N.H., Shaw, R.S., Jones, G., Donnelly, R., 1980, Phys. Lett. 76A, p. 1
13. Aikake, H., in "System identification: advances in case studies" 1979 (Mehra, R.K., Lainiotis, editors), p. 27.

14. Perdang, J., 1981, Astrophys. Space Sci. 74, p. 149.
15. Wolf, A., Swift, J., 1981, Phys. Lett. 83A, p. 184.
16. Blacher, S., Perdang, J., 1981, Physica D 3, p. 512.
17. Lafon, A., Rossi, A., Vidal, C., 1983, J. Physique 44, p. 505.
18. Vidal, C., Bachelart, S., Rossi, A., 1982, J. Physique 43, p.7.

THEORETICAL AND EXPERIMENTAL STUDIES OF CATALYTIC REACTIONS

R.A. Schmitz, G.A. D'Netto, L.F. Razon, J.R. Brown

University of Notre Dame, Indiana 46556

Heterogeneously catalyzed reactions in open systems often exhibit steady-state hysteresis, complex self-generated oscillations and spatial instabilities. These are described by mathematical models which differ in their descriptions of the surface kinetic mechanism and of heat and mass transport processes. While theoretical results qualitatively resemble experimental observations, general agreement as to the cause of the observed behavior has not been reached.

This paper sketches the current state of experimental and theoretical research in this subject area. Progress on experimental methods of obtaining instantaneous spatial temperature distributions on catalytic surfaces by means of infrared thermography are also reported.

1. INTRODUCTION

This general area of the dynamics of heterogeneous catalytic reactions has attracted researchers for two principal reasons. First, in industrial practice, most important chemical reactions involve heterogeneous catalysis. Second, theoretical and experimental research through the past couple of decades has shown that such reactions are rich in fascinating behavioral phenomena including multiple steady states and associated hysteresis, and the main subjects of this publication, complex self-generated oscillations and spatial effects. However, studies involving heterogeneous catalysis meet with insidious difficulties that are not necessarily met in studies of homogeneous systems. For example, there are the omnipresent heat and mass exchange rates

G. Nicolis and F. Baras (eds.), Chemical Instabilities, 33–57.

between the catalytic surface and the surrounding fluid, usually a gas. Further, since reactions take place on the catalytic surface, their rates are sensitive to impurities and to the precise physical and chemical nature of the surface which is often difficult to define. The surface has spatial variation, and it may change during the course of the reaction beyond the control of the experimenter.

Still another difficulty is that the key state variables of concentrations of species on the catalytic surface cannot be measured instantaneously in situ.

All of this means that experiments are often not reproducible -- certainly not in a quantitative sense -- and that close coupling between theory and experiments has not yet been achieved. Still, much has been accomplished and the future promises much more.

An exhaustive review is not possible here and information will have to be given in a rather superficial manner. The intent is to give an overview of the field, pointing out current directions and concepts concerning oscillations and spatial effects in heterogeneous catalysis. Extensive reviews are available elsewhere [1 - 6].

2. THE PHYSICAL PICTURE

Experimental studies of catalytic reaction dynamics have mostly made use of continuous-flow systems, of one of the types shown in Figure 1. The so-called single-particle studies have employed either a flow-through reactor (Figure 1a), in which the particle being studied is suspended in a flowing stream of reactants, or a stirred reactor (Figure 1b) in which the bulk reactant/product mixture surrounding the catalyst is mechanically mixed so that it may be assumed gradientless. A tubular reactor run with a high recycle ratio would have the same effect. Catalyst particles in the form of wires, foils, and dispersions on supports in pellet or wafer form have all been studied.

Other experiments have employed a flow-through tubular reactor packed with catalyst pellets (Figure 1c). Of the three types shown in Figure 1, this is the type which most closely resembles a commercial reactor. It also presents the greatest complexities for analyzing experimental results and coupling them to a mathematical model because of the existence of spatial gradients, at least in the direction of flow. We limit the scope of this paper only to those systems for which it seems reasonable to assume spatial uniformity of the state of the surrounding fluid. This will eliminate tubular reactors -- but not necessarily multiparticle systems, if all particles are exposed to uniform surroundings.

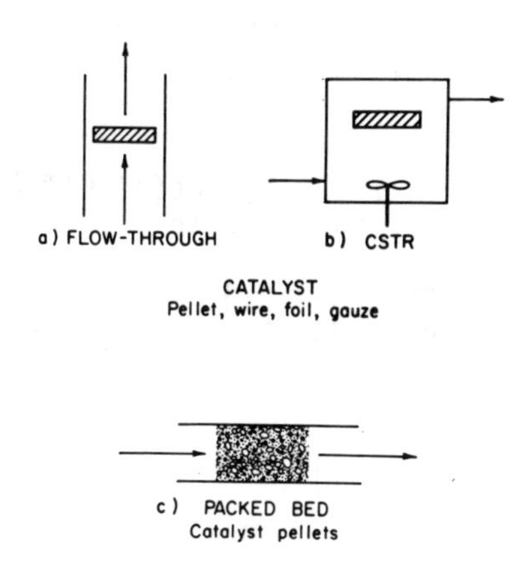

Figure 1. Types of laboratory reactors used in the studies of heterogeneous catalysis.

Table 1 shows the types of reactions and catalysts for which complex dynamics, particularly self-sustained oscillations, have been observed. Notice that most of them are oxidation reactions, the simplest being the first two on the list which are the oxidations of CO and H_2. All the reactions in Table 1 are exothermic. The catalysts most commonly used have been Pt, Pd and Ni.

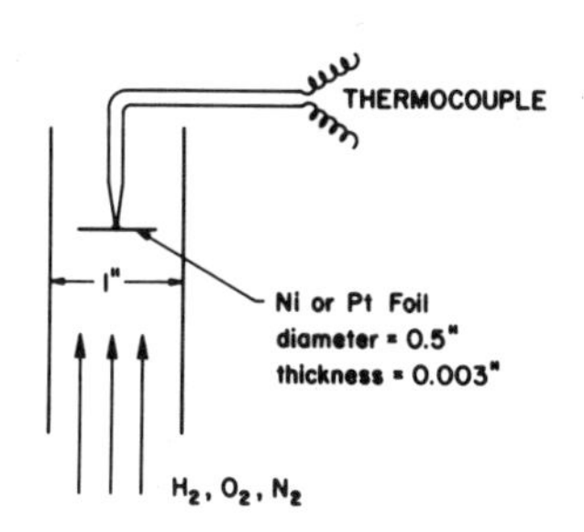

Figure 2. Schematic of the flow-through reactor used in the experiments of Schmitz, Renola, and Garrigan [38].

Experiments conducted in our laboratories a few years ago [38] provide a convenient illustration for purposes of introduction. The first study of this type was done by Keil and Wicke [17]. Our experiments involved the flow of reactants, H_2 and O_2, through a tube under controlled conditions as shown in Figure 2.

TABLE 1

OSCILLATING REACTANT/CATALYST COMBINATIONS

Reactants	Catalysts
$CO + O_2$	Pt [7 - 15], Pt(s) [16 - 26], Pd [27], CuO(s) [28], Ir [27]
$H_2 + O_2$	Pt [29 - 32], Pt(s) [17, 33 - 35], Pd [36], Pd(s) [34], Ni [37 - 39]
$NH_3 + O_2$	Pt [40]
$C_2H_4 + O_2$	Pt [41]
$C_3H_6 + O_2$	Pt [42]
$CH_3OH + O_2$	Pt(s) [43]
$CH_3(CHCH_2)O + O_2$	Ag [44]
$C_6H_{12} + O_2$	NaY(zeolite) [45]
N_2O decomposition	CuO(s) [46]
$CO + NO$	Pt [47]
$NH_3 + NO$	Pt [48]
$C_2H_4 + H_2$	Ni(s) [49]
$CO + NO + O_2 + H_2O$	Pt(s) [50]

A disk of catalytic foil, Ni or Pt, was suspended in the flow. A thermocouple welded to the downstream side of the foil was used to monitor the instantaneous temperature rise resulting from the foil-catalyzed oxidation of hydrogen.

In ranges of experimental conditions, self-sustained oscillations of the thermocouple signal were observed and in some cases, the picture was very complex. Figure 3 shows a case in which the train of oscillations underwent changes, or bifurcations, as time into the experimental test increased. The oscillations settled first into a single-peak pattern, then changed, or bifurcated, to two peaks per cycle after seven hours, then to four peaks per cycle after ten hours. After thirteen hours, the oscillations became chaotic in the sense that there was no longer any apparent periodicity to them. They continued in that fashion for days or as long as we continued the test.

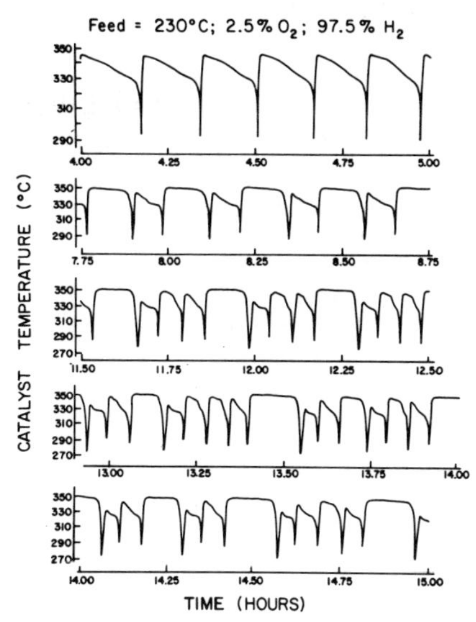

Figure 3. Oscillatory temperature patterns for the oxidation of hydrogen on nickel catalyst [38].

All experimental conditions were held constant during a test of this type. We suspect that some very slow physical or chemical change of the catalyst surface was responsible for the passage through the bifurcations -- changes beyond our capabilities to control and detect. Even disregarding such slow changes, these observations suggest certain obvious lines of inquiry.

First, one might inquire into the underlying cause of oscillations by analyzing mathematical models which incorporate a basic knowledge of physical and chemical rate processes.

Second, one might inquire into spatial effects, asking about what might be taking place at other points on the catalyst surface, when these oscillations were happening at the thermocouple location. This same line of inquiry might lead one to wonder what would happen in a multi-particle assembly, with a number of coupled oscillators.

These two types of inquiry lead to a branching point of this overview, a departure along two paths which recent research in this area has tended to follow. The first branch deals with the subject of mathematical modeling to describe oscillations disregarding the possibility of spatial phenomena. The second considers spatial phenomena.

3. CATALYTIC PROCESSES AND KINETIC MODELS

Catalytic reactions are necessarily multi-step processes as indicated schematically in Figure 4. This particular set of steps involves the reversible adsorption of reactants A and B_2 on active sites. Here the diatomic compound B_2 is shown as being adsorbed in the atomic state. Adsorbed species A and B on adjacent sites react to form product P, which is then also involved in a reversible desorption step. This is the classical Langmuir-Hinshelwood mechanism. This particular illustration depicts a commonly accepted sequence of steps for the oxidation of CO on Pt, for which case A represents CO, B_2 is oxygen, and P is CO_2.

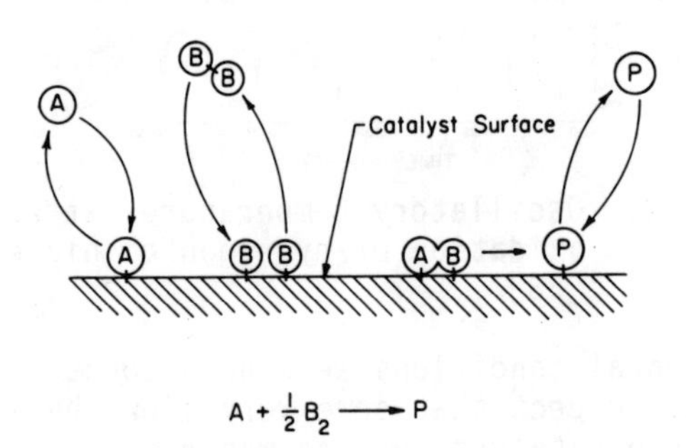

Figure 4. A schematic of the reaction mechanism for a bimolecular gas-solid catalytic reaction.

Such reaction sequences are of an autocatalytic type, generally speaking, and, therefore, it might be suspected that models would yield oscillations. In the case of an exothermic reaction, the autocatalytic agent is the product heat which causes the catalyst temperature to rise and surface reactions to accelerate. Even under isothermal conditions, however, there is autocatalysis, in effect, owing to reactant inhibition. A strongly adsorbed reactant may inhibit the reaction by occupying a large portion of the active surface sites, preventing the other reactant from being adsorbed. Under these conditions, the reaction rate may increase when some of the reactant is converted to product. Analogous effects are found in cases of substrate inhibition in enzyme kinetics [51].

In addition to these kinetic steps, there are also physical processes of heat and mass transfer to be considered. The external transport problem is one of heat and species exchange through the boundary layer between the surrounding bulk fluid and the catalyst surface (Figure 5). Concentration and temperature gradients are necessarily present in this case and would have to be accounted for in the modeling equations. Also, there is often an internal transport problem of heat conduction through the catalytic material -- and in the case of porous catalyst particles, an internal diffusion problem as well. Internal transport problems are beyond the scope of this paper. It must be noted, however, that any model intended to describe real-life systems will have to account for these effects.

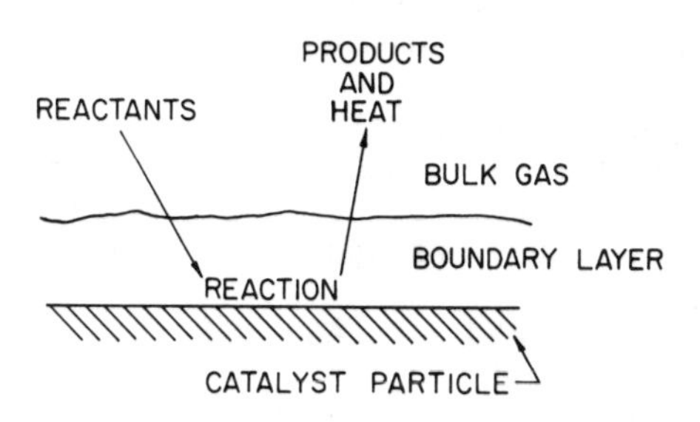

Figure 5. A schematic of the physical processes involved in a catalytic surface reaction.

While transport effects may be eliminated in laboratory reactors, and experiments have shown that certain reactions oscillate under what may be considered isothermal and gradientless conditions, the Langmuir-Hinshelwood mechanism by itself with conventional mass-action kinetics does not give a satisfactory description of them. A number of "extra" features have been added in modeling studies reported in the literature. Among them are an activation energy which depends on the concentration of adsorbed species in one or more of the reaction steps [37, 52 - 54], transition between active and inactive forms of an adsorbed component [7, 17, 55, 56], and periodic switching of the reaction mechanism [16, 18, 40, 57].

More commonly and simply, however, models have been based on an added side reaction which modifies the active sites. For example, steps (1) - (4) below show an added reaction in which an unreactive species D is reversibly adsorbed and thus competes for, and blocks, some sites.

Chemisorption

$$A + S \underset{-1}{\overset{1}{\rightleftharpoons}} AS \qquad (1)$$

$$B_2 + 2S \underset{-2}{\overset{2}{\rightleftharpoons}} 2\,BS \qquad (2)$$

Reaction

$$BS + AS \xrightarrow{3} P + 2S \qquad (3)$$

$$D + S \underset{-4}{\overset{4}{\rightleftharpoons}} DS \qquad (4)$$

This model was first proposed by Eigenberger [58] and further studied by Bykov and co-workers [59] from whom this particular example was taken. If the concentrations of the species in the gas phase are assumed constant, and if the physical rate processes in the external gas phase are not important, then the isothermal dynamics are described by the following set of three ordinary differential equations and a site balance.

$$[AS]' = k_1[S] - k_{-1}[AS] - k_3[AS][BS] \qquad (5)$$

$$[BS]' = 2k_2[S]^2 - 2k_{-2}[BS]^2 - k_3[AS][BS] \qquad (6)$$

$$[DS]' = k_4[S] - k_{-4}[DS] \qquad (7)$$

$$[S] = 1 - [AS] - [BS] - [DS] \qquad (8)$$

Depending on parameter values, there could be multiple states and unique unstable states. Some of the former and all of the latter would lead to sustained oscillations. The usual mathematical methods have been employed in the analysis of oscillatory behavior, including linear stability analysis, Hopf bifurcation analysis and computer simulations.

A more generally accepted model nowadays is one which accounts for a chemical change of the catalytic material such as metal oxidation and reduction as shown below.

The reactions depicted by steps (9) - (11) are again the Langmuir - Hinshelwood steps, i. e. the chemisorption of A and B which then react. Step (12) is a catalyst deactivation reaction in which a site, which has chemisorbed B (presumably oxygen), is converted to a metal oxide SB. In equation (13), the oxidized site is reduced by chemisorbed reducing agent A. It seems reasonable to assume that the first set of reactions is much faster than the second, an assumption which can lead to some mathematical simplifications. If this reaction mechanism, already containing an autocatalytic step, is modeled with heat effects (another autocatalytic step), some very complex oscillations result from simulations. We used a variation of this mechanism in simulations of a nonisothermal process a few years ago [60]. The model accounted for the external resistance of a fluid layer to heat and mass transfer. The major difference was that the reaction occurred on the oxidized sites instead of the reduced ones. The intention of the modeling study was to ask whether a model so formulated could yield strange oscillations of the type observed in aforementioned experimental studies of the oxidation of H_2 on nickel foil.

$$
\begin{array}{llll}
\text{Chemisorption} & A + S \underset{-1}{\overset{1}{\rightleftharpoons}} AS & & (9) \\
 & B_2 + 2S \underset{-2}{\overset{2}{\rightleftharpoons}} 2\,BS & \text{FAST} & (10) \\
\text{Reaction} & BS + AS \xrightarrow{3} P + 2S & & (11) \\
\text{Surface modification} & BS \underset{-4}{\overset{4}{\rightleftharpoons}} SB & \text{SLOW} & (12) \\
 & SB + AS \xrightarrow{5} P + 2S & & (13)
\end{array}
$$

Figure 6 shows a selection of simulations resulting from digital computer solutions of the three ordinary differential equations which comprised the mathematical model after all assumptions were made. The parameters for this set of six curves of dimensionless temperature versus dimensionless time is the dimensionless rate constant for the reaction between chemisorbed reactants. The interesting point is that the model, relatively innocent and simple in appearance, can yield such complex oscillations, including an apparently chaotic state for case (c). Chang and Aluko have shown simulated complex oscillations from a similar model [61].

There are many variations to this theoretical approach and there are as many experimental observations. Such works may be summarized by stating that experimental observations can be described qualitatively by various models which cannot be substantiated at present mainly because of our inability to measure surface species in situ and to observe catalyst changes

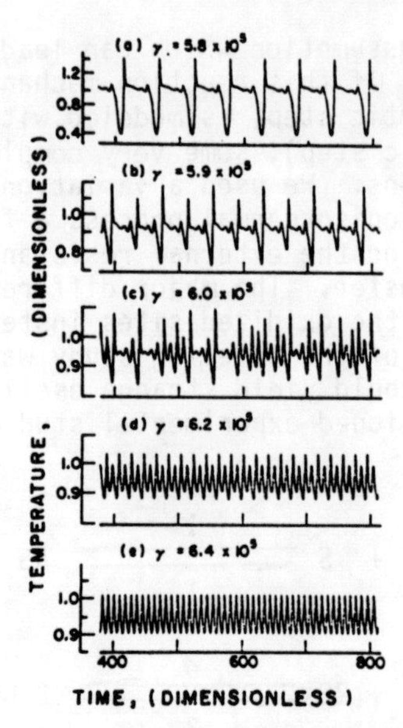

Figure 6. Simulated oscillations showing the changes in the oscillatory patterns as a catalyst activity parameter is changed [60].

in situ. Quantitative agreement between experiment and theory is not expected at this time. All-in-all, theoretical works along this branch have been valuable to our overall understanding. They show, for example, that complex spatial effects are not necessarily responsible for observed behavior. They even satisfy completely those people who are inclined not to expend any more analytical effort than is necessary to predict general qualitative experimental observations. Experimental information at this time does not justify in itself a greater depth of analytical detail. Still, it is not realistic to ignore the spatial identification that is a natural part of catalytic surfaces and systems. The second branch of research to be discussed here considers these spatial phenomena.

4. SPATIAL CONSIDERATIONS

There are at least three different aspects of space-dependent behavior that researchers have adopted. In the first, we retain the assumption that the physical and chemical nature of the catalytic material itself is spatially uniform, but take into account the fact that molecules or atoms of reactants chemisorb on specific surface sites and that reactants can react with each other only when they are on adjacent sites. We might expect, therefore, that clusters or islands of like species will develop and that reaction will occur only at the boundaries of the islands. There would develop, as a natural consequence, a two-dimensional pattern of surface species concentration.

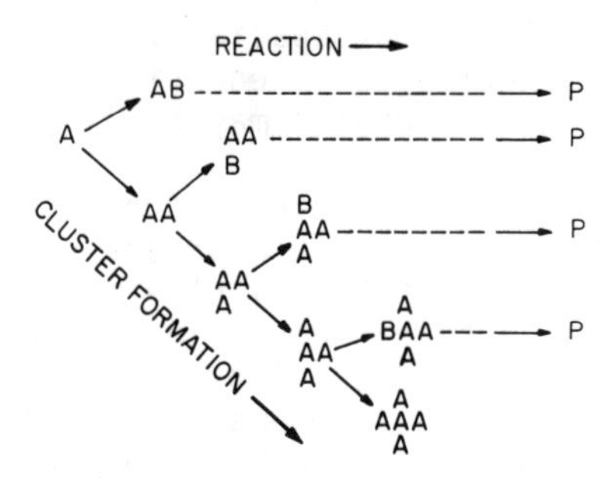

Figure 7. Scheme of a mechanism for cluster formation during the reaction A + B → P [35].

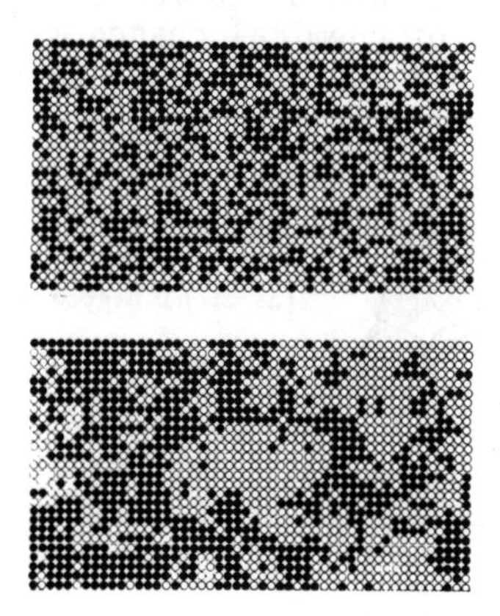

Figure 8. Development of clusters by a Monte Carlo simulation of a reacting catalytic surface [35].

Wicke and co-workers were apparently the first to address this picture [35]. They did so by employing a Monte Carlo simulation in which probabilities for the various events were incorporated including probabilities of molecules striking unoccupied surface sites, of sticking or being adsorbed, and of reacting when on adjacent sites. Figure 7 illustrates the sequence of events. The sequence begins with species 'A' on a certain site. That species may react to form a product only if species 'B' lands on an adjacent site, or a cluster will begin to form if another 'A' sticks on an adjacent site. As this process continues, one can easily see how clusters would grow. The illustration in Figure 8 shows the spatial development, beginning with a random distribution of 'A' (indicated by dark circles) and 'B' (indicated by blank circles) as shown on the top. The picture at the bottom shows the development after the simulation of a large number of steps. Questions about the eventual state, either steady or oscillatory or spatially patterned, have not been answered to our knowledge. Theoretical work is continuing along these lines, and the results may be very enlightening.

In the second aspect, we again picture the catalytic material itself to have spatially invariant properties, but now we ask questions about the stability of a spatially uniform reaction state to spatial perturbations. This stability question is similar to that posed in studies of hydrodynamic stability and of the other reaction-diffusion problems considered by Turing [62], Prigogine [63,64], Nicolis [63], Othmer and Scriven [65,66] and their co-workers. Prigogine and his co-workers labeled this phenomena "symmetry-breaking" instabilities. The key idea is that since there is a finite rate of transport, the complex interactions between the rate of communication by diffusive transport and the rate of chemical change may make it dynamically impossible for a spatially uniform state to be sustained.

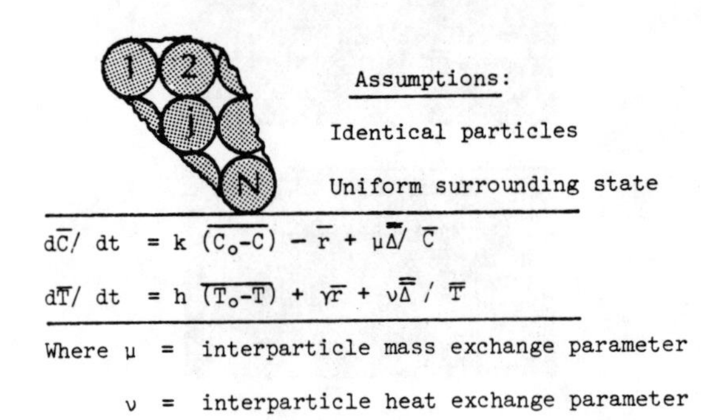

Figure 9. A lumped model to study interactions and spatial patterns in assemblies of catalyst particles [67].

Figure 9, which is taken from a recent theoretical study by Tsotsis and Schmitz [67], serves as an illustration of a particular case of this type of instability in catalytic systems. This situation, reminiscent of problems studied previously in other contexts [62-66], consists of an assemblage of identical catalyst particles in perfectly uniform surroundings. Neglecting gradients over a single particle and lumping all transport resistances at the contact points, we assume that a single-step reaction takes place non-isothermally. Species and energy balances equations for each particle can be written as shown in Figure 9 where $\overline{C}$ and $\overline{T}$ are vectors of concentration and temperature respectively, with each element corresponding to a particular particle. The rate processes on the right in each equation describe exchange with the surrounding milieu at uniform conditions given by C_0 and T_0, chemical conversion and heat generation by chemical reaction, and interchange of mass and energy between neighboring particles. The mathematical details are omitted here, but the essential result is that a steady state in which a single particle alone is globally stable may indeed be unstable to non-uniform perturbations in the interacting assembly. For a particular two-particle example, simulations show that the instability will grow to a standing spatial pattern as shown in Figure 10. The initial state is a uniform state for a bimolecular, endothermic reaction of the Langmuir-Hinshelwood type. The uniform state is unstable to a non-uniform perturbation, but stable to a uniform one. As shown in Figure 10, the instability leads eventually to a state in which the two particles reside stably in different states.

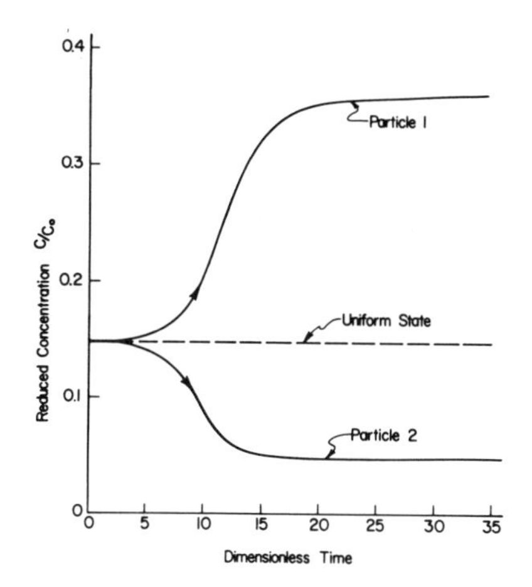

Figure 10. A simulation of the growth of a small non-uniform spatial perturbation from an unstable uniform state in a two-particle system [67].

There are other points which should be made concerning this interesting subject. First, the instabilities may be oscillatory, leading to travelling patterns in place of the standing ones shown in Figure 10. Second, even though the uniform state may be determined to be stable to small perturbations by means of a linear analysis, patterns may arise if perturbations are sufficiently large. Clearly if a single particle has multiple stable states, the various particles in the assembly may be found at any one of the stable states if there is no communication between the particles. The particular set of states would depend on the nature and size of the perturbation from a uniform initial state. As interparticle communication improves, only certain spatial configurations would be allowed. This behavior has been studied in detail by Tsotsis [68].

This general picture is reminiscent of the cell or box models used in other contexts [62-66]. The differences are non-isothermalities and the particular kinetics of catalytic reactions. Other studies have been in biological contexts. The fact that catalyst particles can be modeled and studied in definitive experiments much more easily than is the case with living cells, leads one to the expectation that combined theoretical and experimental studies of catalyst particles will provide useful insights into the behavior of living systems.

As a final point, a continuous catalyst surface could have been adopted rather than a discrete one, and partial differential equations of diffusion and conduction could have been used rather than those shown here. A great deal of interest in recent years has been directed toward studies of such diffusion-reaction systems, prompted strongly by the fascinating spatial patterns of the now-famous Belousov-Zhabotinskii reaction [69,70]. There have been some applications of such studies to catalytic processes (by Pismen [71-73], Sheintuch [73-75] and Hlavacek [22,25,76]); but the possibilities have not been clearly delineated. Concentration and temperature patterns on catalytic surfaces analogous to the colored bands in the B-Z system might be expected.

There is a third aspect of spatial effects in catalytic processes which researchers with a bent toward surface science will claim is the most important. This is the fact that spatial uniformity of a catalyst surface is an idealization. In reality, no surface is truly uniform, and there is reason to believe that the natural spatial variations present in all catalytic systems are important in the understanding of behavioral features. Figure 11 is a view taken from a scanning electron microscope of a polycrystalline foil of the type used in the experiments described previously. Various crystal orientations are evident and the scale of these variations is probably too large to justify a general assumption of spatial uniformity in mathematical models, particularly in the unsteady state. Studies with single crystal surfaces have pointed to the possibility that different crystal planes have different reactions rates and mechanisms, and there is reason to believe that oscillations will occur on some orientations and not on others [14].

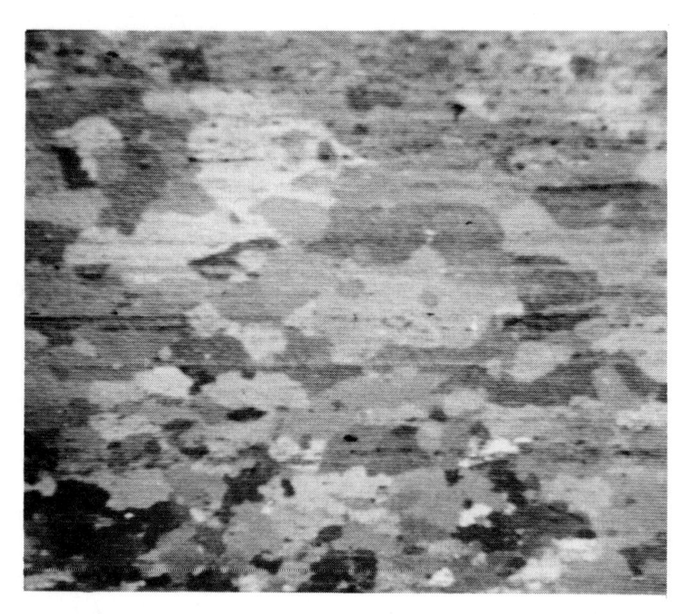

Figure 11. Scanning electron micrograph of a polycrystalline Pt foil taken at a magnification of 250X.

Catalytic wires after being used in a catalytic reaction often show increased roughness and the formation of spatial structures [78]. These observations prompted the so-called fuzzy-wire model in which the protrusions on the roughened wire were modelled by interconnected cylinders [78,79]. This study revealed a plethora of complex oscillations of the average rate of reaction resulting from the coupled protruded oscillators. The model however fails to predict the long period of the oscillations usually seen in catalytic reactions. The conclusion to draw is that rough catalysts may lead to different dynamic behavior than smooth ones; a conclusion not always supported by experimental fact.

There are two other levels of spatial variation of a catalyst. First, within an individual particle, not only is the surface very rough and porous, but the catalyst itself is dispersed on the support material in the form of small crystallites. As suggested in Figure 12, a detailed analysis would have to account for interactions between particles or crystallites. Second, though studies of symmetry-breaking instabilities in the case of exactly identical particles were discussed earlier, no two particles are really identical. They vary slightly in physical and chemical characteristics such as size, porosity, amount of catalyst and its dispersion.

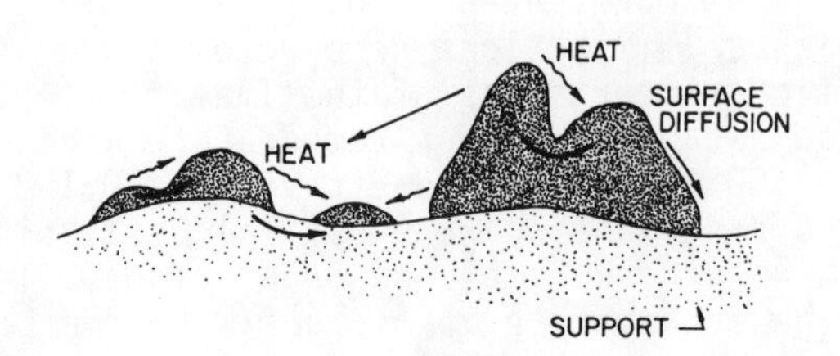

Figure 12. A schematic of the physical transport processes on a supported catalyst.

All these aspects, of course, are part of the composite picture. That is, the complete study of spatial effects must consider the spatial identity resulting from islands of adsorbed species, from spontaneous instabilities due to reaction-diffusion coupling, and from the physical nature and arrangement of the catalyst material. Though there have been some theoretical studies on each of these aspects separately, work remains in each, and studies of the composite picture have not been attempted.

Experimental studies of these spatial effects are lacking because local observations and measurements in situ are not generally possible. In our laboratories we have undertaken an experimental study which we hope will take a large step toward adding experimental information to this conceptual framework and form a basis for further theoretical studies. Although the work is in the early stages and definitive results are not yet available, we can demonstrate the capabilities of the technique and point to current experimental directions.

5. CURRENT EXPERIMENTS

The principal aim in our current experiments is to obtain measurements of local events on a catalytic surface during reaction. The one state variable for which measurements are currently possible is the surface temperature. The local temperature for an exothermic reaction, of course, reflects the local rate of reaction. The technique which we are using is infrared thermography or thermal imaging -- a method in which an instrument measures the infrared thermal radiation from a grid of points on a catalyst surface and produces an image of the temperature patterns. The instrument we use is a Barnes RM50 Infrared Micro Imager. This device scans an entire target area picking up the infrared radiation from a grid of 5376 points over a time span of two seconds. It produces several forms of output, but within the limitations of this overview only one form can be described. That form is a thermogram on which the different temperature levels are represented by different colors on a TV monitor. Owing to space limitations here we omit a description of the operational characteristics of the Imager except to say that (a) it has an adjustable range or temperature window anywhere between 0°C and 600°C, (b) the sensitivity is 0.1°C at best, depending on the range and (c) the spatial resolution in the examples to be shown here is of the order of 0.02 inch. The resolution may be enhanced to 15 microns with the best available lens. Figure 13 shows a schematic of the experimental setup. The reactor is about 30 inches below the Infrared Imager.

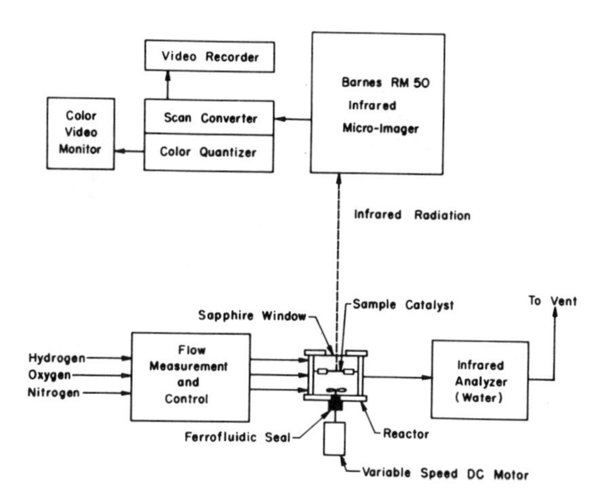

Figure 13. Schematic of the experimental setup for detecting thermal spatial patterns on catalyst surfaces by infrared thermography.

In these experiments we used a well-stirred continuous-flow reactor in which the test specimen was mounted. A sapphire window on the top of the reactor permitted the passage of infrared radiation to the thermal imager. The contents were mechanically stirred with an impeller whose shaft entered the reactor through a gas-tight seal. Hydrogen and oxygen entered the reactor at a rate controlled by electronic mass-flow controllers, and the exit stream was continuously analyzed for product concentration (water vapor in this case) which was recorded on a strip chart.

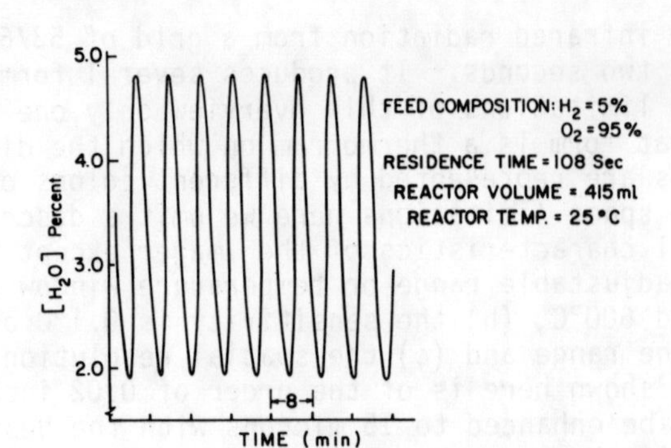

Figure 14. Experimental oscillations observed during the oxidation of hydrogen on Pt foil.

In the first set of thermograms shown here, the catalyst used was a one-inch square polycrystalline platinum foil. A feed to the reactor was a mixture of 5% hydrogen and 95% oxygen at room temperature and atmospheric pressure. The result was an oscillatory state as shown in Figure 14 which is a recording of the water vapor concentration in the bulk gas of the exit stream. Other pertinent experimental details are given in the figure. The Infrared Imager was focused on the foil and a series of thermograms were taken at intervals of four seconds. The range for the thermograms is 68 to 122°C. At the Austin Workshop we presented a continuous series of color slides showing these thermograms during one cycle of an oscillation. The series showed a small area of the foil slowly heating, growing in size and intensity till the entire surface was above 122°C after which the foil cooled till the entire cycle was repeated. Due to technical limitations on this publication, we present selected pictures from this sequence as black-and-white contours Figure 15. The labels numbered 1 to 8 in the figure represent increasing temperature with 1 indicating temperatures below 69°C and 8 indicating those above 122°C. The sequence starts at the lowest point of a cycle and proceeds till virtually the whole foil is above 122°C. After a few seconds at this condition the foil cools, and the sequence of contours in Figure 15 finishes near the end of the cycle. Other cycles are similar although there are minor variations in the ignition point and propagation pattern. The same foil at a non-oscillatory steady state showed a fairly uniform temperature pattern. Different foils may show drastic variations. For example, regions of highest activity are most commonly near the foil edges.

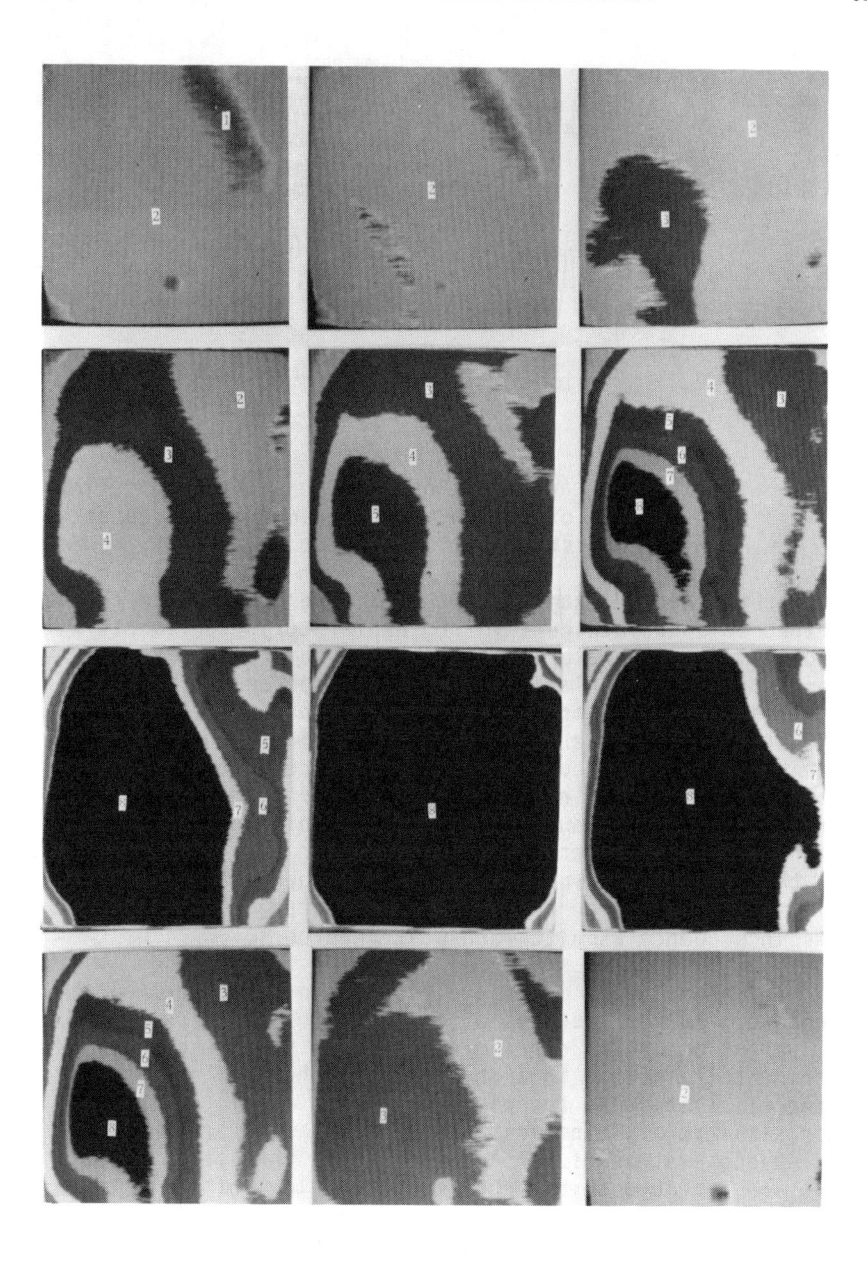

Figure 15. Sequence of thermograms of a Pt foil showing thermal spatial patterns during oscillations. Sequence reads from left to right starting at the top.

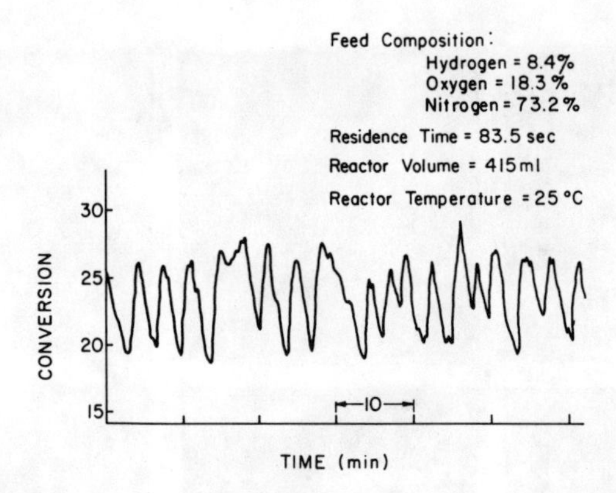

Figure 16. Aperiodic oscillations obtained during the oxidation of hydrogen over a crescent-shaped Pt/SiO_2 wafer.

For the second example we show a more complicated result obtained somewhat accidently from exploratory experiments with a supported catalyst wafer. The wafer was a crescent shaped fragment of a 5 wt% platinum catalyst dispersed on a silica support. The recording from the gas phase analyzer showed complex oscillations which are clearly not periodic (Figure 16). Again, we originally presented a sequence of slides showing the temperature as different colors in a range from 25 to 40°C. The thermograms showed a number of weakly coupled oscillating spots on the surface interacting with each other in a complex way so that the active pattern continuously changed with time in no obvious pattern. Figure 17 shows a selected few of the thermograms in sequence as black-and-white contours.

Many other experiments are planned for this thermal imaging technique. Further studies of catalyst foils and supported catalyst wafers are being carried out, and we hope to couple these experimental pictures to a physical and chemical characterization of the catalyst. The technique will also enable us to study many other situations of interest. For example, we are in the process of studying systems of particles arranged in a string or a ring. These are intended to be experimental analogs of early theoretical studies of Turing's connected particles, and of the systems of cells modeled by Prigogine, Scriven and others. We are also conducting experiments connected with the more recent theoretical studies of interacting catalyst particles reported by Tsotsis [68].

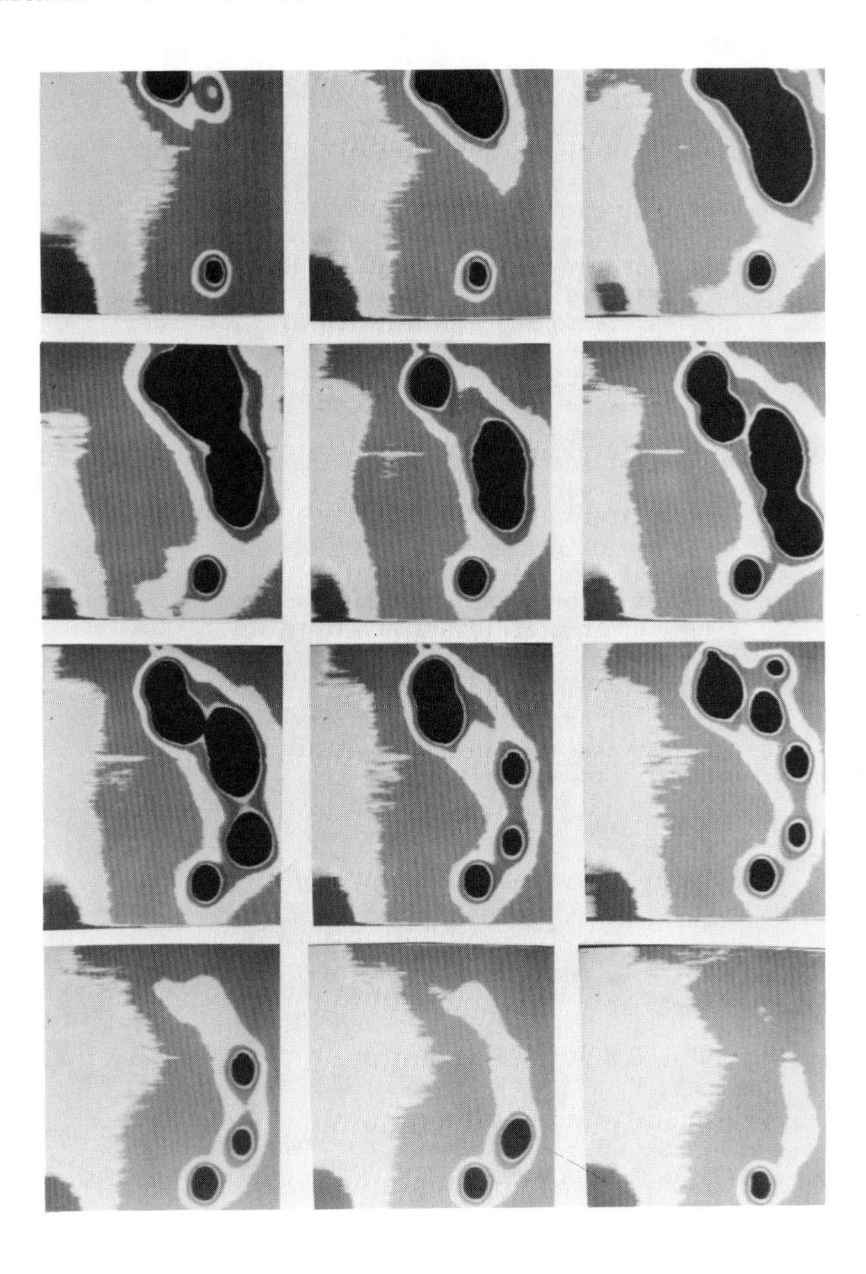

Figure 17. Sequence of thermograms of a crescent-shaped Pt/SiO_2 wafer during aperiodic oscillations. Sequence reads from left to right starting at the top.

TABLE OF NOMENCLATURE

A	A reactive nondissociative species
AS	Species A adsorbed on an active site
B_2	A reactive dissociative species
BS	Species B adsorbed on an active site
$\overline{C}$	Vector of concentrations on catalyst surface
C_o	Concentration of species in surrounding medium
C_p	Heat capacity of the catalyst particle
D	An unreactive species
DS	Species D adsorbed on an active site
h	Heat transfer coefficient to the surrounding medium
ΔH	Heat of reaction
k	Mass transfer coefficient to the surrounding medium
k_i	Rate constant
P	Product of reaction between A and B
$\overline{r}$	Vector of reaction rates
S	Active site
SB	A surface oxide
$\overline{T}$	Vector of temperatures
T_o	Temperature of surrounding medium
V	Volume of the catalyst particle
γ	In Figure 6, a catalyst activity parameter; in Figure 9, $(-\Delta H)/V\rho C_p$
ρ	Density of the catalyst particle
[]	Denotes concentration of a species
'(prime)	Denotes a time derivative

REFERENCES

1. Sheintuch, M. and Schmitz, R.A. 1977, Catal. Rev. - Sci. Eng., 15(1), pp. 107-172.
2. Slin'ko, M.M. and Slin'ko, M.G. 1980, Uspekhii Khimii, 49, pp. 561-587.
3. Bykov, V. and Yablonskii, C. 1981, Int'l Chem. Eng., 21, pp. 142-155 & pp. 716-717.
4. Hlavacek, V. and Vortruba, J. 1978, Adv. Catal., 27, pp. 59-96.
5. Eigenberger, G. 1981, Chem.-Ing.-Tech., 50(12), pp. 924-933.
6. Hlavacek, V. and Rompay, P.V. 1981, Chem. Eng. Sci., 36, pp. 1587-1597.
7. Hugo, P. and Jakubith, M. 1972, Chem.-Ing.-Tech., 44, pp. 383-387.
8. Plichta, R. and Schmitz, R.A. 1979, Chem. Eng. Commun., 3, pp. 387-398.
9. Hetrick, R. and Logothetis, E. 1979, Appl. Phys. Lett., 34(1), pp. 117-119.
10. Gray, P., Griffiths, J. and Rogerson, J. 1979, Joint ASME/AIChE 18th National Heat Transfer Conference, San Diego.

11. Zhang, S. 1980, Ph. D. Thesis, Univ. of Illinois.
12. Turner, J., Sales, B. and Maple, M. 1981, Surf. Sci., 103, pp. 54-74.
13. Barkowski, D., Haul, R. and Kretschmer, U. 1981, Surf. Sci., 107, pp. L329-L333.
14. Ertl, G., Norton, P. and Rustig, J. 1982, Phys. Rev. Lett., 49(2), pp. 177-180.
15. Edelbock, W. and Lintz, H. 1982, Chem. Eng. Sci., 37(9), p. 1435.
16. Dauchot, J. and Van Cakenberghe, J. 1973, Nature Phys. Sci., 245, pp. 61-63.
17. Beusch, H., Fieguth, P. and Wicke, E. 1972, Chem.-Eng.-Tech., 44, pp. 445-451.
18. McCarthy, E., Zahradnik, J., Kuczynski, G. and Carberry, J.J. 1975, J. Catalysis, 39, pp. 29-35.
19. Cutlip, M. and Kenney, C. 1978, ACS Symp. Ser., 65, pp. 475-486.
20. Varghese, P., Carberry, J.J. and Wolf, E. 1978, J. Catalysis, 55, pp. 76-87.
21. Keil, W. and Wicke, E. 1980, Ber. Busenges. Phys. Chem., 84, pp. 377-833.
22. Rathousky, J., Puszynski, J. and Hlavacek, V. 1980, Z. Naturforsch., 35a, pp. 1238-1244.
23. Rathousky, J., Kira, E. and Hlavacek, V. 1981, Chem. Eng. Sci., 36, pp. 776-780.
24. Liao, P. and Wolf, E. 1982, Chem. Eng. Comm., 13, pp. 315-326.
25. Rathousky, J. and Hlavacek, V. 1982, J. Catalysis, 75, pp. 122-133.
26. Elhaderi, A. and Tsotsis, T.T. 1982, ACS Symp. Ser., 196, pp. 77-88.
27. Turner, J., Sales, B. and Maple, M. 1981, Surf. Sci., 109, pp. 591-604.
28. Eckert, E., Hlavacek, V. and Marek, M. 1973, Chem. Eng. Comm., 1, pp. 89-102.
29. Zuniga, J.E. and Luss, D. 1978, J. Catalysis, 53, pp. 312-320.
30. Rajagopalan, K. and Luss, D. 1980, J. Catalysis, 61, pp. 289-290.
31. Belyaev, V., Slin'ko, M.M. and Slin'ko, M.G. 1976, Proc. 6th Int. Cong. on Catalysis, London, pp. 758-767.
32. Volodin, Y., Barelko, V. and Khalzov, P. 1977, DAN SSSR, 234, pp. 1108.
33. Horak, J. and Jiracek, F. 1972, Fifth European/Second International Symposium on Chemical Reaction Engineering, Amsterdam, pp. B8-1.
34. Boudart, M., Hanson, F., and Beagle, B. 1976, 69th Annual AIChE Meeting, Chicago.
35. Wicke, E., Kumman, P., Keil, W., and Scheifler, J. 1980, Ber. Bunsenges. Phys. Chem., 84, pp. 315-323.

36. Rajagopalan, K., Sheintuch, M. and Luss, D., Chem. Eng. Comm., 7, pp. 335-343.
37. Belyaev, V., Slin'ko, M.M., Slin'ko, M.G. and Timoshenko, V. 1974, DAN SSSR, 212, pp. 1098-1100.
38. Schmitz, R.A., Renola, G., and Garrigan, P. 1979, Ann. N. Y. Acad. Sci., 316, pp. 638-651.
39. Kurtanjek, Z., Sheintuch, M., and Luss, D. 1980, J. Catalysis, 66, pp. 11-27.
40. Flytzani-Stephanopoulos, M., Schmidt, L.D. and Caretta, R. 1980, J. Catalysis, 64, pp. 346-355.
41. Vayenas, C.G., Lee, B. and Michaels, J. 1980, J. Catalysis, 66, pp. 36-48.
42. Sheintuch, M. and Luss, D. 1981, J. Catalysis, 68, pp. 245-248.
43. Jaeger, N.I., Plath, P.J. and Van Raaij, E. 1979, Proc. of the Conference on the Kinetics of Physicochemical Oscillations, Aachen, pp. 221.
44. Stoukides, M. and Vayenas, C.G. 1982, J. Catalysis, 74, pp. 266-274.
45. Tsitouskaya, O., Altshuler, V. and Krylov, O.V. 1973, DAN SSSR, 212, pp. 1400-1403.
46. Hugo, P. 1971, Fourth European Symp. on Chem. React. Eng., Brussels, pp. 459-472.
47. Adlhoch, W., Lintz, H., and Weisker, T. 1981, Surf. Sci., 103, pp. 576-585.
48. Takoudis, C. and Schmidt, L. 1983, J. Phys. Chem., 87, pp. 964-968.
49. Niyama and Suzuki 1980, Chem. Eng. Comm., 14, pp. 145-149.
50. Subramaniam, B. and Varma, A. 1983, Chem. Eng. Comm., 20, pp. 81-91.
51. Bunow, B. and Colton, C.K. 1975, Biosystems, 7, pp. 160-171.
52. Sheintuch, M. and Schmitz, R.A. 1978, ACS Symp. Ser., 65, pp. 487-497.
53. Pikios, C. and Luss, D. 1977, Chem. Eng. Sci., 32, pp. 191-194.
54. Ivanov, E., Chumakov, G., Slin'ko, M.G., Bruns, D. and Luss, D. 1980, Chem. Eng. Sci., 35, pp. 795-803.
55. Suhl, H. 1981, Surf. Sci., 107, pp. 88-100.
56. Sheintuch, M. 1980, Chem. Eng. Sci., 35, pp. 877-881.
57. Berman, A. and Elinek, A. 1979, DAN SSSR, 248, pp. 643-647.
58. Eigenberger, G. 1978, Chem. Eng. Sci., 33, pp. 1263-1268.
59. Bykov, V., Yablonskii, G., and Kim, V. 1978, DAN SSSR, 242(3), pp. 637-640.
60. Schmitz, R.A., Renola, G. and Zioudas, A. 1980, "Dynamics and Modelling of Reactive Systems", pp. 177-193, Academic Press, NY.
61. Chang, H. and Aluko, M., Chem. Eng. Sci., submitted.
62. Turing, A. 1952, Phil. Trans. of the Royal Society, B237, pp. 37-72.

63. Prigogine, I. and Nicolis, G. 1967, J. Chem. Phys., 46, pp. 3542-3550.
64. Prigogine, I. and Lefever, R. 1968, J. Chem. Phys., 48, pp. 1695-1700.
65. Othmer, H. and Scriven, L. 1971, J. Theor. Biol., 32, pp. 507-537.
66. Othmer, H. and Scriven, L. 1974, J. Theor. Biol., 43, pp. 83-112.
67. Tsotsis, T.T. and Schmitz, R.A. 1979, 72nd Annual Meeting of the AIChE, San Francisco, Paper No 109d.
68. Tsotsis, T.T. 1981, Chem. Eng. Commun., 11, pp. 27-58.
69. Winfree, A.T. 1974, Sci Am., 230, pp. 82-95.
70. Walker, J. 1978, Sci. Am., 239, pp. 152-160.
71. Pismen, L. 1982, Chem. Eng. Sci., 36, pp. 1950-1978.
72. Pismen, L. 1979, Chem. Eng. Sci., 34, pp. 563-570.
73. Sheintuch, M. and Pismen, L. 1981, Chem. Eng. Sci., 36, pp. 489-497.
74. Sheintuch, M. 1981, Chem. Eng. Sci., 36, pp. 893-900.
75. Sheintuch, M. 1981, Chem. Eng. Sci., 37, pp. 591-599.
76. Hlavacek, V., Puzynski, J. and Rompay, P.V. 1982, ACS Symp. Ser 196, Paper No. 8, pp. 89-96.
77. Schmidt, L. and Luss, D. 1971, J. Catalysis, 22, pp. 269.
78. Jensen, K. and Ray, W.H. 1980, Chem. Eng. Sci., 35, pp. 2439-2457.
79. Jensen, K. and Ray, W.H. 1982, Chem. Eng. Sci., 37, pp. 1387-1410.

SENSITIVITY ANALYSIS: A NUMERICAL TOOL FOR THE STUDY OF PARAMETER VARIATIONS IN OSCILLATING REACTION MODELS

Raima Larter

Department of Chemistry
Indiana University-Purdue University at Indianapolis
Indianapolis, Indiana 46223 U.S.A.

A numerical tool, sensitivity analysis, which can be used to study the effects of parameter perturbations on systems of dynamical equations is briefly described. A straightforward application of the methods of sensitivity analysis to ordinary differential equation models for oscillating reactions is found to yield results which are difficult to physically interpret. In this work it is shown that the standard sensitivity analysis of equations with periodic solutions yields an expansion that contains secular terms. A Lindstedt-Poincare approach is taken, instead, and it is found that physically meaningful sensitivity information can be extracted from the straightforward sensitivity analysis results, in some cases. In the other cases, it is found that structural stability/instability can be assessed with this modification of sensitivity analysis. Illustration is given for the Lotka-Volterra oscillator.

INTRODUCTION

Analytical and computational methods have proven to be useful mathematical tools for the elucidation of mechanisms of oscillating chemical reactions [1,2]. The ordinary or partial differential equations which describe the oscillating system are parameterized by rate constants, initial conditions, boundary conditions, etc. The successful modeling of an oscillating system depends both on 1) an appropriate choice of the form of the differential equations and 2) an appropriate choice of values for each of the parameters. The usual mathematical methods for studying the properties of models of oscillating reactions

G. Nicolis and F. Baras (eds.), Chemical Instabilities, 59–66.

(linear stability analysis, bifurcation theory [3], numerical simulation) do not address these points explicitly. Recently, a new computational method known as sensitivity analysis has been proposed as a tool which will address the questions of structural and parameter sensitivity [4, 5, 6]. Sensitivity analysis promises to be a fast, inexpensive means for studying the dependence of simulations on parameter values and on the algebraic structure of the model.

Several works describe the general mathematical and computational aspects of sensitivity analysis [7,8] and its application to chemical kinetic systems [9]. In this work I will give a general description of sensitivity analysis, referring the interested reader to the above references for details. I will also discuss the use of sensitivity analysis in determining and characterizing the structural stability of multi-parameter models. The more standard topological or bifurcation-theoretic methods for assessing structural stability have generally been restricted to single parameter systems of equations, a restriction which limits their application to models of real systems. However, as will be seen below, sensitivity analysis is related to the linear stability/bifurcation theory approach to analyzing an oscillator.

METHOD

Models of oscillating chemical reactions consist of systems of ordinary differential equations which describe the temporal (or temporal and spatial) behavior of the chemical species' concentrations. These equations are parameterized by rate constants, concentrations of constant species, initial conditions, etc. and the solution to the equations depends on the values chosen for each of these various parameters. If the parameters are not themselves functions of space or time, the solution components (the chemical species' concentrations and the temperature, if nonisothermal) can be thought of as hypersurfaces in parameter space where one numerical solution of the differential equations for a given set of parameter values yields a point (which is a function of time and space, in general) on this hypersurface. The gradients of the solution components in parameter space thus give a measure of the sensitivity of the solution to perturbations in the chosen parameter values. As such, the gradients also give a measure of the structural stability of the model. As will be shown below, it is also possible to use sensitivity analysis to determine <u>which</u> parameter variations will cause structural instabilities.

Standard Theory

To quantify these ideas, consider a general model of an isothermal oscillating reaction without spatial variations,

$$\frac{dC_i}{dt} = R_i(C_1, \ldots, C_N, \alpha_{N+1}, \ldots, \alpha_M), \quad i=1, \ldots, N \tag{1}$$

subject to the initial conditions,

$$C_i(t=0) = \alpha_i \qquad , \; i=1, \ldots, N \tag{2}$$

where C_i is the concentration of one of the N species, R_i describes the chemical reaction kinetics of species i and $\underline{\alpha}=\{\alpha_1,\ldots,\alpha_N,\alpha_{N+1},\ldots,\alpha_M\}$ is a vector containing all the parameters in the problem. The first N components of $\underline{\alpha}$ are the initial conditions while the last N+1 to M are all those parameters which appear explicitly in the differential equations, such as the rate constants. A typical computer simulation would be performed by choosing values for the parameters (let us denote this reference point $\overline{\underline{\alpha}}$) and numerically solving the system of equations (1) and (2), which yields a solution which we shall denote $\overline{\underline{C}}$.

The more widely used sensitivity analysis method involves an expansion about the reference point $\overline{\underline{\alpha}}$ in parameter space; keeping only the linear terms in the expansion results in

$$C_i(t) = \overline{C}_i(t) + \sum_{j=1}^{M} \overline{\frac{\partial C_i}{\partial \alpha_j}}(t)\, d\alpha_j + \ldots \quad , \; i=1, \ldots, N \; . \tag{3}$$

The gradients $\overline{\partial C_i(t)/\partial \alpha_j}$ are evaluated at the reference point in parameter space and are obtained by solving the linearized differential equations

$$\frac{d}{dt}\left(\overline{\frac{\partial C_i}{\partial \alpha_j}}\right) - \sum_{k=1}^{N} \overline{\frac{\partial R_i}{\partial C_k}}\;\overline{\frac{\partial C_k}{\partial \alpha_j}}(t) = \overline{\frac{\partial R_i}{\partial \alpha_j}} \quad , \quad \begin{matrix} i=1, \ldots, N \\ j=1, \ldots, M \end{matrix} \quad , \tag{4}$$

subject to the initial conditions

$$\overline{\frac{\partial C_i}{\partial \alpha_j}}(t=0) = \begin{cases} \delta_{ij} & j=1, \ldots, N \\ 0 & j=N+1, \ldots, M \end{cases} \quad , \quad i=1, \ldots, N \; . \tag{5}$$

(See ref.[9] for details.) Sensitivity information for many independent parameter variations can thus be obtained with a single additional numerical calculation rather than a large number of simulations for each chosen set of parameters.

Modified Theory

A straightforward application of the above ideas to oscillating systems yields numerical results which are very difficult to interpret physically. The problem lies in the inherent secular terms which always appear in these types of expansions[6]. It can be shown that the sensitivity coefficients obtained by the method described above always have the form

$$\overline{\frac{\partial C_i}{\partial \alpha_j}}(t) = \left(\overline{\frac{\partial C_i}{\partial \alpha_j}}\right)_\tau - \frac{t}{\tau}\,\overline{\frac{\partial \tau}{\partial \alpha_j}}\,\overline{\frac{\partial C_i}{\partial t}} \tag{6}$$

where τ is the period of the oscillation and $\overline{(\partial C_i/\partial \alpha_j)_\tau}$ is a periodic function describing the change in C_i if τ is held fixed. If the period depends on the parameter α_j, the second term is nonzero and is seen to grow without bound as t becomes large, meaning that the second term is a secular term and equation (3) is a nonuniformly valid expansion.

By applying the method of Lindstedt and Poincare [10] which essentially renormalizes the time variable t so that it depends on the α_j, one can recast the straightforward expansion as one in which the expansion coefficients contain no secular terms. The linear expansion coefficient in this modified approach is found to be identical with the first term of equation (6) (see ref.[6]for details). The modified expansion in equation (7) is now uniformly valid for all times t and the expansion coefficients are found to be periodic functions with easily interpretable physical meaning.

The modified sensitivity coefficients $\overline{(\partial C_i/\partial \alpha_j)_\tau}$ can be extracted from the standard sensitivity coefficients $\overline{\partial C_i/\partial \alpha_j}$ via equation (6) if the quantity $\overline{\partial \tau/\partial \alpha_j}$ can be determined, since all other quantities in equation (6) are known. In reference 6, one method is proposed which involves the evaluation of integrals of the standard sensitivity coefficients and yields the quantity $\overline{\partial \tau/\partial \alpha_j}$ via

$$\overline{\frac{\partial \tau}{\partial \alpha_j}} = - \frac{\int_{t_1}^{t_1+\tau} dt' \, \overline{\frac{\partial C_i}{\partial \alpha_j}}(t') - \int_{t_2}^{t_2+\tau} dt' \, \overline{\frac{\partial C_i}{\partial \alpha_j}}(t')}{\bar{C}_i(t_1) - \bar{C}_i(t_2)} \ . \tag{7}$$

The times t_1 and t_2 are chosen such that the denominator in (7) is not too small, i.e. by picking t_1 and t_2 to correspond to different points along the cycle. Tomovic and Vukobratovic, 1972, discuss a similar approach in which the modified sensitivity coefficients are determined directly from a linearized differential equation; the method given here and in reference 6 has the advantage of yielding additional information about the sensitivity of the period via equation (7).

RESULTS

More complete results concerning the following examples can be found in ref. 5 and 6. Here some examples are given which illustrate the main points of the theory covered in the preceding section.

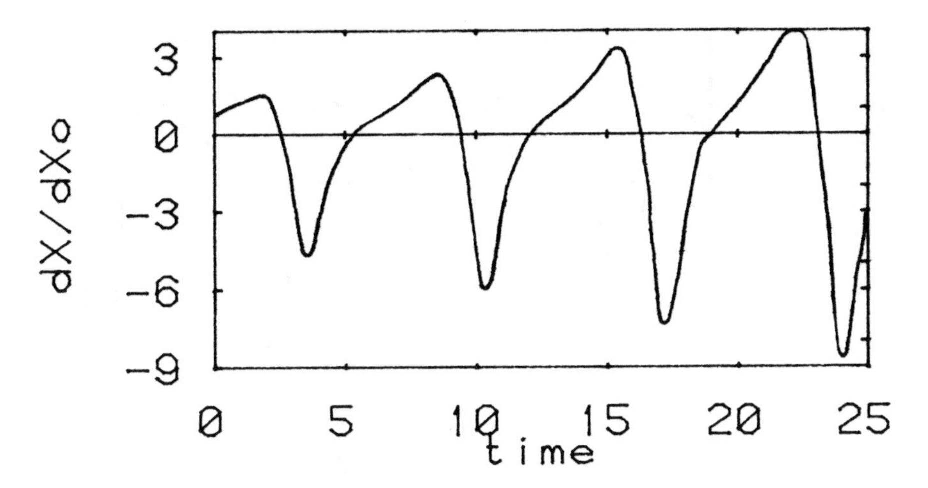

Fig. 2a) The initial condition sensitivity coefficient $\partial X/\partial X_0$ for the Lotka-Volterra oscillator evaluated over four cycles of the reference solution.

Initial Conditions

The initial condition parameters play different roles in the two well-known model oscillators shown in the Table, the Lotka-Volterra oscillator and the Brusselator.

Model Oscillators Studied Via Sensitivity Analysis

Brusselator	Lotka-Volterra
$\dot{X}=k_1+k_2X^2Y-k_3X-k_4X$	$\dot{X}=k_1X-k_4X^2-k_2XY+k_5Y^2$
$\dot{Y}=k_3X-k_2X^2Y$	$\dot{Y}=k_2XY-k_5Y^2-k_3Y$
Reference Parameter Values	
$k_1=k_2=k_4=1.0$	$k_1=k_2=k_3=1.0$
$k_3=3.0$	$k_4=k_5=0.0$
$X(0)=1.1,\ Y(0)=3.0$	$X(0)=Y(0)=0.5$

In the latter, the same cycle is always attained, regardless of the choice of initial conditions, meaning that $\overline{\partial\tau/\partial\alpha}=0$ for α_j an initial condition. Hence, the initial condition sensitivity coefficients are periodic functions for the Brusselator and will yield a physical interpretation immediately. The Lotka-Volterra oscillator is not a limit cycle oscillator and for initial conditions sufficiently far from the unstable focus in the phase plane, the initial condition sensitivity coefficients shown in Figure 2(a) are qualitatively different from those of the Brusselator. The method described in the previous section can be

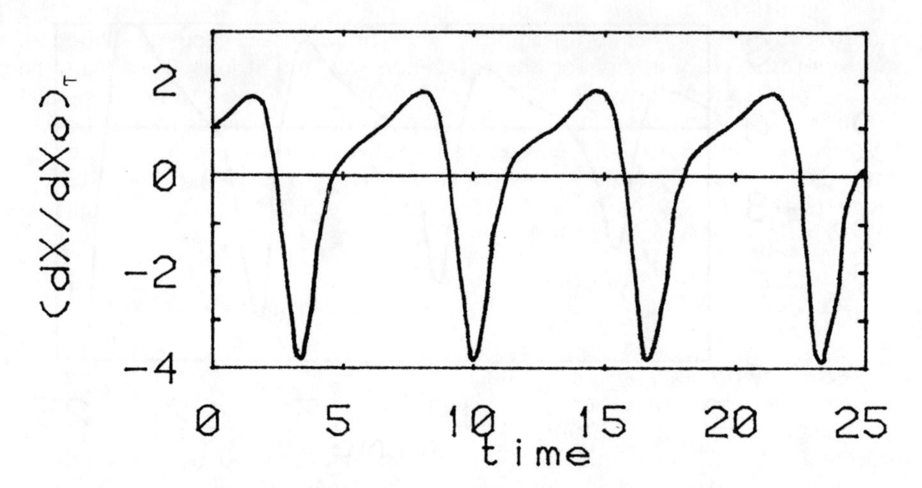

Fig. 2(b) The modified sensitivity coefficient $(\partial X/\partial X_o)_\tau$ evaluated from 2(a) and equations (6)-(7). The modified coefficient is periodic and we find that $\partial\tau/\partial X_o = -1.1$

implemented here and used to extract two physically meaningful terms, the period sensitivity $\overline{\partial\tau/\partial\alpha_j}$ (equation (7)) and the modified sensitivity coefficient $(\partial C_i/\partial\alpha_j)_\tau$ shown in Figure 2(b). For both the Lotka-Volterra and the Brüsselator it is found that the largest variations in concentration, which result from a perturbation of initial conditions, occur when the reference solution is at a maximum or minimum. See reference for details.

Rate Constants

The sensitivity coefficients for parameters, such as the rate constants, are qualitatively like those shown in Figure 2(a) for both oscillators, i.e. undamped oscillatory functions of time. The reference solution for the Lotka-Volterra oscillator is taken to be the periodic solution valid for $k_1=k_2=k_3=1.0$ and $k_4=k_5=0.0$. If k_4 and k_5 are different from zero, the oscillations will damp out or completely disappear. Hence, this reference point is a bifurcation point in the two directions k_4 and k_5 in parameter space. We would expect the sensitivity coefficients $\partial C_i/\partial k_4$, $\partial C_i/\partial k_5$ to be qualitatively different from the sensitivity coefficients for the parameters k_1-k_3, but no such qualitative differences can be observed. However, when using the method described in the previous section of this paper to extract the period sensitivities $\partial\tau/\partial k_4$ and $\partial\tau/\partial k_5$, numerical difficulties were encountered which were not encountered during the calculations of $\partial\tau/\partial k_1$, $\partial\tau/\partial k_2$ and $\partial\tau/\partial k_3$. In fact, it is impossible to find values for $\partial\tau/\partial k_4$ and $\partial\tau/\partial k_5$ by this method. These quantities are, in fact, not well-defined for this problem since variations in k_4 and k_5 away from the reference point of zero cause the periodic function to become unstable.

Sensitivity analysis thus provides a method by which the structural stability of multi-parameter models can be assessed and described in more detail. We can say for the reference solution studied here that the Lotka-Volterra oscillator is structurally unstable to variations in k_4 and k_5 but not to variations in k_1, k_2 and k_3. These properties of the Lotka-Volterra oscillator are, of course, well-known. The success of sensitivity analysis in unambiguously (and quantitatively) verifying these facts suggests that it will be a useful tool for the study of models which are not so well-understood.

MORE COMPLEX SITUATIONS

Sensitivity analysis has been extended to include spatial effects in reaction/diffusion systems [11,12] and has been applied to quantum mechanical scattering problems [13,14]. Sensitivity analysis is, in principal, applicable to any problem which is described by a system of dynamical equations. It is also possible to consider problems in which the parameters are themselves functions of space and/or time. This leads to linearized equations for sensitivity densities, e.g. $(\delta C_i(t,t')/\delta\alpha_j)$ for a purely temporal problem, which are functional derivatives and give the effect on C_i at time t due to a perturbation of α_j at an earlier time t' (see ref. [12] for a discussion of reaction/diffusion functional sensitivity densities).

The modified sensitivity analysis described in this work is likely to be necessary for problems in the spatial domain which are unstable towards patterning. A Lindstedt-Poincare approach will probably need to be taken before sensitivity analysis can be used to study the effect of parameters on spatially patterned solutions.

REFERENCES

1. D. Edelson, R.M. Noyes and R.J. Field, Intl. J. Chem. Kin. 11, 155 (1979)
2. D. Edelson and H. Rabitz, "Numerical Techniques for Modeling and Analysis of Oscillating Reactions", manuscript in preparation for Oscillations and Traveling Waves in Chemical Systems, R. J. Field and M. Burger, eds.
3. D. Sattinger, Topics in Stability and Bifurcation Theory, Springer-Verlag, New York (1973)
4. D. Edelson and V.M. Thomas, "Sensitivity Analysis of Oscillating Reactions. 1. The Period of the Oregonator", J. Phys. Chem. 85, 1555 (1981)
5. R. Larter, H. Rabitz and M. Kramer, "Sensitivity Analysis of Limit Cycles", submitted to J. Chem. Phys., (1983)
6. R. Larter, "Sensitivity Analysis of Autonomous Oscillators: Separation of Secular Terms and Determination of Structural Stability", in press, J. Phys. Chem. (1983)

7. P.M. Frank, Introduction to System Sensitivity Theory, Academic Press, New York (1978)
8. R. Tomovic and M. Vukobratovic, General Sensitivity Theory, Elsevier, New York (1972)
9. J.T. Hwang, E.P. Dougherty, S. Rabitz and H. Rabitz, "The Green's Function Method of Sensitivity Analysis in Chemical Kinetics", J. Chem. Phys. 69, 5180 (1978)
10. A.H. Nayfeh, Introduction to Perturbation Techniques, Wiley, New York (1981)
11. M. Demiralp and H. Rabitz, "Chemical Kinetic functional sensitivity analysis: Elementary sensitivities", J. Chem. Phys. 74, 3362 (1981)
12. R. Larter, H. Rabitz and M. Kobayashi, "Derived sensitivity densities in chemical kinetics: A new computational approach with applications", in press, J. Chem. Phys. (1983)
13. L. Eno and H. Rabitz, "Generalized sensitivity analysis in quantum collision theory", J. Chem. Phys. 71, 4824 (1979)
14. R. Larter and H. Rabitz, "Scattering Theory Sensitivity Analysis for Spatial Hamiltonian Variations", submitted to J. Chem. Phys. (1983)

PART II

COMBUSTION

ISOTHERMAL AUTOCATALYSIS IN THE CSTR: EXOTIC STATIONARY-STATE PATTERNS (ISOLAS AND MUSHROOMS) AND SUSTAINED OSCILLATIONS

P. Gray and S.K. Scott

School of Chemistry, University of Leeds, Leeds LS2 9JT

The prototype, cubic autocatalytic reaction (A + 2B → 3B) forms the basis of a simple homogeneous system displaying a rich variety of complex behaviour. Even under well-stirred, isothermal open conditions (the CSTR) we may find multistability, hysteresis, extinction and ignition. Allowing for the finite lifetime of the catalyst (B → inert products) adds another dimension. The dependence of the stationary-states on residence-time now yields isolas and mushrooms. Sustained oscillations (stable limit cycles) are also possible. There are strong analogies between this simple system and the exothermic, first-order reaction in a CSTR.

1. INTRODUCTION

There is a widespread belief that exotic patterns of behaviour in chemical systems require either very complex kinetic mechanisms or non-isothermal influences. There have been many investigations of the single, irreversible, exothermic reaction [see e.g. 1-5] proceeding under well-stirred, open conditions (in a CSTR). By contrast, the isothermal systems [6] covered have tended to be rather specific enzyme rate-laws or reactions at surfaces. Models proposed for homogeneous, isothermal reactions include complicated schemes [7,8] such as the "Brusselator" and "Oregonator". Table 1 lists some of the important historical landmarks of this subject.

The dependence of the stationary-states on residence time for such examples typically displays S-shaped curves, raising questions of unique or multiple solutions and rationalizing the

G. Nicolis and F. Baras (eds.), Chemical Instabilities, 69–90.

phenomena of ignition, extinction and hysteresis. More varied stationary-state patterns include closed curves (isolas) and mushrooms (which also have Z-shaped sections). Later work has focussed on stability, and on the possibilities of sustained oscillations.

period	Chemical Engineering		Combustion		Other	
pre-1940	Liljenroth	1918	Taffanel & le Floch	1913	Bredig	1904
			Semenov	1928	Bodenstein & Wolgast	1908
	Damköhler	1936	Frank-Kamenetskii	1938+	Lotka	1910, 1920
1950			Zel'dovich	1941		
			Salnikov	1948	Denbigh	1944+
			Longwell	1952	C. Wagner	1945
	van Heerden	1953	Vulis (book)	1954		
	Bilous & Amundson	1955			Prigogine	1945+
	Aris & Amundson	1958				
1960			B. Gray & Yang I	1965		
			Sarkosov	1967		
	Nishimura & Furusawa	1968	B. Gray & Yang II	1969	Denbigh & Dutton	1968
	Root & Schmitz	1969				
					Catastrophe theory	
1970	Hlavácek	1970				
			Lignola	1973+		
			JF Griffiths	1974+		
			B. Gray & Felton	1975		
	Uppal, Ray & Poore	1974 1976			Chaos 1973	
			Merzhanov	1976		
			Abramov	1978		
1980	Luss & Balakotaiah	1981				

Table 1. Some key papers related to multistability, isolas, mushrooms and oscillations

There has been, by comparison, almost no study of the simplest isothermal systems where feedback is not thermal but autocatalytic. Investigations by Lin [9] and by Heinrichs and Schneider [10] present some special features, but a comprehensive introduction has only recently been produced [11-15].

The simplicity of these systems permits a complete algebraic analysis of the stationary-state solutions and their stability. They yield an unexpected richness of behaviour, and provide striking insights into the physical origins of isolas and mushrooms. Strong analogies may be drawn between isothermal and non-isothermal examples. Isothermal autocatalysis with a stable catalyst resembles non-isothermal reaction under adiabatic conditions; isothermal autocatalysis with an unstable catalyst resembles non-isothermal, non-adiabatic behaviour. Recognizing the finite lifetime of the catalyst adds another dimension to the problem just as does finite heat loss.

2. KINETIC MODEL

We shall consider throughout one of the simplest representations of an autocatalytic reaction, coupled with a first-order decay (or poisoning) of the catalyst:

$$A + 2B \rightarrow 3B \tag{1a}$$

$$B \rightarrow C. \tag{1b}$$

For simplicity we at first assume these steps to be irreversible, and we neglect any parallel, uncatalysed conversion of A to B. These features may be included [12,13] at the cost of algebraic simplicity, but they do not cause significant qualitative changes to the phenomena chiefly of interest here.

The mass balance equations for (1) are

$$\frac{da}{dt} = -k_1 a b^2 + k_f (a_o - a) \tag{2a}$$

$$\frac{db}{dt} = k_1 a b^2 - k_2 b + k_f (b_o - b) \tag{2b}$$

where a and b are the concentrations of A and B, a_o and b_o are their concentrations in the inlet and k_f is the inverse of the mean residence time. It is convenient to reduce equation (2) to dimensionless forms. We choose the inlet concentration of the reactant a_o as the base for extent of reaction, etc.:

$$\alpha = a/a_o, \quad \beta = b/a_o, \quad \beta_o = b_o/a_o.$$

For our time-scale, the most sensible choice is the characteristic reaction time t_{ch} for the autocatalytic step (1a)

$$t_{ch} = 1/k_1 a_o^2 ;$$

hence

$$\tau = t/t_{ch} = k_1 a_o^2 t, \quad \text{dimensionless time}$$

$$\tau_{res} = t_{res}/t_{ch} = k_1 a_o^2 t_{res}, \quad \text{reduced residence time}$$

$$\tau_2 = t_2/t_{ch} = k_1 a_o^2/k_2. \quad \text{reduced decay time.}$$

This leads to compact, expressive forms for the dimensionless equations that also allow us to see most clearly the effects of varying the residence-time or the decay time independently. The limiting case of a completely stable catalyst is readily derived by allowing k_2 to tend to zero or τ_2 to become infinite.

Thus (2) becomes

$$\frac{da}{d\tau} = - \alpha\beta^2 + \frac{(1 - \alpha)}{\tau_{res}} \qquad (3a)$$

$$\frac{d\beta}{d\tau} = \alpha\beta^2 - \frac{1}{\tau_2}\beta + \frac{1}{\tau_{res}}(\beta_o - \beta). \qquad (3b)$$

In general this system has two independent variables α and β. Much of our investigation, however, will be concerned with determining the stationary-states of (3). Under these conditions, α and β are not independent but linked. For the stationary-states it is convenient to introduce an additional dimensionless measure of the concentration, the extent of reactant conversion (degree of advancement)

$$\gamma = (a_o - a)/a_o = 1 - \alpha.$$

We shall investigate how the stationary-state extent of conversion γ_{ss} (or equivalently $1 - \alpha_{ss}$ and β_{ss}) varies with the residence-time, and how this dependence is affected by the stability of the catalyst (i.e. by the value of τ_2) and by its inlet concentration (β_o). Variations in τ_2 and β_o affect the stability and character of the stationary-state as well as its location.

3. FLOW DIAGRAMS AND STATIONARY-STATES (ILLUSTRATED FOR SYSTEMS WITH A STABLE CATALYST)

Stationary-states arise when $d\alpha/d\tau$ and $d\beta/d\tau$ go simultaneously to zero, i.e. when

$$-\alpha_{ss}\beta_{ss}^2 + (1 - \alpha_{ss})/\tau_{res} = 0 \tag{4a}$$

$$\alpha_{ss}\beta_{ss}^2 - \beta_{ss}/\tau_2 + (\beta_o - \beta_{ss})/\tau_{res} = 0. \tag{4b}$$

Equation (4) readily provides the relationship between α and β in the stationary-state

$$\begin{aligned}[1 + (\tau_{res}/\tau_2)]\,\beta_{ss} &= (1 + \beta_o - \alpha_{ss}) \\ &= (\beta_o + \gamma_{ss}).\end{aligned} \tag{5}$$

Although we may now substitute (5) into (4) to obtain and manipulate a cubic expression for α_{ss}, there is much to be gained by first examining the equations graphically. We shall illustrate this at first with reference to the system with a perfectly stable catalyst, i.e. for which $\tau_2 \to \infty$. From (5) β and α are even more simply related

$$\beta = (1 + \beta_o - \alpha) = \beta_o + \gamma. \tag{6}$$

Furthermore, this relationship holds at all times, not just in the stationary-state. The problem has been reduced to one independent variable.

3.1 Inflow contains no catalyst ($\beta_o = 0$)

The simplest case to be considered is that in which the inflow is of pure A, so that $\beta_o = 0$. Substituting (6) into (4a) we obtain, in terms of the extent of conversion γ:

$$\begin{aligned}\gamma^2(1 - \gamma) &= \gamma/\tau_{res} \\ R &= L.\end{aligned} \tag{7}$$

The group on the left-hand side of this equation, $R = \gamma^2(1 - \gamma)$ has been plotted as a function of γ in figure 1a. It is a cubic curve, with a point of inflexion at $\gamma = 1/3$ and a maximum at $\gamma = 2/3$. R represents the dependence of the rate of chemical production of B from A on the extent of reaction. The loss-term L, represents the net rate of removal of B by flow. L varies linearly with γ and the gradient of the line is given by the inverse of the residence time.

The stationary-state solutions of (7) are given by the intersections of R and L. When the inflow contains no catalyst, there is always one intersection at the origin, $\gamma_{ss} = \gamma_1 = 0$. This solution clearly satisfies (7). For long residence-times, the loss-line has a low gradient; there are two additional intersections. These correspond to non-zero extents of conversion

$$0 < \gamma_2 \leqq 1/2 \leqq \gamma_3 < 1.$$

As the residence time is decreased, so L becomes steeper and γ_2 and γ_3 move closer together; they merge at $\gamma_2 = \gamma_3 = 1/2$ when R and L become tangential. The condition for tangency is $\tau_{res} = 4$, and multiple solutions exist only if

$$\tau_{res} \geqq 4. \tag{8}$$

For residence-time shorter than this, the loss-line has no intersection other than that at the origin. There is a unique stationary-state.

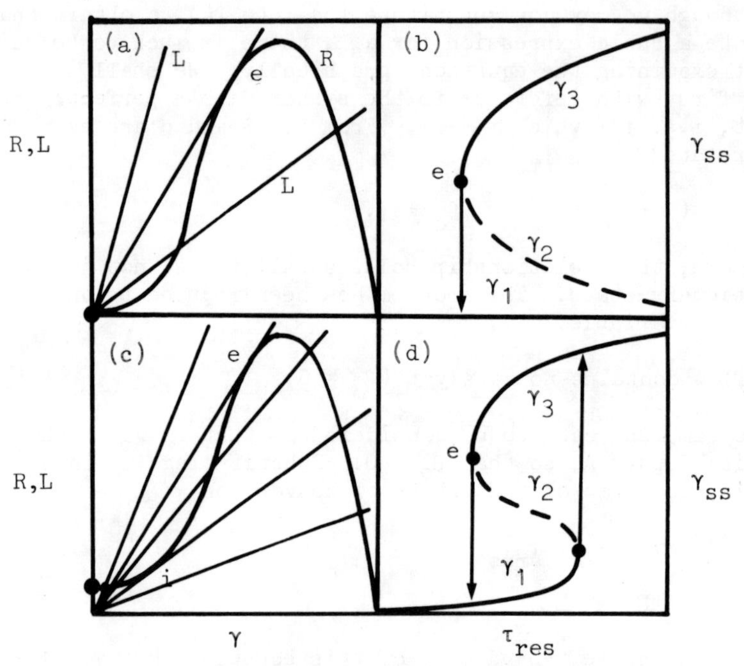

Figure 1. Flow-diagram for cubic autocatalysis A + 2B → 3B:
(a) Dependence of rate of chemical production R on extent of conversion with no catalyst inflow. Also shown are three representative loss or outflow lines corresponding to high flow-rates (steep line) $\tau_{res} < 4$ with one intersection only at $\gamma = 0$; intermediate flow $\tau_{res} = 4$ leading to tangency of R and L and; low flow-rates (low gradient) $\tau_{res} > 4$ with three intersections;
(b) Variation of stationary-state extent of conversion γ_{ss} with residence time swept out in (a);
(c) and (d) Effect of non-zero catalyst inlet concentration.

The dependence of the stationary-state extent of conversion on τ_{res} swept out in this way is shown in figure 1b. As the residence-time becomes very long, $\tau_{res} \to \infty$, the upper solution γ_3 tends to unity (complete conversion), whilst the middle solution γ_2 tends to zero.

3.2 Inflow contains some catalyst ($\beta_o > 0$)

If the concentration of B in the inflow is not zero, the stationary-state condition can be expressed as

$$(\gamma + \beta_o)^2 (1 - \gamma) = \gamma/\tau_{res} \tag{9}$$

generation R = outflow L.

Figure (1c) shows the dependence of R on γ for a small value of β_o. The chemical rate of production of B is now non-zero when $\gamma = 0$; the inflexion occurs at $\gamma \cong (1/3) - (2/3)\,\beta_o$ and the maximum in R occurs at $\gamma \cong (2/3) - (1/3)\,\beta_o$. The main effect, however, is near to the origin. There is no longer an intersection at $\gamma = 0$. There may now be two loss-lines which are tangential to some portion of the production curve, and there will be either one or three intersections of R and L in the positive quadrant depending on the residence time.

For the longest residence times (lowest slopes of L), there is one intersection, lying at high γ (the "ignited" state). As τ_{res} increases, the gradient increases. The lower (ignition) tangency is reached, and there are now three solutions altogether. This tangency occurs when $\gamma_{ss} \cong \beta_o$ (for $\beta_o \ll 1$). The upper tangent is encountered when $\gamma_{ss} \cong \frac{1}{2} - \beta_o$. Here, the two highest stationary-states merge. This corresponds to extinction or wash-out. For the lowest residence times and higher gradients, there is a unique solution close to, but not at, $\gamma = 0$.

The variation of the stationary-states with τ_{res} corresponding to this sequence is shown in figure 1d. There is a region of hysteresis and multistability. This region ends with an abrupt jump or "ignition" from a state of low conversion (γ_1) to one of high conversion (γ_3) as τ_{res} is raised, and with an abrupt jump (an "extinction" or "washout") from γ_3 to γ_1 as the residence-time is decreased.

As the concentration of the autocatalytic species in the inflow increases, the two tangents move closer together. The range of τ_{res} over which hysteresis occurs decreases. When $\beta_o = 1/8$ (i.e. when $b_o = a_o/8$), the ignition and extinction points merge and multiplicity disappears completely.

4. ISOLA FORMATION: THE EFFECTS OF CATALYST DECAY

We now consider the variation of the stationary-state solutions with the residence-time when the catalyst is not indefinitely stable, but decays at a finite rate. The system now has two independent concentrations, but these are linked in the stationary-state by equation (5). Substituting this relationship into 4(a), with $\beta_o = 0$, yields

$$\gamma_{ss}^2 (1 - \gamma_{ss}) = \frac{1}{\tau_{res}} (\frac{\tau_{res}}{\tau_2} + 1)^2 \gamma_{ss} \qquad (10)$$

$$R = L$$

Again R is a cubic curve, and L is linear with γ. The gradient of L depends upon τ_{res} in a more complex manner than found above. It is now a quadratic function. For short residence-times ($\tau_{res}/\tau_2 \ll 1$), the gradient is proportional to $1/\tau_{res}$; L starts with a high slope but becomes less steep as τ_{res} increases. At long residence-times ($\tau_{res}/\tau_2 \gg 1$), the gradient is proportional to τ_{res}; L becomes steeper as τ_{res} increases. In between these extremes, the slope of L passes through a minimum. This occurs when $\tau_{res} = \tau_2$. The minimum gradient, therefore, has the value $4/\tau_2$.

Figure 2 shows the different patterns of response possible for the dependence of γ_{ss} on the residence-time. If the catalyst is not very stable, τ_2 will be short. This means that the minimum slope of the loss-line will be quite high, e.g. L_1 in figure 2a. For such a system, there can be no intersections between R and L except that at the origin, corresponding to zero conversion. This response is shown in figure 2b, for which the $\gamma_{ss} = 0$ axis is a solution for all τ_{res}, and arises when $\tau_2 < 16$.

For the special case $\tau_2 = 16$, the loss-line of minimum slope is given by L_2 (figure 2a), the tangent to R. The zero solution, $\gamma_{ss} = 0$ exists for all τ_{res} again, but for $\tau_{res} = \tau_2$ there is a second, non-zero intersection at the point of tangency, corresponding to $\tau_{ss} = 1/2$. This represents the birth of an isola, or closed curve (figure 2c). The isola grows from a point in the $\gamma_{ss} - \tau_{res}$ (or $\gamma_{ss} - k_f$)plane, its size increases as τ_2 increases as shown in figure 2d. For all other residence-times, L lies above L_2 and does not intersect with R.

If the catalyst is more stable, so that $\tau_2 > 16$, the loss-line can penetrate further into the production curve. L_3 represents the minimum slope for such a system. For the shortest residence-times (or highest flow-rates k_f), the loss-line is steep, lying well above L_3 and also above the tangent line L_2. The gradient decreases as τ_{res} is increased, and when (7) is satisfied L is tangential to R. We then have a non-zero intersection $\gamma_{ss} = \frac{1}{2}$. The slope of L continues to decrease as the residence-time gets longer, and two non-zero intersections γ_2 and γ_3 emerge: γ_2 decreases and γ_3 increases as τ_{res} increases. These solutions move further apart until $\tau_{res} = \tau_2$. Further lengthening of the residence-time causes L to become steeper: γ_2 and γ_3 move closer together, and merge again at the tangency of L and R. Thus the branches of γ_2 and γ_3 form a closed island (figures 2c and d). The ranges of τ_{res} and γ_{ss} over which this isola exists vary with τ_2, becoming larger as the catalyst becomes more stable. In the limit of complete stability ($\tau_2 \rightarrow \infty$), the loss-line has no minimum.

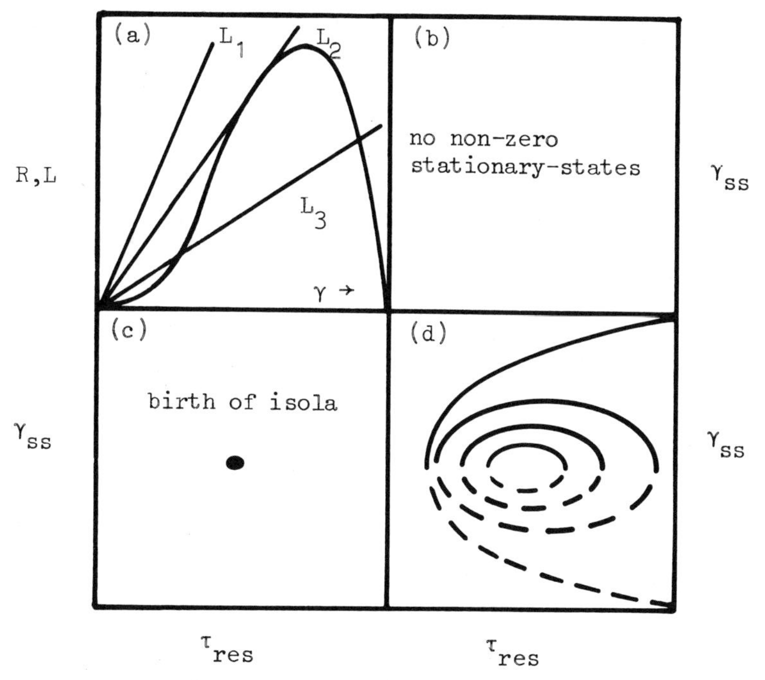

Figure 2. Flow-diagram for cubic autocatalysis A + 2B → 3B with catalyst decay B → C:
(a) dependence of R and L on extent of conversion γ_{ss};
(b) variation of γ_{ss} with τ_{res} for highly unstable catalyst $\tau_2 < 16$, no non-zero intersections;
(c) variation of γ_{ss} with τ_{res} showing birth of isola, $\tau_2 = 16$;
(d) isola in $\gamma_{ss} - \tau_{res}$ plane for various values of $\tau_2 > 16$.

The isola then "opens-up" at its right-hand end; γ_3 tends to unity (complete conversion) and γ_2 tends to zero as τ_{res} becomes infinite ($k_f \rightarrow 0$) and we recover the pattern on figure 1b.

The origin of the isola can also be seen from the stationary-state condition [6]. Noting that $\gamma_{ss} \neq 0$ on the closed curve, we have for γ_2 and γ_3:

$$\gamma_{ss}(1 - \gamma_{ss}) = \frac{1}{\tau_{res}}\left(\frac{\tau_{res}}{\tau_2} + 1\right)^2. \quad (11)$$

This is quadratic both in γ_{ss} and in τ_{res}. Thus for any given residence-time there may be two stationary-state extents of conversion, and for any stationary-state there may be two residence-times. The quadratic dependence on τ_{res} disappears as τ_2 becomes infinite (stable catalyst). Note also that the left-hand side of (9) has a maximum value of 1/4 (when $\gamma_{ss} = 1/2$). Hence, if the minimum value of the right-hand side $4/\tau_2$, is not less than or equal to this (i.e. if τ_2 is less than 16), the solutions of the quadratic cannot be real.

5. DEPENDENCE OF STATIONARY-STATES ON RESIDENCE-TIME: CATALYST IN INFLOW; ISOLA AND MUSHROOM FORMATION

When the concentration of the autocatalyst in the inflow is not zero, the stationary-state relationship becomes

$$(\gamma_{ss} + \beta_o)^2 (1 - \gamma_{ss}) = \frac{1}{\tau_{res}}\left(\frac{\tau_{res}}{\tau_2} + 1\right)^2 \gamma_{ss}. \quad (12)$$

This equation is cubic in γ_{ss}, indicating three or one stationary-state extents of conversion for any given residence-time. It is still a quadratic in τ_{res}, so there exist either two or no residence-times corresponding to any stationary-state solution for γ.

The different patterns of response for this system, with a small value of β_o (i.e. $\beta_o < 1/8$) are displayed in figure 3. The loss-line L has two tangencies (L_2 and L_4) with the production curve R occurring at

$$\gamma_{ss} \cong \beta_o \quad \text{and} \quad \tfrac{1}{2} - \beta_o \qquad (\beta_o \ll 1)$$

with slopes of

$$4\,\beta_o(1 - 2\,\beta_o) \quad \text{and} \quad 1/4 + \beta_o$$

respectively. It is the relative positions of these two tangents and the loss-line of minimum slope that dictates the nature of the $\gamma_{ss} - \tau_{res}$ diagram. There are five possibilities. The minimum slope of the removal line is again $4/\tau_2$, occurring at $\tau_{res} = \tau_2$. The effects of varying τ_2 and the inlet concentration of the catalyst β_o are described below: (i)-(v) correspond to a small value of β_o ($\beta_o \ll 1$); (vi) corresponds to a relatively stable catalyst (high τ_2).

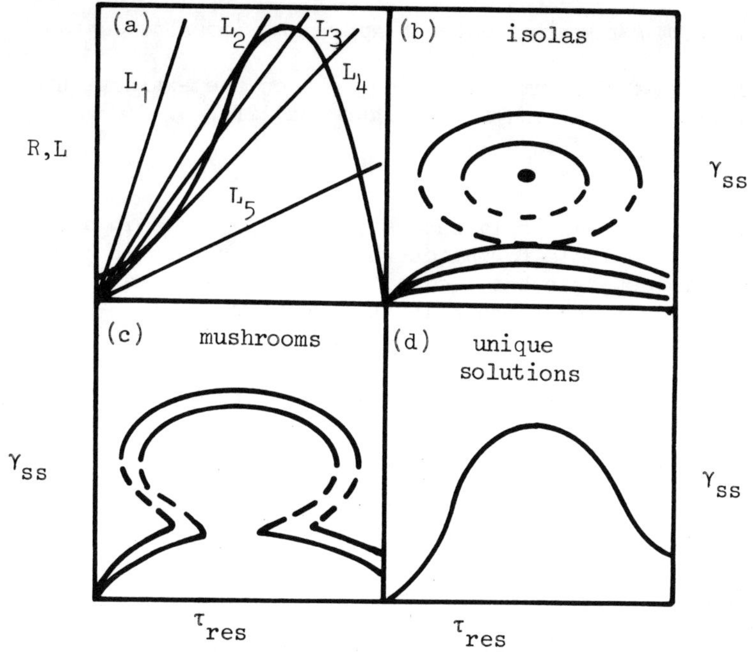

Figure 3. Flow-diagram for cubic autocatalysis A + 2B → 3B with catalyst decay B → C and non-zero inlet concentration $b_o > 0$:
(a) dependence of R and L on extent of conversion γ_{ss};
(b) variation of γ_{ss} with τ_{res} for systems with loss-lines of minimum slope corresponding to L_2, L_3 and L_4 leading to isolas;
(c) mushroom formation, systems with loss-lines of minimum slope represented by L_5;
(d) the effect of large inlet concentration of catalyst when mushrooms open up giving solutions unique for all τ_{res}.

(i) *minimum slope lies above upper (extinction) tangent*

The condition for the loss-line of minimum slope to lie above the upper tangent (e.g. L_1 in figure 3a) is

$$\frac{4}{\tau_2} > \frac{1}{4} + \beta_o.$$

There is a unique intersection γ_1, close to the origin corresponding to low extents of conversion, for all τ_{res}. No other intersections can exist.

(ii) *minimum slope coincides with upper (extinction) tangent*

This corresponds to the birth of an isola, when (11) becomes an equality. This nascent island lies above a branch of solutions γ_1 which arises from the lowest intersection as in (i) above.

(iii) *minimum slope lies between upper and lower tangents*

The loss-line of minimum slope lies between the extinction and ignition tangents (L_2 and L_4 in figure 3a) if

$$4\ \beta_o\ (1 - 2\ \beta_o) < \frac{4}{\tau_2} < \frac{1}{4} + \gamma_o.$$

This gives rise to an isola in the $\gamma_{ss} - \tau_{res}$ plane in an analogous way to isola formation when there is no B in the inflow (§4). The lowest branch of solutions γ_1, lying beneath the isola, now corresponds to non-zero extents of conversion ($\gamma_1 = 0$ for all τ_{res} when $\beta_o = 0$).

The three cases (i)-(iii) are represented in figure (3b).

(iv) *minimum slope coincides with lower (ignition) tangent*

The loss-line of minimum slope and the lower tangent to R are superimposed if

$$4/\tau_2 = 4\ \beta_o\ (1 - 2\ \beta_o).$$

The bottom of the isola, corresponding to the middle intersection γ_2, just touches the lower branch of solutions γ_1. This represents the lowest value of τ_2 for which an isola can exist, and is shown in figure 3(c).

(v) *minimum slope lies beneath lower (ignition) tangent*

The loss-line may lie below the lower tangent if

$$4/\tau_2 < 4\ \beta_o\ (1 - 2\ \beta_o).$$

The variation of γ_{ss} with τ_{res} in this situation is shown in figure 3(c). The isola has now merged with the lower branch to give a mushroom. This has two "extinction" points at which the highest stationary-state γ_3 merges with γ_2 and two "ignition" points at which the lowest solution γ_1 merges with γ_2. At the centre of the mushroom lies a range of residence-time for which γ_3 is the unique stationary-state. The upper branch of solutions is now accessible: it exists but cannot be reached for systems displaying isolas. There are two regions of hysteresis on the mushroom, for each of which multistability occurs.

(vi) *effect of catalyst inlet concentration*

The systems with and without catalyst inflow have many features in common. We readily recover equation (10) from (12) as β_o tends to zero. The main different between the two cases is seen in the lower branch of stationary-states γ_1. Instead of corresponding to no conversion, γ_1 rises from zero to a maximum (when $\tau_{res} = \tau_2$) and then falls as τ_{res} increases. For a given catalyst decay rate τ_2, the maximum value of γ_1 increases as the concentration of B in the inlet is increased. Once the lower branch merges with the bottom of the isola we obtain a mushroom. Further increases in β_o thicken the stem of the mushroom and the regions of hysteresis become smaller. Hysteresis disappears completely when $\beta_o = 1/8$ (i.e. when $b_o = a_o/8$). There then exists a unique stationary-state solution for any residence-time (figure 3d).

6. CHARACTER AND LOCAL STABILITY OF STATIONARY-STATES

The stability and character of the individual stationary-states is reflected by and determines how the system responds to small perturbations. We are, therefore, interested in the time-dependent solutions $\alpha(\tau)$ and $\beta(\tau)$. For small departures from the stationary-state (α_{ss}, β_{ss}), the time-dependence is governed by the linearized equations

$$\frac{d\Delta\alpha}{d\tau} = \frac{\partial}{\partial\alpha}\left(\frac{d\alpha}{d\tau}\right)\Delta\alpha + \frac{\partial}{\partial\beta}\left(\frac{d\alpha}{d\tau}\right)\Delta\beta$$

$$\frac{d\Delta\beta}{d\tau} = \frac{\partial}{\partial\alpha}\left(\frac{d\beta}{d\tau}\right)\Delta\alpha + \frac{\partial}{\partial\beta}\left(\frac{d\beta}{d\tau}\right)\Delta\beta$$

where $\Delta\alpha = \alpha(\tau) - \alpha_{ss}$ and $\Delta\beta = \beta(\tau) - \beta_{ss}$. The solutions of these coupled differential equations has the form

$$\Delta\alpha(\tau) = A \exp \lambda_1\tau + B \exp \lambda_2\tau$$

$$\Delta\beta(\tau) = C \exp \lambda_1\tau + D \exp \lambda_2\tau$$

where the coefficients A,B,C and D are related to the initial perturbation. The decay or growth of the departure from the stationary-state thus depends upon the sign and nature (real or complex) of the exponents λ_1 and λ_2, which are given by

$$\left| \begin{matrix} \frac{\partial}{\partial\alpha}\left(\frac{d\alpha}{d\tau}\right) - \lambda & \frac{\partial}{\partial\beta}\left(\frac{d\alpha}{d\tau}\right) \\ \frac{\partial}{\partial\alpha}\left(\frac{d\beta}{d\tau}\right) & \frac{\partial}{\partial\beta}\left(\frac{d\beta}{d\tau}\right) - \lambda \end{matrix} \right|_{ss} = 0 \qquad (13)$$

This yields a quadratic equation

$$\lambda^2 - P\lambda + Q = 0 \qquad (14)$$

where the coefficients P and Q are given by

$$P = \frac{\partial}{\partial\alpha}\cdot\left(\frac{d\alpha}{d\tau}\right) + \frac{\partial}{\partial\beta}\left(\frac{d\beta}{d\tau}\right)$$

$$Q = \frac{\partial}{\partial\alpha}\left(\frac{d\alpha}{d\tau}\right)\frac{\partial}{\partial\beta}\left(\frac{d\beta}{d\tau}\right) - \frac{\partial}{\partial\alpha}\left(\frac{d\beta}{d\tau}\right)\frac{\partial}{\partial\beta}\left(\frac{d\alpha}{d\tau}\right).$$

The sign of P and Q, and the group $P^2 - 4Q$ determine the stability and character of the stationary-state as summarized in Table 2. If both roots of (14) are negative or have negative real parts, the terms $\exp \lambda_1\tau$ and $\exp \lambda_2\tau$ become vanishingly small as $\tau \to \infty$. Thus the perturbation decays and the system moves back to the stationary-state. If λ_1 and λ_2 are real this decay is monotonic (nodal behaviour), if they are complex it is via a series of damped oscillations (focal behaviour). If either one or both roots λ_1 and λ_2 are positive or have positive real parts, the perturbations grow corresponding to an unstable stationary-state. Again monotonic (nodal) or divergent oscillatory (focal) responses are possible.

P	Q	$P^2 - 4Q$	stability	character	λ_1, λ_2
−	+	+	stable	node	real, negative
−	+	−	stable	focus	complex, negative
+	+	−	unstable	focus	complex, positive
+	+	+	unstable	node	real, positive
+ or −	−	+	unstable	saddle	opposite sign

Table 2 Character and stability of stationary-state extents of conversion.

Figure 4 shows how the stability and character of the stationary-state extent of conversion vary with the residence time for three different combinations of the catalyst lifetime τ_2 and inlet concentration β_o.

(i) *lowest stationary-state*, γ_1

This solution is always stable: λ_1 and λ_2 are both negative. The exponents are also both real. Hence small perturbations from γ_1 decay monontically back to the stationary-state. This behaviour is characteristic of a stable node (sn). No other type of singularity is found for γ_1.

(ii) *middle stationary-state*, γ_2

This solution is always unstable: λ_1 and λ_2 have opposite signs.

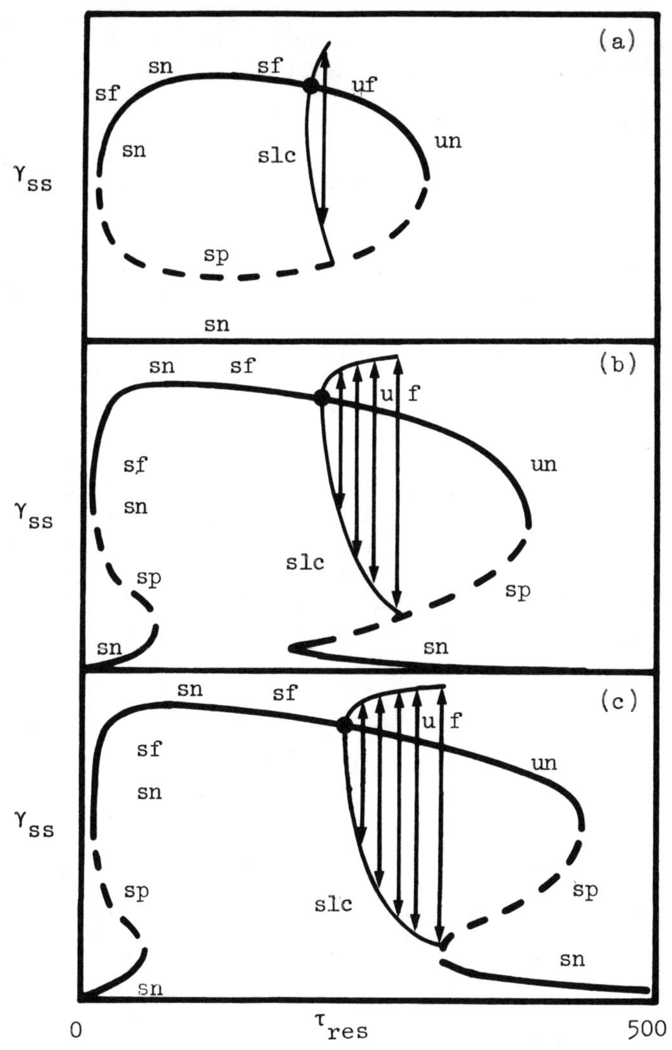

Figure 4. Stability of stationary-state extents of conversion for different catalyst stability and inlet concentrations:
(a) $\tau_2 = 40$, $\beta_o = 0$; gives isola;
(b) $\tau_2 = 40$, $\beta_o = 1/30$; gives mushroom;
(c) $\tau_2 = 40$, $\beta_o = 1/15$; gives mushroom.
Different stabilities and characters are represented by: sn, stable node; sf, stable focus; un, unstable node; uf, unstable focus; and sp, saddle point. Also shown are the envelopes of stable limit cycles (slc) surrounding unstable solutions on upper branch.

The smallest perturbation grows in time, and the system initially near γ_2 immediately diverges either to γ_1 or γ_3. Stationary-states of this type cannot be realized during an experiment. This behaviour is characteristic of a saddle point (sp).

(iii) *highest stationary-state*, γ_3

The stability and character of the highest stationary-state extent of conversion varies with the residence-time and with τ_2 and β_o. The exponents λ_1 and λ_2 may be real or complex and may have either positive or negative real parts. The conditions, in terms of τ_{res}, τ_2 and β_o, required for each of these four possibilities has been given [11] elsewhere. These allow us to distinguish different regions along the upper branch in figure 4(a)-(c). At the shortest residence-times, at the left-hand end of the isola or mushroom, γ_3 is a stable node: λ_1 and λ_2 are real and negative. As τ_{res} increases, the exponents become complex, with negative real parts. Small perturbations now decay back to γ_3 via a series of damped, oscillatory overshoots. This behaviour characterizes a stable focus (sf). For high values of τ_2 (e.g. figures 4b and c, $\tau_2 = 40$), the region of stable foci may be divided by a seconded region of stable nodes. As the residence-time is increased beyond τ_2, the stable focus may become an unstable focus (uf). This occurs when the real parts of λ_1 and λ_2 become positive. At the highest residence-times along the γ_3 branch, the exponents become purely real but remain positive. The stationary-state is an unstable node (un).

7. SUSTAINED OSCILLATIONS AND LIMIT CYCLES

The analysis of the previous section provides detailed information only of the local stability of the individual stationary-states. The overall, global stability of the system is also of interest, especially when the highest extent of conversion is unstable. In particular, there is the possibility of sustained oscillatory responses. These correspond to movement around closed curves or limit cycles in the α-β phase-plane (as opposed to isolas which are closed curves in the $\alpha - \tau_{res}$, or $\beta - \tau_{res}$, plane, and which are not related to time-dependent phenomena).

When the highest stationary-state extent of conversion γ_3 changes from a stable focus to an unstable focus at $\tau_{res} = \tau^*_{res}$, sustained oscillations set in. This is a Hopf bifurcation [16]. It leads [13] to stable limit cycles surrounding the unstable focus for some range of residence-time $\tau_{res} > \tau^*_{res}$. Stable limit cycles can arise even when γ_1 and γ_2 exist and γ_1 is stable.

The amplitude and period of the oscillations can be evaluated analytically in the vicinity of the bifurcation (i.e. for τ_{res}

close to τ^*_{res}). In general, however, it is more convenient to compute these, by integrating the differential equations (3) numerically. Typical oscillatory traces are shown in figures 5 and 6.

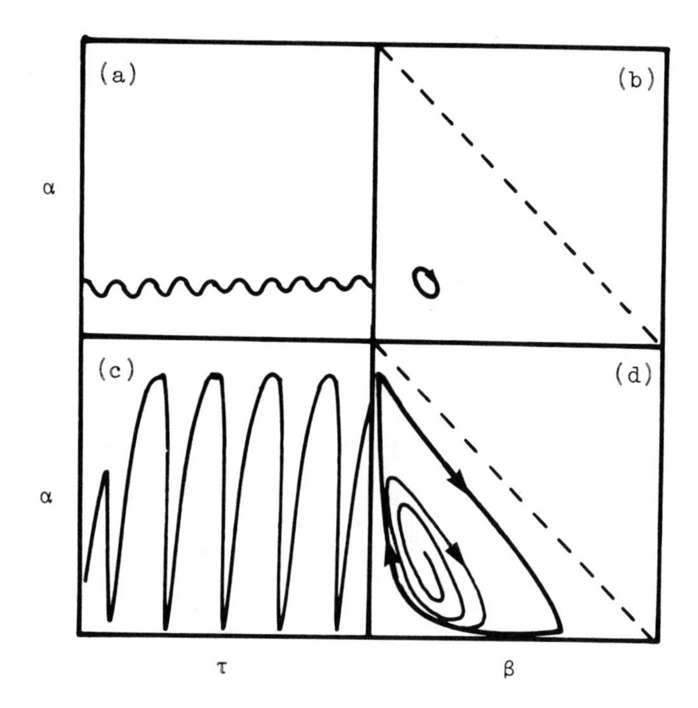

Figure 5. Concentration-time histories $\alpha(\tau)$ and trajectories in α-β phase-plane for sustained oscillations at different residence times for system in figure 4(c), $\tau_2 = 40$ and $\beta_o = 1/30$:
(a) and (b) $\tau_{res} = 205$;
(c) and (d) $\tau_{res} = 250$.

As the residence-time is increased, the pulses get larger and are separated by a longer period (compare 5a and c). The increasing amplitude leads to larger and larger limit cycles surrounding the unstable focus in the α-β phase-plane (5b and d). Figure 5 corresponds to systems for which the unstable focus is a unique solution. The only limitation on the size of the oscillatory pulses is that α and β remain between zero and unity and between zero and $1 + \beta_o$, respectively.

The concentration-time histories in figure 6 correspond to the system whose stationary-states are given by the isola in 4(a). The upper solution is never unique, and the phase-plane always contains at least one stable singularity (the lowest solution γ_1

is a stable node). With τ_{res} = 190, a sequence of sustained oscillations is observed (figure 6a). The phase-plane (figure 6b) contains a stable limit cycle. The separatrices of the saddle point divide the phase-plane into two sections. Trajectories cannot cross from the vicinity of the limit cycle to the vicinity of the stable node. As τ_{res} is increased, the oscillations plunge deeper and the limit cycle grows in the phase plane. The limit cycle gets nearer and nearer the phase plane to the separatrices of the saddle point. Contact annihilates it; the cycle "evaporates". All computed trajectories now ultimately lead to the lower, stable node (figure 6d). Sustained oscillations cannot be achieved. Instead a system starting near the unstable focus shows a short series of divergent oscillations before settling at steady concentrations α and β. This crossing of the separatrices may arise because of numerical instability which, in turn, may mirror the experimental fluctuations likely in, for instance, the flow-rate.

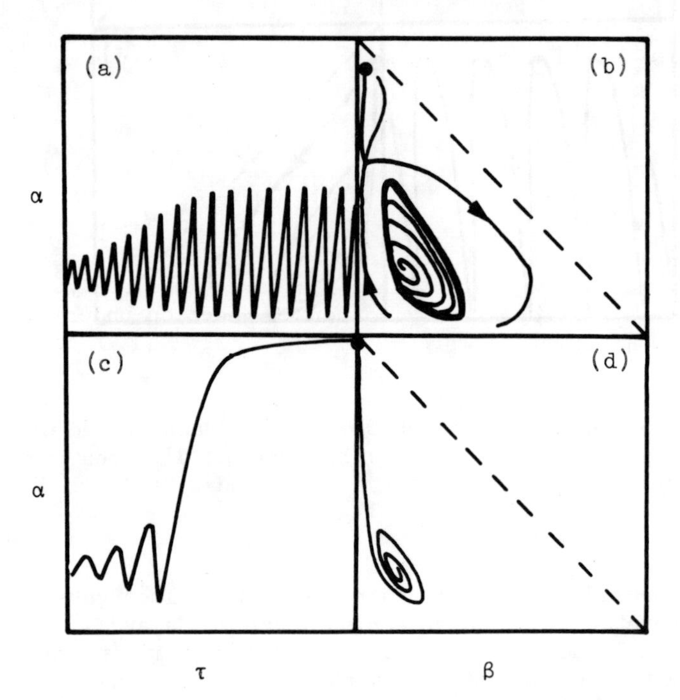

Figure 6. Concentration-time histories $\alpha(\tau)$ and trajectories in α-β phase-plane at different residence times for system in figure 4(a), τ_2 = 40 and β_o = 0:
(a) and (b) τ_{res} = 190 - sustained oscillations and stable limit cycle lying within separatrices of saddle point;
(c) and (d) τ_{res} = 200 - divergent oscillations as system crosses from vicinity of highest stationary-state conversion (lowest α) to lowest extent (highest α).

8. DISCUSSION

Despite its simplicity, our model displays an unexpected variety of stationary-state behaviour, with isolas and mushrooms for the dependence of the extent of conversion on residence time or flow-rate. There is also a rich variety of stability and character, with stable and unstable nodes and foci. It is also the simplest isothermal scheme to show sustained oscillatory responses.

This richness arises primarily from the non-linear dependence of the reaction rate on the concentration of the autocatalyst B. The interaction provides a chemical route for feedback and gives rise to the cubic shape of the production curve R (figure 1). Similar production curves are also displayed by the first-order, irreversible, exothermic reaction $A \rightarrow B$ proceeding non-isothermally in a CSTR, for which feedback is thermal via the heat released; by some enzyme systems; and by heterogeneous reaction at a catalyst surface obeying Langmuir-Hinshelwood kinetics. The cubic rate-law has also been shown to approximate the behaviour observed in the autocatalytic reactions between arsenite and iodate [17] ions and between permanganate and oxalate [18]. Thus our simple system may be regarded as the prototype of a wide range of complicated schemes.

The analogies between our autocatalytic sequence and the non-isothermal system may be taken further. Autocatalysis with a stable catalyst resembles non-isothermal reaction under adiabatic conditions. In these cases, the amount of reactant A consumed is uniquely related to and completely fixed by the amount of catalyst B or heat produced, <u>at all times</u>. These are one-variable systems, the concentration of A cannot be varied independently of the concentration of B or the temperature-excess ΔT, respectively.

Autocatalysis with an unstable catalyst resembles non-isothermal reaction under non-adiabatic operation. In each case the species responsible for feedback (B or the heat released) can be removed from the system by a route independent of the reactant A. These extra channels for removal are chemical decay and Newtonian cooling at the walls respectively. These circumstances lead to two independent variables. Only under stationary-state conditions is the concentration of A directly linked to the concentration of the catalyst or the temperature-excess. In our system we find that the quotient t_{res}/t_2, where $t_2 = 1/k_2$ is a characteristic half-life for the catalyst, plays an important role in determining the stationary-states and their stability. For the non-isothermal reaction, the quotient t_{res}/t_N, where $t_N = V c_p \rho/\chi S$ is the Newtonian cooling time, arises in the same way. Each of these represents the ratio of the rate of decay of B or of the heat evolved to the rate of their removal via the outflow.

Throughout our studies we have found the pictorial representation of flow-diagrams to be particularly helpful. They are of great value, illustrating all aspects of the stationary-state behaviour: unique and multiple solution, hysteresis and jumps between different branches (ignition and extinction or washout), and the effects of reversibility and of non-zero inlet concentration of the autocatalyst. The algebraic analyses are, by comparison, far less transparent, although their forms can also be expressive.

The flow diagrams come into their own in explanations of the occurrence of isolas and mushrooms and the general dependence of the stationary-state extent of conversion on residence-time. Isolas have two extinction points. At either, a state of high conversion changes abruptly to one of low conversion. Both these jumps lead to an increase in the Gibbs free energy of the system; one extinction accompanies an increase in the flow-rate (or a decrease in the residence-time), the other accompanies a decrease in k_f (or an increase in t_{res}).

When we have sufficient B in the inflow - but not too much - the isola opens up to give a mushroom. This pattern displays two extinctions as above. It also has two points of ignition at each of which a state of low conversion changes abruptly to one of high conversion. These transitions lead to an overall decrease in the Gibbs free energy of the system. One ignition accompanies an increase in the residence time (or a decrease in the flow-rate), the other is achieved by decreasing t_{res} (or increasing k_f).

The possibility of extinction accompanying an increase in the residence-time and ignition accompanying a decrease, are associated with the right-hand end (long t_{res}) of the isola and mushrooms shown in figure 4. They provide an explicit and unequivocal counter-example to a proposed but erroneous "rule" [19] concerning the possible responses of the stationary-states to variations in the flow-rate or residence-time.

There is one significant distinction between the extinction occurring at short residence-times (left-hand end of isola or mushrooms in figure 4) and that at long residence times. For the former, the transition is from one branch of stable and steady stationary-states to another. This may be envisaged as the system suddenly sliding off the end of the upper branch as t_{res} is reduced. For the extinction at the long-residence-time end, the stationary-states on the upper branch have already become unstable and are surrounded by stable limit cycles. Thus the extinction now arises experimentally as the abrupt quenching of an oscillatory sequence. Furthermore this quenching occurs when the limit cycle grows large enough for numerically computed trajectories to cross the separatrices of the saddle point corresponding to the middle solution. This extinction occurs before the upper branch of stationary-states vanishes.

9. CONCLUSIONS

1) In a CSTR achieving "exotic" patterns of behaviour requires neither self-heating nor non-uniformity nor elaborate kinetics: rather elementary autocatalysis ($A + 2B \rightarrow 3B$) under isothermal, homogeneous conditions can suffice.

2) This scheme is a simple prototype of numerous more complex systems (e.g. large families of enzyme reactions or surface catalysis), and lies at the heart of the Brusselator and Oregonator.

3) Isolas and mushrooms arise in the dependence of the extent of conversion upon residence time as soon as the finite lifetime of the catalyst is recognized - most simply by the first order decay ($B \rightarrow$ inert).

4) When there are multiple solutions, the lowest branch of stationary-states, corresponding to small extents of reaction, is always stable.

5) When there are multiple solutions, the middle branch is always unstable.

6) When there are multiple solutions, the character and stability of the highest branch varies with the residence-time and with the catalyst stability and inflow concentration.

7) When there is only one solution, equally varied patterns are possible.

8) Unique, unstable stationary-states or those on the highest branch may be surrounded by stable limit cycles. These give rise to sustained oscillations in the concentrations of A and B.

9) Stable limit cycles are found for both isolas and mushrooms: the stationary-state need not be unique and no catalyst is needed in the inflow.

REFERENCES

1. Zeldovich, Ya. B. 1941 Zh. Tekh. Fiz. 11, pp. 493-508; Zeldovich, Ya. B. and Zysin, Y.A. 1941 Zh. Tekh. Fiz. 11, pp. 509-524.
2. van Heerden, C. 1953 Ind. Engng. Chem. 45, pp. 1242-1247.
3. Uppal, A., Ray, W.H. and Poore, A.B. 1974 Chem. Engng. Sci.

29, pp. 967-985; 1976 Chem. Engng. Sci. 31, pp. 205-214.
4. Vaganov, D.A., Samoilenko, N.G. and Abramov, V.G. 1978 Chem. Engng. Sci. 33, pp. 1133-1140.
5. Balakotaiah, V. and Luss, D. 1982 Chem. Engng. Sci. 37, pp. 1611-1623.
6. Aarons, L.J. and Gray, B.F. 1976 Chem. Soc. Rev. 5, pp. 359-375.
7. Prigogine, I. and Lefever, R. 1968 J. Chem. Phys. 48, pp. 1695-1700.
8. Field, R.J. and Noyes, R.M. 1974 J. Chem. Phys. 60, pp. 1877-1884.
9. Lin, K.F. 1979 Can. J. Chem. Engng. 54, pp. 476-480; 1981 Chem. Engng. Sci. 36, pp. 1447-1452.
10. Heinrichs, M. and Schneider, F.W. 1981 J. Phys. Chem. 85, pp. 2112-2116; 1980 Ber. dts. Bunsenges. phys. Chem. 84, pp. 857-865.
11. Gray, P. and Scott, S.K. 1983 Chem. Engng. Sci. 38, pp. 29-43.
12. Scott, S.K. 1983 Chem. Engng. Sci. 38
13. Gray, P. and Scott, S.K. 1983 J. phys. Chem.
14. Gray, P. and Scott, S.K. 1983 Ber. dts. Bunsenges. phys. Chem.
15. Gray, P. and Scott, S.K. 1983 Chem. Engng. Sci.
16. Hassard, B.D., Kazarinoff, N.D. and Wan, Y.-H. 1981 Theory and Applications of Hoph Bifurcation (London Mathematical Society Lecture Note Series no. 41, I.M. James ed.), Cambridge University Press.
17. Papsin, G.A. Hanna, A. and Showalter, K. 1981 J. Phys. Chem. 85, pp. 2575-2582.
18. Reckley, J.S. and Showalter, K. 1981 J. Amer. Chem. Soc. 103, pp. 7012-3.
19. Noyes, R.M. 1981 Proc. Natl. Sci. USA 78, pp. 7248-7250.

THERMOKINETIC OSCILLATIONS AND MULTISTABILITY IN GAS-PHASE OXIDATIONS

J.F. Griffiths

Department of Physical Chemistry, The University,
Leeds LS2 9JT, England.

The roots of spontaneous combustion studies are deeply seated in physico-chemical interactions due to non-linear kinetics, heat release and heat dissipation. Discontinuous responses to slowly varying conditions are features that are most familiar as ignition and extinction, but the transitions can also be between stationary or oscillatory states. This presentation is concerned with oscillations and multiple stabilities that owe their existence to complex thermokinetic interactions.

The realm of interest is the part of gas-phase, thermal oxidation of organic compounds described classically as the "cool-flame" region. Reaction is exothermic and self-heating occurs. But the kinetic response is an unusual one; there are competitive modes of reaction, and the higher temperatures favour a displacement from chain branching to non-branching. The overall effect is to generate a non-Arrhenius dependence of rate on temperature, including a region exhibiting a negative temperature coefficient. Oscillatory modes of reaction may result and the most rewarding method for their investigation is that of non-adiabatic, well-stirred flow (cstr).

Very rich patterns of behaviour are found which, at constant flow, fit into well-defined regions of the pressure-temperature diagram. The regions include two types of stationary states and two types of oscillatory modes in which there are sub-divisions. These phenomena are described. A background to the kinetic framework is presented and the present state of understanding is summarized.

G. Nicolis and F. Baras (eds.), Chemical Instabilities, 91–107.

INTRODUCTION

This presentation is concerned with the steady and oscillatory states, and multiple stabilities that are associated with the gas-phase, thermal oxidation of organic compounds. These phenomena result from an interaction between complex, homogeneous chemical kinetics and internal temperature change due to heat release. They are widely encountered in the oxidation of gaseous or volatile hydrocarbons (natural gas - predominantly methane, and gasoline - predominantly C_6-C_8 containing hydrocarbons, are the most common examples) but they are also a feature of the oxidation of other classes of organic compounds, such as alcohols, aldehydes, ethers and ketones.

The system that has been subjected to the most detailed modern study is the oxidation of acetaldehyde at sub-atmospheric pressures in the temperature range 500-700 K, and it is to be discussed here in order to exemplify the kinetic background of oscillatory phenomena and the principal experimental observations. In chemical terms acetaldehyde oxidation is more simple than most organic oxidations yet it exhibits all of the thermokinetic features so far discovered. It has been a natural choice for study not only because of the relative ease of understanding kinetic mechanisms but also for convenience of experimental investigation.

The existence of thermokinetic oscillations is demonstrated most convincingly under well-stirred flowing conditions [1-4] and the deepest insights have been obtained, within the last decade, using continuously-stirred tank reactors (cstr). But flowing conditions are not a pre-requisite for the existence of thermo-kinetic oscillations even if their existence in closed conditions is ephemeral, and there is a very substantial history of the study of oscillatory "cool flame" phenomena in closed, unstirred vessels [5-8].

The principal common feature between flow-through and non-flow systems is that each is very far from adiabatic. Moreover, heat dissipation rates are very similar since flow rates adopted are sufficiently slow that most of the heat generated is transferred to the surroundings via the walls; only a small proportion is carried by outflowing gases. The major distinction between flowing and non-flowing conditions is that, with flow, the continuous replenishment of reactants permits stationary and oscillatory states to be maintained indefinitely. In a batch mode the reaction proceeds to a state of completion, and the best that can be achieved is a quasi-stationary or quasi-oscillatory approach to the final state. Mechanical mixing ensures spatial uniformity of temperature and concentration throughout the reactants, and so quantitative comparisons between theory and experiment became possible.

There are four parts in this paper. A general chemical background is presented in the first. The second deals with the most simple kinetic foundation that is capable of providing a conceptual framework for interpretation of thermokinetic phenomena. Modern experimental procedures and the principal experimental observations are discussed in the third part. In conclusion, there is a summary of the present state of development and understanding of this topic.

CHEMICAL BACKGROUND

The fundamental property of any combustion process is that heat is released during chemical conversion and that some of this energy (or, in adiabatic conditions, all of it) is retained by the system, causing an internal temperature rise. This constitutes the thermal feedback that is essential to self-sustained combustion. Chemical feedback through atoms and radicals produced in chain branching processes is also a common feature, though not necessarily pre-requisite. (It is, however, a pre-requisite for thermokinetic oscillatory phenomena.)

The mechanisms of combustion propagated by free radicals and atoms are always complex: there is no oxidation process of such simplicity that conversion of the reactant to the final products occurs by a single, elementary step. The complete combustion of acetaldehyde is given by the overall stoichiometry

$$CH_3CHO + 2.5\ O_2 \longrightarrow 2CO_2 + 2H_2O \quad ;$$

$$\Delta H^{\theta}_{298} = -1100\ \text{kJ mol}^{-1} \qquad (1)$$

and the associated adiabatic temperature in the absence of dissociation is approximately 4000 K. But even by the shortest conceivable route, this conversion is accomplished only via very many consecutive and competitive elementary steps (certainly in excess of 50). A number of molecular intermediates are possible and a variety of free radicals are involved in chain propagation (Table 1). In fact, complete oxidation (Equation 1) is achieved only when the system is virtually adiabatic and when the initial reactant ratio $[CH_3CHO]/[O_2] \leqq 2.5$.

In other circumstances some of the molecular intermediates of complete combustion appear as final products and the enthalpy of reaction is much diminished. For example, at temperatures between 1000 and 1100 K the stoichiometry

$$CH_3CHO + 0.50\ O_2 \longrightarrow 1.35\ CO + 0.5\ CH_4 + 0.05\ C_2H_4 + 0.02\ CH_2O + 0.01\ CO_2 + 0.20\ H_2 + 0.65\ H_2O \ ;$$

$$\Delta H^{\theta}_{298} = -180\ \text{kJ mol}^{-1} \qquad (2)$$

Table 1: Selected elementary steps yielding molecular products of acetaldehyde oxidation

$$CH_3CHO + O_2 \longrightarrow CH_3CO + HO_2$$
$$CH_3CHO + HO_2 \longrightarrow CH_3CO + \underline{H_2O_2}$$
$$CH_3CO + O_2 \longrightarrow CH_3CO_3$$
$$CH_3CO_3 + CH_3CHO \longrightarrow \underline{CH_3CO_3H} + CH_3CO$$
$$CH_3CO + M \longrightarrow CH_3 + \underline{CO} + M$$
$$CH_3 + O_2 + M \longrightarrow CH_3O_2 + M$$
$$CH_3 + CH_3CHO \longrightarrow \underline{CH_4} + CH_3CO$$
$$CH_3 + CH_3 + M \longrightarrow \underline{C_2H_6} + M$$
$$CH_3CO_3H \longrightarrow CH_3 + \underline{CO_2} + OH$$
$$CH_3CO_3H \longrightarrow CH_3CO_2 + OH$$
$$CH_3CO_2 + CH_3CHO \longrightarrow \underline{CH_3CO_2H} + CH_3CO$$
$$CH_3CHO + OH \longrightarrow CH_3CO + \underline{H_2O}$$
$$CH_3O_2 + CH_3O_2 \longrightarrow CH_3O + CH_3O + O_2$$
$$CH_3O + CH_3O \longrightarrow \underline{CH_2O} + \underline{CH_3OH}$$
$$CH_3O + CH_3CHO \longrightarrow \underline{CH_3OH} + CH_3CO$$

(molecular products underlined)

seems most appropriate [9] , and at still lower temperatures (*ca.* 600 K)

$$CH_3CHO + 1.10\ O_2 \longrightarrow 0.50\ CO_2 + 0.50\ CH_3CO_2H + 0.15\ CH_3OH + 0.25\ CH_2O + 0.10\ CH_4 + 0.25\ H_2O$$

$$\Delta H^{\theta}_{298} = -370 \text{ kJ mol}^{-1} \qquad (3)$$

These isolated examples cannot convey a detailed picture of the changing networks of elementary steps but they do serve to show that different mechanistic routes predominate at different temperatures. It is not the aim to discuss comprehensive kinetic mechanisms here, but to provide an illustration only of the kinetic features that are of greatest importance in the realm of thermokinetic oscillations.

The key is that degenerate chain branching (via molecular intermediates) should be possible but that competitive processes, favoured by rising temperature, divert this main course from the predominantly branching to a non-branching mode of reaction. In acetaldehyde oxidation at temperatures in the range 500-700 K this competition is attributed to steps involving the acetyl radical (CH_3CO), as follows:

Oxidation route

$$CH_3CHO \dashrightarrow CH_3CO \xrightarrow{O_2} CH_3CO_3 \dashrightarrow CH_3CO_3H \text{ (peracetic acid)} \dashrightarrow \text{free radicals (branching)} \quad (4)$$

Acetaldehyde

$$CH_3CO \xrightarrow{M} CH_3 + CO \dashrightarrow \text{propagation (non-branching)} \quad (5)$$

Decomposition route

Because the activation energy for acetyl radical decomposition is greater than that for its oxidation there is a shift in the balance of the ratio of reaction rates (v_4/v_5) towards decomposition as the temperature is raised.

In isothermal circumstances the sole consequence of this type of chemistry is to generate an abnormal dependence of overall reaction rate on temperature: there is a negative temperature - coefficient of reaction rate (ntcr) over a limited range of temperature. In non-isothermal circumstances, not only is the ntcr a feature but when these kinetic properties are coupled to self-heating, oscillatory states become a possibility.

SKELETON KINETIC SCHEMES: THERMOKINETIC OSCILLATIONS AND MULTISTABILITY

In order to amplify these points and to proceed in a more general context it is appropriate to introduce here the most simple, skeleton, thermokinetic model from which it is possible to predict many of the phenomena observed during the non-isothermal oxidation of organic compounds. The model was devised by B.F. Gray and C.H. Yang about 15 years ago and its properties investigated by them [10,11]. Its representation was reduced to the essential kinetic features determining the behaviour of an intermediate species x, as follows:

		Rate Constant	Activation Energy	Exothermicity
Initiation	Fuel → x	k_i	E_i	h_i
Branching	x → 2x	k_b	E_b	h_b
Termination	x → product	k_1	E_1	h_1
Termination	x → product	k_2	E_2	h_2

The scheme is subject to the condition $E_1 < E_b < E_2$, and the fuel concentration is assumed to be constant. Under non-isothermal conditions a two-dimensional analysis in terms of the temperature T and concentration of x follows from the pair of equations

$$\frac{d[x]}{dt} = k_i[F] - (k_1 + k_2 - k_b)[x] \tag{6}$$

$$c\frac{dT}{dt} = k_ih_i[F] + (k_1h_1 + k_2h_2 + k_bh_b)[x] - \frac{ls}{V}(T-T_a) \tag{7}$$

c is a volumetric heat capacity and s/V is a surface to volume ratio for the reaction vessel. l is a Newtonian heat-transfer coefficient. On the right-hand side of equation (7),the first and second terms combined represent the rate of heat release. The last term represents the rate of heat loss, proportional to the temperature excess within the system. These equations are coupled via the linear concentration term [x] and the non-linear (Arrhenius) temperature-dependence of the rate constants.

The linear dependences of branching and termination on the concentration of the intermediate x make a detailed stability analysis of this system relatively straight-forward. If the scheme were to be elaborated to match real experimental systems or established kinetic features, there would be a consequent loss of flexibility. The full richness of behaviour that is predicted in such a system is revealed by investigating the properties of singularities in the T-[x] phase-plane. However, some elementary features can be deduced in terms of the thermal properties originating from the stationary state solutions of the equation

$$k_ih_i[F] + (k_1h_1 + k_2h_2 + k_bh_b)[x]_{ss} - \frac{ls}{cV}(T-T_a) = 0 \tag{8}$$

$$\text{or} \quad k_ih_i[F] + \frac{(k_1h_1 + k_2h_2 + k_bh_b)}{(k_1 + k_2 - k_b)}k_i[F] - \frac{ls}{cV}(T - T_a) = 0$$

The thermal diagram offers the most simple representation of the location of and access to stationary states and it paves the way to a description of events as they are commonly observed experimentally. Depicted in Figure 1 are qualitative representations of the way in which the heat-release rate (R) and the heat-loss rate (L), described by Equations (6) and (7), vary with temperature over a moderate range. The main feature of the heat-release rate is its rise to a maximum and a fall (in the realm of the ntcr) followed by a further growth as the temperature increases. (If we were to trace R to very much higher temperatures it would eventually reach a limiting rate determined by the condition $E_{overall} < RT$.) L is a linear function of T and its intersections with R determine the number of singularities that

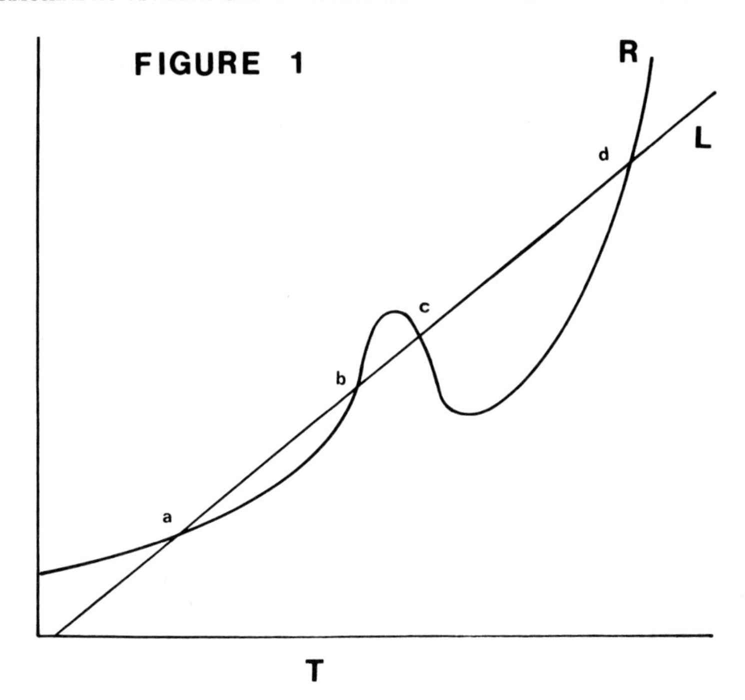

are possible and the temperature at which each is located. The maximum number of singularities in the present scheme is five; four of them are represented in Figure 1, the fifth is located at much higher temperatures. Of those displayed in Figure 1, b and d are unstable states, a and c are stable and may be realized experimentally.

Suppose that the vessel temperature is sufficiently low (T_1) and the heat-transfer coefficient (the gradient of L) sufficiently high that the only accessible state is a. The observed heat-release rate will correspond to that at the single intersection of R and L and the measured temperature excess will correspond to that displayed in Figure 2a. There are three ways in which this stationary reaction rate may be increased; (i) by raising the vessel temperature ($T_1 \rightarrow T_2$ in Figure 2b), (ii) by raising the reactant pressure ($R_1 \rightarrow R_2$ in Figure 2c), (iii) by diminishing the heat-transfer coefficient ($l_1 \rightarrow l_2$ in Figure 2d). A smooth variation in each of these control parameters will eventually bring about a discontinuous transition from state a to state c (here illustrated only by the change of vessel temperature, $T_1 \rightarrow T_3$ in Figure 2e). Within the range of temperatures $T_3 \rightarrow T_4$ (Figure 2f) the observed behaviour will correspond to that of state c. Another discontinuity occurs when the temperature is raised beyond T_4 and the expected response would be a transiton to high temperature ignition. If the vessel temperature is diminished from T_4 the pathway of states observed during the upward

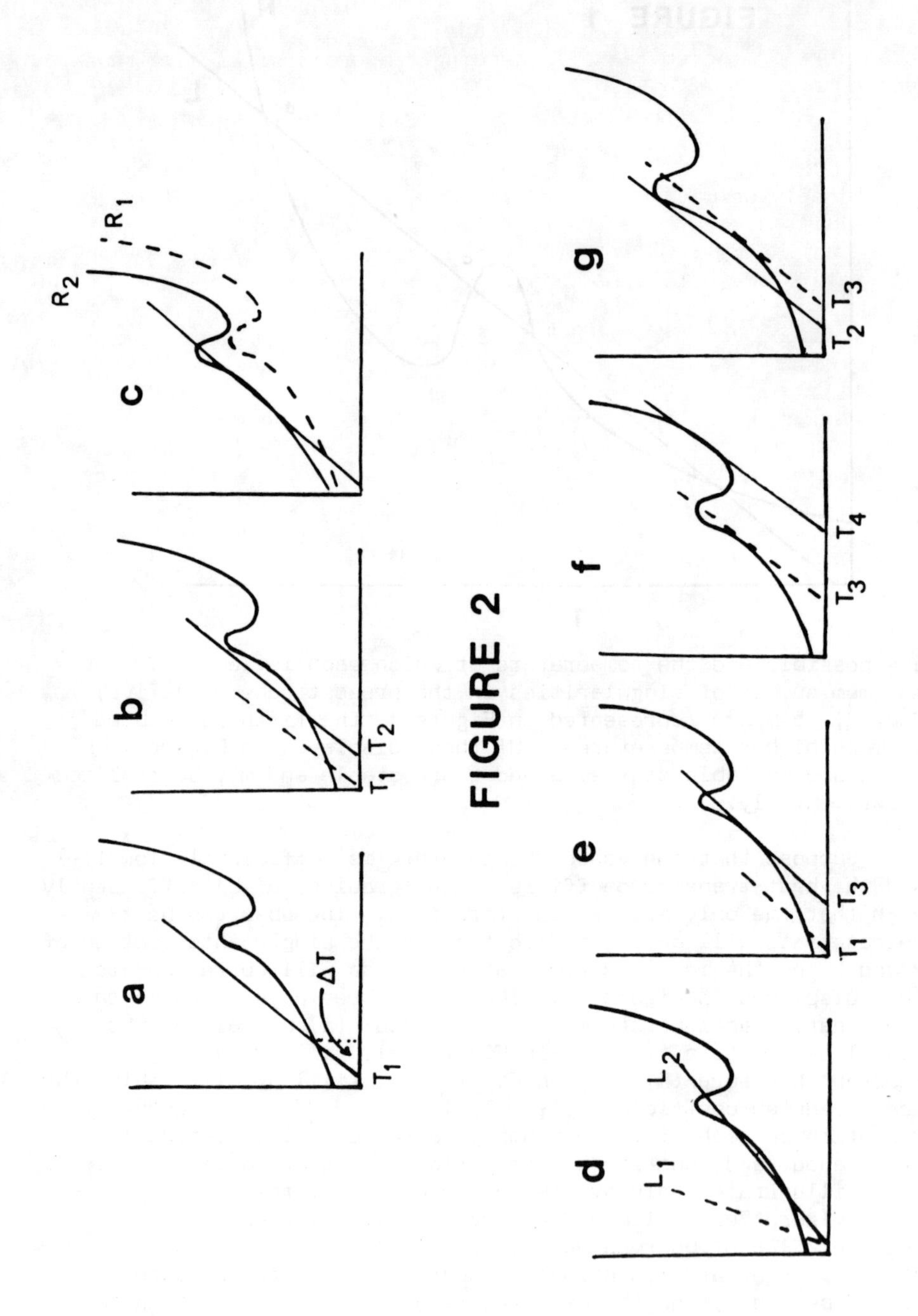

FIGURE 2

traverse is retraced as far as T_3. In the range of temperatures $T_3 \rightarrow T_2$ the system now exhibits behaviour that was not accessible in the course of ascending temperature: we have thus located a realm of bistability and hysteresis (Figure 2g). Extinction occurs from state c to state a, at a temperature marginally lower than T_2.

Finally we may note that (i) there is no criticality and no bistability associated with isothermal reaction ($l_1 \rightarrow \infty$ in Figure 2d), and (ii) only high temperature ignition occurs when the system is virtually adiabatic ($l_2 \rightarrow 0$ in Figure 2d).

This analysis, based on the single variable T, shows that multiple critical transitions ("ignition" and extinction) and regions of bistability are possible in non-isothermal reaction when complex kinetics occur. It cannot reveal the types of behaviour associated with the singularities of the Gray and Yang scheme, but the proper two-dimensional (T-[x]) stability analysis establishes the following:

state a: stable node (sn)

state b: saddle point

state c: stable focus (sf) or unstable focus surrounded by limit cycle (uf)

state d: saddle point

When this information is patched into the pattern of events predicted by the thermal diagrams, a picture emerges illustrating the mode of behaviour in terms of their location in a (p-T_a) ignition diagram (Figure 3). All of this is possible from a qualitative algebraic basis. Still richer patterns of behaviour are found when the Gray and Yang model is analysed numerically [11,12].

MODERN EXPERIMENTAL METHODS AND RESULTS

Apparatus and Procedures

At the heart of present-day experiments is the cstr; a typical example is shown in Figure 4. It is a Pyrex-glass sphere (0.5 dm^3) containing a stainless-steel, mechanical stirrer driven by an external rotating magnet. Reactants are pre-heated separately and mixed at entry to the vessel. Probes for monitoring the rate and extent of reaction are inserted through entry arms so that they protrude into the vessel. The whole is mounted in an oven to maintain a uniform, controlled surface temperature. This type of vessel is very satisfactory for the

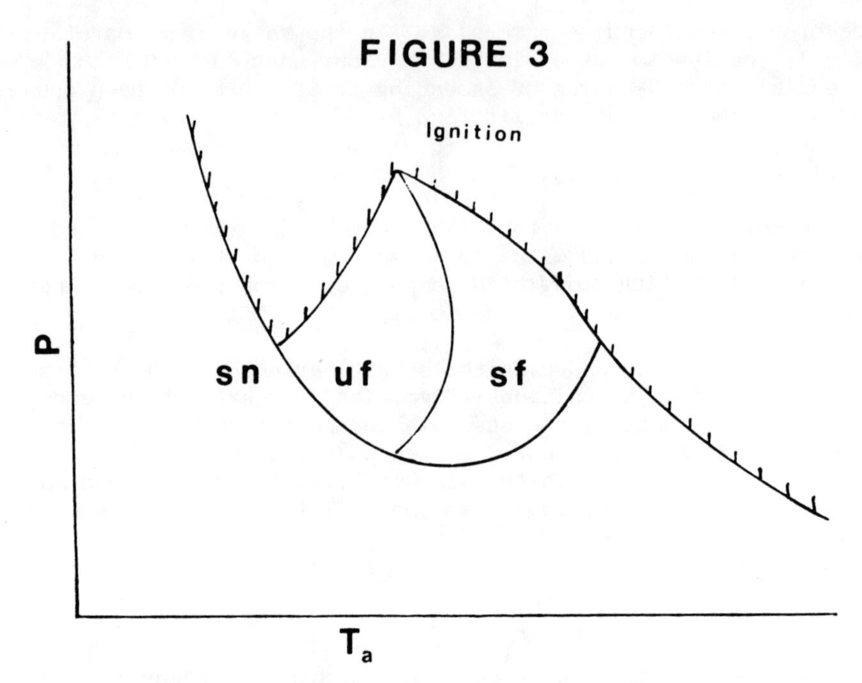
FIGURE 3
Ignition
P
sn
uf
sf
T a

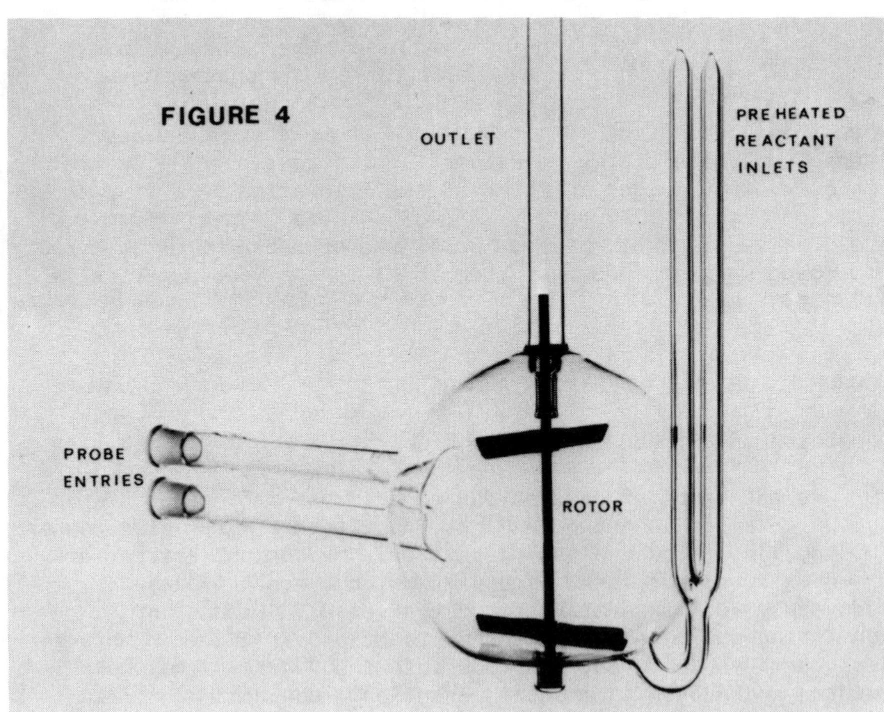
FIGURE 4
OUTLET
PRE HEATED
RE ACTANT
INLETS
PROBE
ENTRIES
ROTOR

study of reaction at sub-atmospheric pressure and mean residence times in excess of 1 s. Reactant compositions, flow rates and the total pressure are controlled carefully by auxiliary, external apparatus [3].

In order to investigate modes of behaviour and transitions between them it is most convenient to establish the reactant pressure, composition and residence time and to vary the vessel temperature either very slowly and continuously or in a stepped fashion, permitting sufficient time to elapse for stability to be re-established after each step.

The temperature within the system above that of the vessel wall is measured using a very fine, responsive thermocouple (25 μm, Pt-Pt/13% Rh wire, coated with a thin layer of silica). The extent of conversion of reactants and the product composition is measured by continuous withdrawal of samples via a very fine probe to a mass spectrometer. Light output associated with oscillatory "cool flames" (a chemiluminescent emission from excited formaldehyde) is detected by a photomultiplier through a window in the wall of the oven. Continuous and simultaneous measurements are thus made, and since the system is "well-stirred", what is measured at one location by the thermocouple, or by withdrawal to the mass spectrometer, is the same as that at any other point. These measurements are time-dependent and so at a stationary state (sn or sf in the phase-plane) they will be invariant; at an oscillatory state (uf in the phase-plane) periodic phenomena will be observed.

Results

We may assemble the information gained from the time-dependent measurements in terms of modes of behaviour in the (p-T_a) ignition diagram, and it is most convenient to present the experimental ignition diagram first and to describe the behaviour within each region subsequently. The (p-T_a) ignition diagram for $CH_3CHO + O_2$ at a mean residence time of 3 s in a spherical (0.5 dm^3) cstr is shown in Figure 5. It is divided into five principal regions of which two are stationary states (I and V) and three are oscillatory modes (II, III and IV). The main features of each of the stationary states are summarized in Table 2. There are marked contrasts between them, the most important of which are (i) the responses to small disturbances which signify the nature and (ii) the temperature dependences of each state. Heat-release rates, derived from temperature excesses, are displayed in Figure 6. Also shown is a region of bistability and hysteresis at the upper temperature bound of region I. "Ignition" occurs here to the oscillatory mode IV and extinction occurs from it. There is no experimentally accessible stationary-state within IV. The present measurements do not extend to sufficiently high vessel

Table 2: Stationary states (I and V) compared

	I	V
Pressure range/kPa	0-20	5-20
Temperature range T_a/K	430-500	600-650
Self-heating T/K (as T_a is increased)	0 → 30	100-60
Heat release rate W dm^{-3}	0-20 (rising)	80-50 (falling)
Extent of reaction	<20%	>60%
Abundant carbon oxide	CO_2	CO
$(-\Delta H)$/kJ mol^{-1}	≈370	≈350
Nature of stationary states (response to disturbance)	stable nodal point (monotonic)	stable nodal focus (oscillatory)
Light emission	none	steady weak blue glow

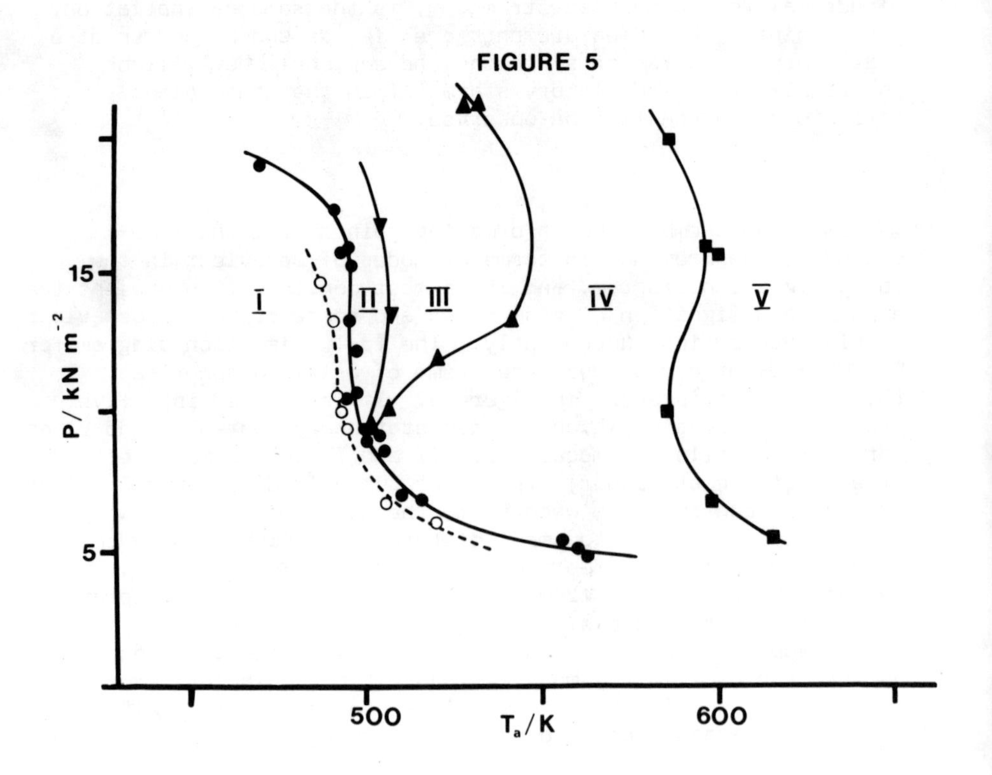

FIGURE 5

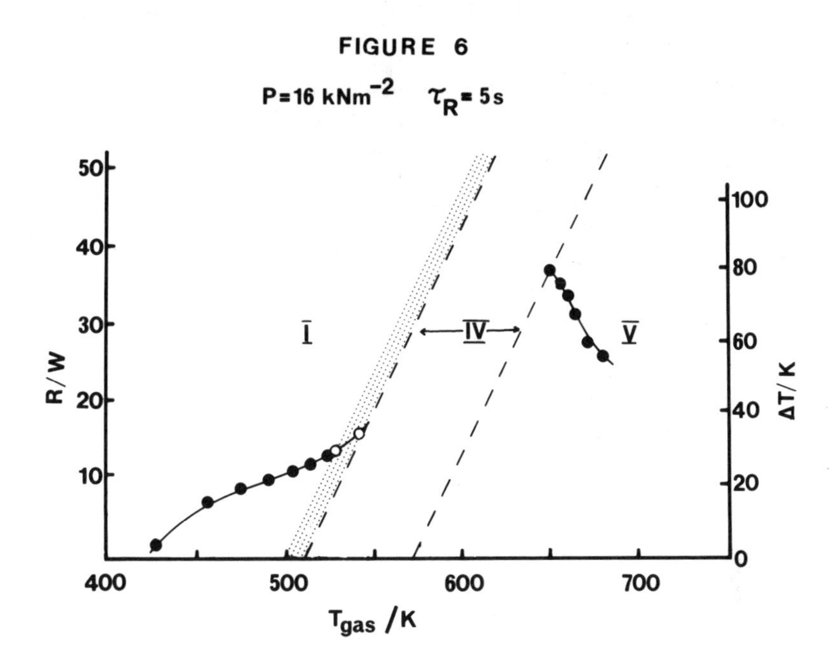
FIGURE 6
P=16 kNm-2 τR= 5s
R/W
ΔT/K
I
IV
V
Tgas /K
400
500
600
700
10
20
30
40
50
0
20
40
60
80
100

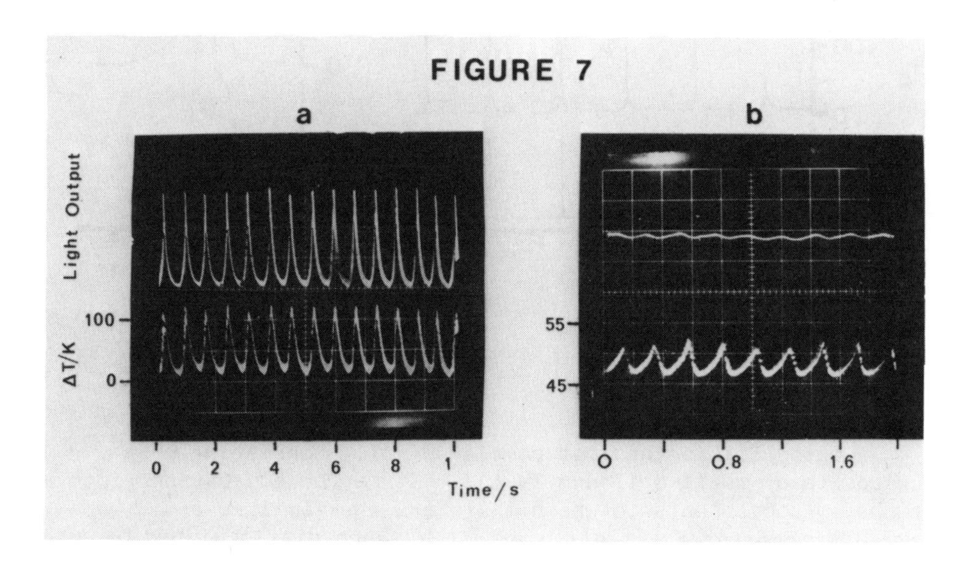
FIGURE 7
a
b
Light Output
ΔT/K
100
0
55
45
0
2
4
6
8
1
Time / s
O
O.8
1.6

temperature to locate a minimum heat-release rate and a subsequent increase.

Region IV of the ignition diagram is occupied by oscillatory cool flames. Their periods are longest and amplitudes highest ($\Delta T \leqq 200$ K) at the lowest vessel temperatures (Figure 7a). As the vessel temperature is raised both the periods and amplitudes diminish (Figure 7b) and the location of the boundary between IV and V is marked only by the disappearance of oscillations in the light output and temperature change.

Hot ignition is observed in region II and has two outstanding features. The first is that it occurs in the present circumstances as an oscillatory relaxation phenomenon (Figure 8a). The residence time would have to be diminished by several orders of magnitude if a stationary flame were to be stabilized in the vessel. The second is that each repetitive ignition has a complex two-stage structure (Figure 8b), the first stage of which has all of the physical and chemical features of the cool flame [4].

FIGURE 8

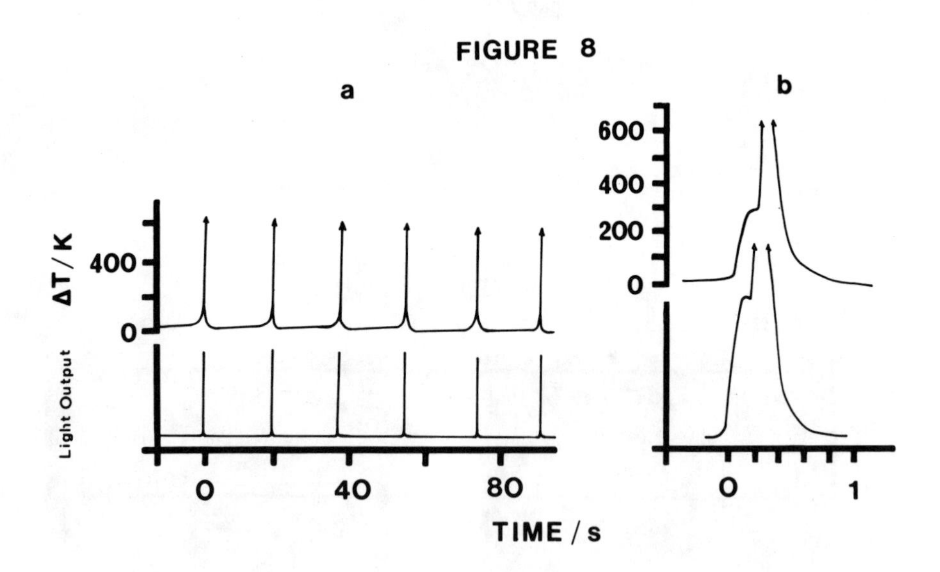

Region III is the most complex of all, consisting of a composite of regions II and IV [4]. It may be sub-divided into a number of components in which there are the families of oscillations (n cf + ign + n cf + ...) each distinguished by a different magnitude of n (Figure 9). n = 1 at the boundary between regions II and III and rises to 5 at the boundary between regions III and IV. As n increases the range of temperatures within which a particular family is stable diminishes and is the order of the experimental control of T_a when n = 5. With improved experimental technique, it may be possible to distinguish still longer trains of cool flame oscillations within each family. The cycles of events (n cf + 1 ign + (n+1) cf + 1 ign + n cf + ...) have also been established, and their location in the (p-T_a) ignition diagram is close to the confluence of the boundaries for region II, III and IV. Their existence as a facet of experimental control cannot yet be discounted. Sequences in which the multiplicities are reversed, namely (n ign + 1 cf + n ign + ...) have not been observed, nor have period doubling or chaos yet been seen.

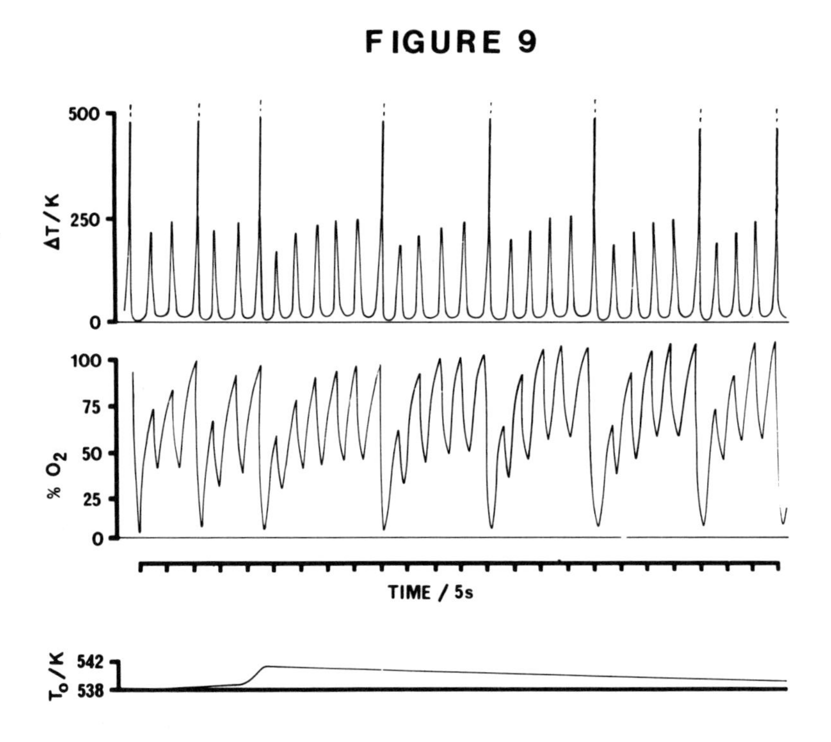

Figure 9 displays the transition through families of hybrid oscillatory waveforms as the vessel temperature is raised rapidly by 4 K and then allowed to diminish slowly. Temperature excesses and oxygen consumption are displayed.

CONCLUSIONS

This brief overview sets out the main observations from present-day experiments, drawing particularly on studies of acetaldehyde oxidation. Within the class of organic oxidations, the phenomena are believed to be quite general, and other systems are being similarly investigated. The academic stimulus in the context of chemical instabilities is to provide a general thermo-kinetic framework to explain the known features and to predict others. Practical needs, concerning auto-ignition and "knock" in engines, might suggest that a greater emphasis on investigations of "higher" hydrocarbons is required but it is more simple chemical systems that are most likely to reveal their secrets.

The broad principles underlying the existence of stationary states (both nodal and focal stabilities), simple oscillatory states and multiple stabilities, involving competitive chain branching and non-branching modes and the interaction with heat release and heat loss, are clear. The more complex oscillatory waveforms do not emerge naturally from this simple recipe, and greater chemical complexity, not yet taken into account, is probably their cause.

Gray and Yang were prudent in their quest to discover the most simple skeleton kinetic scheme and to confine their attention to a two-variable model; investigation of the properties of this system is still proving to be instructive [12]. But the match with the kinetic features of organic oxidations is not perfect, a quadratic termination step being one major omission. There is a need for similar theoretical attacks on more detailed schemes, although the price to be paid may be loss of clarity, tractability and generality as the complexity increased and the number of variables escalates to meet the demands of realism. Numerical analysis is an extremely valuable tool and useful applications have already been made [13,14]. However, its success demands not only confidence in detailed kinetic interpretations and certainty of the kinetic data but also comprehensive quantitative experimental data for its validation. The present cstr studies are able to satisfy that need.

Finally, since this contribution forms part of a broad-based discussion of chemical instabilities, it may be of interest to summarize the very long history of cool flame studies and to pinpoint some of its landmarks, as follows:

1816 Discovery by Davy

1880 First experimental investigations (Perkin)

ca. 1928 Luminescence related to CH_2O^* (Kondratiev and Emeleus)
Engine "knock" linked to cool flame chemistry
ntcr discovered in hydrocarbon oxidation (Pease)

1938	Periodicity of propane oxidation found (Newitt and Thomas) Qualitative links established between periodicity and ntcr (Pease)
1940	Primitive thermokinetic theories (Salnikov)
1965-69	Unified chain-thermal theories (Gray and Yang)
1971	CSTR experiments begin Detailed thermokinetic modelling inaugurated

ACKNOWLEDGEMENT

The author wishes to thank NATO for a travel grant.

REFERENCES

1. Felton, P.G. and Gray, B.F., 1974, Combust. Flame 23, pp.295-310.
2. Caprio, V. Insola, A. and Lignola, P.G., *Sixteenth Symposium (International) on Combustion*, The Combustion Institute 1976, pp.1155-1164.
3. Gray, P., Griffiths, J.F., Hasko, S.M. and Lignola, P.G., 1981, Proc. Roy. Soc. A374, pp.313-339.
4. Gray, P., Griffiths, J.F., Hasko, S.M. and Lignola, P.G., 1977, Combust. Flame 43, pp.175-185.
5. Lewis, B. and Von Elke, G., *Combustion, Flames and Explosions of Gases*, Academic Press, 1961, pp.90-198.
6. Shtern, V. Ya, *The Gas-Phase Oxidation of Hydrocarbons*, Pergamon, 1964.
7. Minkoff, G.J. and Tipper, C.F.H., *Chemistry of Combustion Reactions*, Butterworths, 1962, pp.200-230.
8. Ben-Aim, R. and Lucquin, M., *Oxidation and Combustion Reviews, Volume 1*, Editor Tipper, C.F.H., Elsevier, 1967, pp.1-46.
9. Colkett, M.B., Naegeli, D.W. and Glassman, I., *Sixteenth Symposium (International) on Combustion*, The Combustion Institute 1976, pp.1023-1039.
10. Gray, B.F. and Yang, C.H., 1969, Trans. Farad. Soc., 65, pp.1614-1622.
11. Yang, C.H. and Gray, B.F., 1969, J. Phys. Chem., 73, pp.3395-3406.
12. Gray, B.F. and Gonda, I., 1983, Proc. Roy. Soc., to be published.
13. Halstead, M.P., Prothero, A. and Quinn, C.P., 1971, Proc. Roy. Soc., A322, pp.377-403.
14. Felton, P.G., Gray, B.F. and Shank, N., 1976, Combust. Flame, 27, pp.363-376.

DYNAMICS OF PREMIXED FLAMES

Paul CLAVIN

Département de Combustion du LA 72
Université de Provence. Centre St Jérôme
rue H.Poincaré, 13397 Marseille cedex 13, France.

1. INTRODUCTION

Let us consider the motion of flame fronts in flows that can be unsteady and non-uniform. When the length and time scales of the initial flow are larger than those associated with the planar flame (the flame thickness d and the transit time d/u_o where u_o is the laminar flame speed), the flame front can be considered, in first approximation, as a surface of discontinuity whose the motion is controlled by two distinct factors : the normal burning velocity u_n associated with the mass flux of fresh mixture crossing the front and the value at the front of the flow velocity field. Each point of the front moves with a velocity equal to the difference between the values (at this point) of the upstream flow velocity and of the normal burning velocity oriented toward the fresh mixture in the direction normal to the front.

As soon as one is concerned with wrinkled fronts and/or inhomogeneous flows, the flame **cannot be described by a pure reaction-diffusion model**. Because of the gas expansion produced by the temperature increase in the preheated zone, the streamlines are deflected across the tilted front and a strong coupling with hydrodynamics is developed. When the size Λ of the wrinkles of the front is large compared to the flame thickness d, the corresponding fluid mechanical effects can be splitted in two distinct parts :
-i) The flame structure is locally influenced by the convective transfert produced in the preheated zone by the deflection of the streamlines
-ii) Outside the flame where the gas density ρ and the temperature T are uniform, the flow field is also modified (from its ini-

G. Nicolis and F. Baras (eds.), Chemical Instabilities, 109–115.

tial value without combustion) upstream and downstream the flame on a distance Λ from the front.

The first effect i) results in a change of the normal burning velocity $u_n \neq u_o$. It must be noticed that, in addition to the convective transfert produced by the gas expansion, the diffusion fluxes of heat and mass play also a great part in this modification to flame structure of wrinkled fronts.

The second effects ii) results in a modification to the gas flow at the front.

Both of these effects i) & ii) influences the motion of the front, but the second is the stronger one.

2. THE HYDRODYNAMICAL INSTABILITY

When the modification to flame structure is neglected $u_n = u_o$, the flame could be considered as a passive surface in the sense that the motion is completely prescribed by the value of the flow at the front. Even in this case, the second effect ii) produces a strong hydrodynamical feedback in the motion of the front. This effect was first described in the pioneering works of Darrieus (1938) and Landau (1944) who computed the flow field induced by the front wrinkling when $u_n = u_o$ and $d=0$. The analysis was carried out at the linear approximation in the amplitude of the front corrugations and the induced flow velocity was found to be in phase with the front wrinkles. The resulting motion of the front reveals a strong instability mechanism of planar flames in uniform flows. The least fluctuation around the planar steady state solution (described by Clavin and Linan 1983) is amplified under the hydrodynamical effect ii). As the wavelength becomes shorter, the front is more unstable and Darrieus and Landau concluded that planar fronts freely propagating in uniform mixture cannot exist. The instability mechanism appeared to be so strong that they conclude also that the combustion must be a self turbulizing phenomena. In fact, only two parameters being involved in this theory, the gas expansiòn parameter γ, $\gamma = (\rho_u - \rho_b)/\rho_u$ (the subscripts u and b refer to the unburnt and burnt gases respectively), and the laminar flame speed u_o, the growth rate σ of the instability is found to be proportional to the modulus k of the wave vector of the front wrinkles:

$$\sigma = \sigma_1(\gamma)\ u_o k \qquad (1)$$

Where $\sigma_1(\gamma)$ is a positive adimentional quantity vanishing only in the irrealistic limit $\gamma \to 0$ (zero gas expansion, $\rho_u = \rho_b$).

3. THE EFFECTS OF MODIFICATION TO FLAME STRUCTURE

The first attempt to take into account the effect of the modification to flame structure was performed in the fiftyth by Markstein (see his review paper 1965) who assumed the following phenomenological relation between u_n and the mean radius of curvature* of the front R.

$$\frac{u_n - u_o}{u_o} = \frac{\mathcal{L}}{R} \tag{2}$$

Where $\mathcal{L}$ is a phenomenological length (called Markstein length) that was assumed to be proportional to the flame thickness d and whose the expression is a characteristic of the reactive mixture. According to the semi analytical theory of Eckhauss (1961), Markstein modified the eq(2) to try to take into account the fact that the flame structure must be modified also by the inhomogeneity of the local flow. Such an effect is well known for planar fronts stabilized in stagnation point flows. It can easily be anticipated from eq(2) that the effects associated with the modification to flame structure can only change the dispersion relation (1) through ak^2 term. And thus the large wavelength can never be stabilized by such a mechanism. This is easily understood for the diffusion processes that are well known to be associated with a relaxation time of the order $(Dk^2)^{-1}$ where D is a diffusivity. Concerning the effect i), it is sufficient to notice that, according to (1), the amplitude of the induced flow field (computed by Darrieus and Landau) is proportional to k ; but it is the gradient of the flow velocity that is the relevant quantity involved in the modification to flame structure by convective fluxes ; thus the effect i) induces also ak^2 term.

The first analysis of the wrinkled flame structure was carried out by Barenblatt, Zeldovich and Istratov (1962) but in the diffusive-thermal model where the gas expansion effects i) and ii) are neglected. This model was extensively used these ten last years to culminate in the derivation by G. Sivashinsky (1977) of a non linear differential equation for the flame motion describing a self turbulizing behavior of the cellular structures (Michelson & Sivashinsky 1977). The main interest of this model is to provide us with a simple framework for studying systematically all the dynamical effects that can possibely be produced by the diffusion of heat and mass. The asymptotic technique applied to solve this model in the limit of large values of the Zeldovich number $\beta \to \infty$, is presented in the paper of Joulin & Clavin (1979) that is devoted to the dynamical effects in the presence of heat losses that can produce the thermal extinction. Travelling and spinning waves as well as oscillatory fronts have been predicted

---

* R is positive when the front is concave toward the unburnt gases.

by this model (for a review see the look of Buckmaster and Ludford 1982 and the review article of Sivashinsky 1983). But even as modified by Sivashinsky to take into account a weak gas expansion, this model underestimates the hydrodynamical effects that has been proved to play a dominant role (see eq(1)).

4. STABILITY LIMITS OF FLAMES PROPAGATING DOWNWARD

Recently the coupling between diffusion and hydrodynamics has been properly taken into account for describing the wrinkled flame structure in an anlytical work by Clavin & Williams (1982). The asymptotic expansion $\beta \to \infty$ is used together with a multiscale method based on the assumption $\varepsilon = d/\Lambda$ smaller than unity. The corresponding result was used by Pelcé & Clavin (1982) to study the stability limits of planar fronts propagating downward. The results can be summarized as follows :

a) In the approximation of a one step overall chemical reaction, the modification to flame structure by wrinkling is predicted to be a stabilizing mechanism for most of the ordinary hydrocarbon-mixtures whatever the equivalence ratio may be. The only exception could be mixture of very light reactive components (as hydrogen) diluted in nitrogen. This conclusion contradicts the result obtained by the diffusive-thermal model and appears as a typical effect of the mechanism i).

b) The gas viscosity has a neutral effect on the stability properties of planar fronts.

c) The acceleration of gravity g associated with the effects of the modification to flame structure by wrinkling can counterbalance the hydrodynamical instability for all the wave numbers when the flame velocity is low enough. The cellular threshold is predicted to be observable for flame velocity u_o varying between 5cm/s and 17cm/s with rich mixture of ordinary fuel for low flame velocities ($u_o < 12$cm/s) and with lean composition for higher velocities.

d) The cell size at the threshold can be expressed in terms of only the variables g, u_o and γ ; the detailed properties (chemical kinetics, transport processes...) do not enter into the final expression.

These predictions are in good agreements with the recent experiments of Quinard & al (1983). Furthermore the induced velocity field has been recently recorded in the unburnt mixture by Searby & al (1983) in the case of stable fronts stabilized in weakly "turbulent" flows. As predicted by the theory for planar stable flames, the induced velocity field is found to be out of phase with the front corrugations leading to a blocking of the low frequencies in the turbulence approaching the front.

5. NON LINEAR ANALYSIS

The analysis of Clavin & Williams (1982) concerning the flame structure of wrinkled fronts in a non homogeneous flows has been extended independently by Matalon & Matkowsky (1982) and by Clavin & Joulin (1983) to the nonlinear case of finite amplitudes of the front corrugations. As anticipated by the early phenomenological analysis of Karlowitz & al (1953), the modification to the normal burning velocity u_n produced by the front curvature and by the flow inhomogeneities can be expressed in terms of only **one geometrical scalar** i.e. the total flame stretch experienced by the front.

$$u_n - u_o = - \mathcal{L} \left(\frac{1}{\sigma} \frac{d\sigma}{dt}\right) \quad (3)$$

where σ is the surface element of the front and $\frac{d}{dt}$ its time derivative when each point of the flame surface moves as described at the beginning of the introduction. $\mathcal{L}$ is the Markstein length that depends on the laminar flame velocity u_o, on the physico-chemical properties of the reactive mixture (transfert properties chemical kinetics, etc...). The corresponding expression of has been obtained for different cases ; the effects of the dependence on temperature of the transport coefficients has been studied by Clavin & Garcia (1983) ; the non adiabatic case and the proximity of the limits of inflamability have been studied by Clavin & Nicoli (1983). Some aspects of the multiple step chemistry has been investigated by Linan & Clavin (1984).

Except its limitation to the weak stretch, this surprinsingly simple result is general and can be used in any flow configuration : stagnation point flow, spherical flames, turbulent flames...etc... The effects of the strong stretch has been recently investigated by Libby, Linan and Williams (1983) in the particular case of a planar front in a stagnation point flow. Once again, the effect of the gas expansion is proved to be important ; for example the flame extinction under strong stretch predicted by the "diffusive-thermal model"(γ=0) for one overall chemical reaction is no more accessible when the effects i) are properly taken into account.

It is withwhile to express the result (3) in terms of the mean radius of curvature of the front R. At the same order, the modification to normal burning velocity can be expressed as (see Clavin & Joulin 1983) :

$$u_n - u_o = \mathcal{L} u_o \left\{ \frac{1}{R} + \frac{1}{u_o} \underline{n} \cdot \underline{\underline{\nabla u}} \cdot \underline{n} \right\} \quad (4)$$

Where $\underline{n}$ is the unit vector normal to the front and $\underline{\underline{\nabla u}}$ is the "rate of straintensor" of the upstream flow evaluated at the flame position. Each of the two terms in r.h.s. of eq(4) represents a contribution to the total flame stretch. The term $-u_o/R$ represents the stretch of the front moving in a uniform flow with a constant normal velocity u_o. This term is the effect of the non planar geometry of the front; The term $\underline{\nabla}.\underline{u} - \underline{n}.\underline{\underline{\nabla u}}.\underline{n}$ is known to represent the stretch of a surface convected by the flow field $\underline{u}$. (Here $\underline{\nabla}.\underline{u}=0$ outside the flame). This second term is the effect of the non homogeneity of the flow. Eq(3) and (4) reconcile the point of view of Karlowitz (1953) and Markstein (1965). As noticed by Frankel & Sivashinsky (1983), when eq(3) or (4) are written in the burned gases, $\mathcal{L}$ has not the same value as in the unburnt gases. This result is associated to the fact that, the normal mass flux is no more constant across the flame when the flame thickness cannot be neglected. This result explains the difference in the behavior of converging and diverging spherical front.

Finally let us mentioned that the eq(3) and (4) provide a non linear equation for the front in a non uniform and/or unsteady flow. But, as already mentionned, the value of the flow field appearing in this equation is not a given quantity and, because of the hydrodynamics effects ii), this flow is, in fact, a functionnal of the flame surface. This aspect of the problem of the flame dynamics has been solved only in the linear approximation (see Pelcé & Clavin 1982, Searby & Clavin 1984). Nevertheless, for turbulent wrinkled flames, and when the corresponding random process is assumed to be stationary and homogeneous, the local equations (3) and (4) for the front evolution provide an expression for the turbulent flame speed. The time average of the modification to normal burning velocity is found to be zero and the turbulent flame speed is proved to be given by simply the laminar flame speed times the mean area increase of the front. This result was known front a long time (see the book of Lewis & Von Elbe 1961) but was associated with the assumption of a constant normal burning velocity that is known to produce cusps on the front surface. The above analysis shows that this simple result concerning the turbulent flame speed holds also for smooth wrindled flame fronts and is not limited to the case of a constant normal burning velocity.

More details concerning flame dynamics can be found in the recent review article by Clavin (1984).

REFERENCES

Barenblat, G.I., Zeldovich, Y.B., and Istratov, A.G. 1962, Prikl. Mekh. Tekh. 2, pp.21-26.
Buckmaster, J.D. and Ludford, G.S.S. 1982, Theory of laminar flames, Cambridge Univ. Press.
Clavin, P. and Williams, F.A. 1982, J.Fluid Mech. 116, pp.251-282
Clavin, P. and Garcia, P. 1983, J.Méc.Théorique et Appliquée 2, pp.245-263.
Clavin, P. and Joulin, G. 1983, J.Physique Lettres 44, pp.L1-L12.
Clavin, P. and Nicoli, C. 1983, Combust. and Flame, to appear.
Clavin, P. 1984, Prog. Energy Combust. Sci., to appear.
Clavin, P. and Linan, A. 1983, NATO ASI Series ed. M.G. Velarde, Plenum Press, to appear.
Darrieus, G. 1938, unpublished work.
Eckhauss, W. 1961, J. Fluid Mech. 10, pp.80-100.
Frankel, M.L. and Sivashinsky, G.I. 1983, J. Combust. Sci. Technol. to appear.
Joulin, G. and Clavin, P. 1979, Combust. and Flame 35, pp.139153.
Karlowitz, B., Denniston, J.R., Knapschaefer, D.H. and Wells,F.E. 1953, IVth Symp.(International) on Combustion, The Combustion Institute, pp.613-620.
Landau, L.D. 1944, Acta Physico-chimica URSS XIX, 1 pp.77-85.
Lewis, B. and Von Elbe, G. 1967, Combustion Flames and Explosions of Gases Academic 2nd ed.
Libby, P.A., Linan, A. and Williams, F.A. 1983, J.Combust.Sci. Technol. to appear.
Linan, A. and Clavin, P. 1984, J. Chem. Phys. Submitted
Markstein, G.H. 1964, Non steady flame propagation Pergamon Press.
Matalon, M. and Matkowsky, B.J. 1983, J. Fluid Mech.124, pp.239-259.
Michelson, D.M. and Sivashinsky, G.I. 1977, Acata Astronautica 4 pp.1207-1221.
Pelcé, P. and Clavin, P. 1982, J.Fluid Mech. 124, pp.219-237.
Quinard, J., Searby, G. and Boyer, L. 1983, IX ICODERS Progress in Astronautics and Aeronautics , to appear.
Searby, G., Sabathier, F., Clavin, P. and Boyer, L. 1983, Phys. Rev. Letter, to appear.
Searby, G. and Clavin, P. 1984, J. Combust.Sci.Technol. Submitted.
Sivashinsky, G.I. 1977, Acta Astronautica 4, pp.1177-1206.
Sivashinsky, G.I. 1983, Ann. Rev. Fluid Mech. 15, pp.179-199.

LINEAR STABILITY OF A PLANAR REVERSE COMBUSTION FRONT PROPAGATING THROUGH A POROUS MEDIUM: GAS-SOLID COMBUSTION MODEL

J. A. Britten and W. B. Krantz

Department of Chemical Engineering
University of Colorado
Boulder, Colorado 80309

The method of activation energy asymptotics has been applied to analyze the zeroth-order dynamics of a planar combustion wave traveling through a porous medium in a direction opposite to that of the forced oxygen flux. Such a process finds practical application in the "reverse combustion" of coal seams in-situ. This two-phase oxygen-limited combustion process assumes an infinite effective Lewis number. The adiabatic front temperature is the eigenvalue for the basic-state problem, and is related to the front velocity by the integral energy balance.

Darcy's law describes the gas flow through the porous medium. Combustion of the medium increases its permeability to gas flow. Thus, the combustion front can be visualized as an unstable displacement front. The linear stability of the basic state was investigated, considering non-oscillatory normal modes, and it was found that for a zero permeability change across the front, all waves are stable, with heat conduction being the stabilizing influence, while for infinite permeability increase, all waves are unstable. For intermediate changes in permeability, a most highly amplified mode exists, which depends on this permeability change and the characteristics of the basic-state solution.

INTRODUCTION

Reverse combustion (RC) is a process in which a combustion wave passes through a porous medium in a direction opposite to the direction of the forced gas flow, which supplies oxygen for

G. Nicolis and F. Baras (eds.), Chemical Instabilities, 117–135.

the combustion. This process is important in underground coal gasification (UCG), where the coal is ignited at one well and an oxygen containing gas is forced down another well, drawing the combustion front to the latter well. Reverse combustion has been widely used as a preliminary permeability enhancement step prior to actual gasification of the coal. This is because RC is known to be an unstable process, in that a broad RC front will evolve to form relatively narrow channels through the coal connecting the wells. These channels of partially or totally combusted coal have a much higher permeability to gas flow than the virgin coal and act as "pipelines" to conduct the gas generated during subsequent forward combustion, in which the flow and the combustion wave travel in the same direction, resulting in a large volume of coal gasified. Thus, RC is employed to insure high gasification rates.

Combustion tube experiments and indications from field tests dealing with Wyoming subbituminous coals [1,2] and tar sands [3] have shown that RC is a relatively low temperature carbonization process in which volatile matter released from the medium is combusted, leaving behind char and ash. However, recent experimental results of RC in solid blocks of a high-quality German subbituminous coal [4] show that char combustion can also occur to a large extent. The dynamics of channel formation during RC is not as yet well understood. This, along with the difficulty of performing laboratory-scale experiments which faithfully reproduce in-situ conditions, make it highly desirable to develop simple models that describe the process. It is the aim of this study to develop analytical equations for the temperature and the propagation speed of an RC front by use of activation energy asymptotics (AEA) for the limiting case of only gas-solid heterogeneous combustion, and to analyze the linear stability of this front to determine the effect of the permeability changes caused by combustion. This work will complement a similar study recently completed [5] which considers only gas phase combustion of volatile matter released from the solid.

Modeling studies of RC in the past have dealt only with the gas phase combustion of volatile matter. Kotowsky and Gunn [6] developed a one-dimensional numerical model, assuming quasi-steady-state conditions in a moving coordinate system. The model of Amr [7] relaxed this assumption. Corlett and Brandenburg [8] employed phenomenological arguments after the fashion of the Frank-Kamenetskii approximation of gas-phase flames [9], to describe analytically the steady propagation of an RC front.

The above models do not address the question of the stability of the planar RC front or the evolution of channels. Physically an RC front can be viewed as an unstable displacement front, where the thin combustion zone separates a region of low permeability (the undisturbed medium) from a high permeability medium created by devolatilization and combustion. That is, if the mobility ratio (defined as the ratio of $\frac{\kappa}{\mu}$ in the displaced phase to that of the displacing phase, where κ is the permeability and μ the gas viscosity) is less than unity, channel formation will be favored. A situation similar to RC occurs in secondary oil recovery, where a fluid with a lower viscosity than oil is pumped into an oil reservoir to force oil out. A more detailed description of the physics of channel formation during RC is given by Gunn and Krantz [10]. These authors analyzed the linear stability of a planar RC front represented as a discontinuity in temperature and pressure gradients. They assumed a linear relationship between the front temperature (T_f) and the gas flux N_g obtained by fitting a region of the numerical data of ref. [6] in lieu of solving the nonlinear reactive zone equations directly. Mui and Miller [11] recently obtained a dispersion relation from the same model, using a quadratic relationship of the front speed S_f-vs-N_g fitted to the same data of [6], but obtained a different result. The dispersion relation of [10] predicted that for a zero mobility ratio (infinite permeability increase from unburned to burned zones), a fastest growing mode of finite wave number exists, while the result of [11] predicted that the fastest growing wave number was infinitely large for this case. The analysis of [5] found that to leading order in the activation energy expansion, no finite fastest growing mode exists for a zero mobility ratio, but does when effects at the next order are included.

Recent advances in the theory of AEA have allowed for the analytical description of gas-phase flames. These methods use the fact that in the limit of a very large activation energy (approached in combustion processes) the zone of active combustion becomes negligibly thin, separating the combustion space into two macroscopic regions, a preheat zone and an afterburn zone, in which no reaction occurs. Simplified equations in each of the three zones are solved, and matched according to the formalism of matched asymptotic expansions in the activation energy, to obtain the unknowns of the system. This theory has been used to study the stability of gas-phase flames with respect to heat losses [12], kinetic mechanisms [13] and hydrodynamic effects [14], for example, and also the dynamics of non-porous carbon particle combustion [15]. A concise description of the basics of AEA is given by Ludford [16], and details of the inner reactive zone structure of gas-phase flames can be found in [12]. These studies of gas-phase flames are restricted to Lewis numbers (Le) of order unity. The effective Le in the presence of a low-porosity solid phase is typically very much larger than one. To the authors' knowledge, only this work and its companion work [5] treat rigorously the asymptotics of high Le combustion.

The following model describes to leading order gas-solid

reverse combustion of a combustible porous medium for infinite Le. Such a process is probably not important during RC of tar sands, but can occur along with volatile combustion during RC of coal. The basic-state solution can also be applied to describe the combustion zone of a countercurrent packed-bed oil-shale retort, and possibly also to describe the combustion of coke deposited on cracking catalyst particles during catalyst reactivation. The model grossly oversimplifies the combustion kinetics, but it is felt adequate to describe the overall behavior of the process and to show the primary effect of process parameters on the system.

MODEL FORMULATION

The model envisions a planar combustion front propagating through an infinite isotropic combustible porous medium. Figure 1 depicts the situation and the assumed temperature, O_2 and pressure profiles. Important assumptions in the model are described below, and are discussed in more detail in a later section.

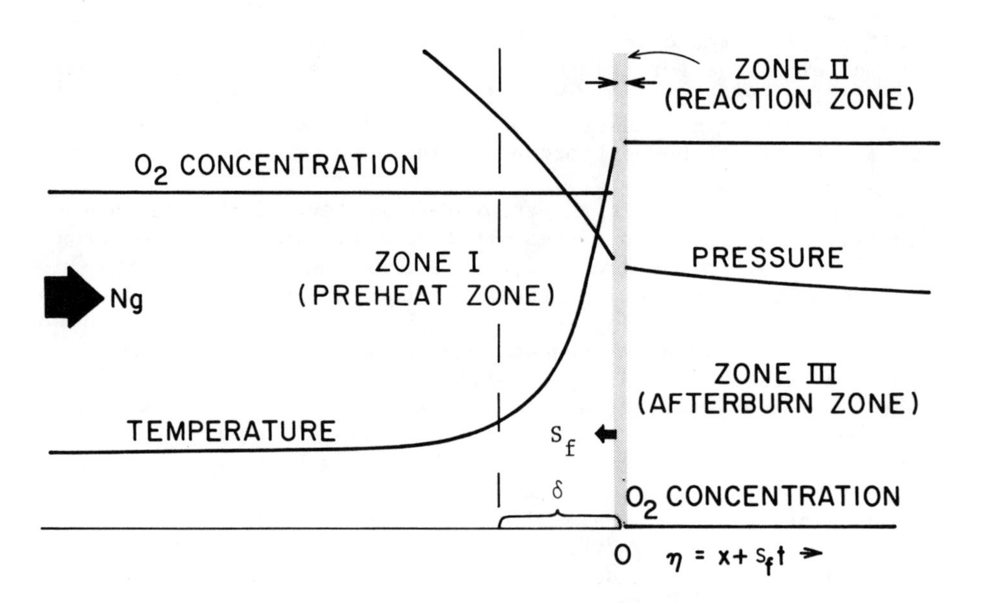

Figure 1. Schematic of planar RC front.

The front moves with constant speed S_f relative to a stationary observer in a direction opposite to a uniform upstream gas flux N_g with respect to stationary coordinates. The gas flow is assumed to be incompressible and governed by Darcy's law. The reactive zone is very thin compared to the characteristic length of the system δ, the length of the preheat zone. The gas and solid phases are assumed to be locally in thermal equilibrium. Physical properties are assumed constant and evaluated at T_f, except the heat capacities, which are evaluated at an average temperature. Heat conduction in the gas phase is not considered, nor is radiative heat transfer. Oxygen reacts with an excess of solid fuel according to single-step Arrhenius kinetics, assumed first-order in O_2. The permeability change due to the devolatilization is represented by a discontinuity in the pressure gradient across the reactive front.

In general dimensional (*) form, the energy, O_2, total mass and momentum balances are:

$$C_s\rho_s^* \frac{\partial T^*}{\partial t^*} + C_g \underline{N}_g^* \cdot \nabla T^* = \lambda \nabla^2 T^* + q r^* \quad , \tag{1}$$

$$\phi\rho_g^* \frac{\partial Y^*}{\partial t^*} + \nabla \cdot (\underline{N}_g^* Y^*) = D_{eff}\rho_g^* \nabla^2 Y^* - r^* \quad , \tag{2}$$

$$\nabla \cdot \underline{N}_g^* = - \frac{\partial \rho_s^*}{\partial t^*} = \Delta\nu r^* \quad , \tag{3}$$

$$\nabla \cdot \underline{N}_g^* = - \frac{\kappa \rho_g^*}{\mu} \nabla^2 P^* \quad . \tag{4}$$

In equations (1) and (3) terms involving ρ_g^* are neglected in comparison with ρ_s^*. Although ρ_s^* and $\underline{N}_g^*$ are not constant, the terms involving $\partial\rho_s^*/\partial t$ and $\nabla \cdot \underline{N}_g^*$ in the energy balance (1) are incorporated into q, which is defined for a gasification reaction as the enthalpy of the reactants in the gas phase minus the enthalpy of the products in their respective phases. The reaction rate expression is:

$$r^* = A\rho_g^* Y^* e^{\frac{-E}{RT^*}} \quad . \tag{5}$$

We define a coordinate system with its origin at the front:

$$\eta^* = x + S_f t^* - \beta f^*(z,t) \quad , \tag{6}$$

where f represents the corrugations of the front in the stability analysis and β is a dimensionless order parameter ($\beta \ll 1$). For brevity we will consider only one lateral dimension.

When this coordinate system and the following scale factors and definitions

$$T = \frac{T^*}{T_f} \; ; \; Y = \frac{Y^*}{Y_{-\infty}} \; ; \; P = \frac{P^*}{P_f} \; ; \; \rho_s = \frac{\rho_s^*}{\rho_{s-\infty}} ; \; (\eta, z, f)$$

$$= (\eta, z, f)^* \left(\frac{\Sigma_t}{\lambda}\right) \; ; \; \rho_g = \frac{\rho_g^*}{\rho_{gf}} \tag{7}$$

$$t = t^*\left(\frac{\Sigma_t^2}{\lambda \rho_{s-\infty} C_s}\right) \; ; \; \Sigma_t = \rho_{s-\infty} C_s S_f + N_{g-\infty} C_g \; ; \; \alpha = \frac{T_{-\infty}}{T_f} \; ;$$

$$E_{-\infty} = \frac{E}{RT_{-\infty}} \, , \tag{8}$$

are introduced into eqns. (1), (2), (3) and (5) and these equations are manipulated to give a form convenient for the asymptotic analysis, we obtain:

$$\rho_s \frac{\partial T}{\partial t} + (F_t N_g + (1 - F_t)\rho_s - \beta(\rho_s \frac{\partial f}{\partial t} - \frac{\partial^2 f}{\partial x_2^2})) \frac{\partial T}{\partial \eta}$$

$$= \nabla^2 T + D(1 - \alpha)^2 Y e^{-E_{-\infty}\alpha(\frac{1}{T} - 1)} \; ; \tag{9}$$

$$\frac{d(N_g Y)}{d\eta} = -D(1 - \alpha) Y e^{-E_{-\infty}\alpha(\frac{1}{T} - 1)} \; ; \tag{10}$$

$$(-N_3 + \beta \frac{\gamma}{F_t} \frac{\partial f}{\partial t}) \frac{d\rho_s}{d\eta} = \frac{dN_g}{d\eta} = \Delta\nu Y_{-\infty} D(1 - \alpha) Y e^{-E_{-\infty}\alpha(\frac{1}{T} - 1)} \; ; \tag{11}$$

where,

$$D = \frac{\lambda A T_{-\infty}}{N_{g-\infty}^2 Y_{-\infty} q} \left(\frac{P_f PM}{RT_{-\infty}}\right) e^{-E_{-\infty}\alpha} \; . \tag{12}$$

Equation (2) has been simplified in arriving at (10) by noting that the groups N_1 (multiplying $\partial Y/\partial t$) and N_2 (multiplying $\partial Y/\partial \eta$, see Nomenclature), are of $O(\rho_{gf}/\rho_{s-\infty})$, and are negligibly small at low to moderate pressures, and by assuming an infinite Le.

Equation (12) was derived using the integral energy balance for the basic state, which relates the unknowns T_f and S_f. In dimensionless form, this is written:

$$\frac{N_{g-\infty} q Y_{-\infty}}{\Sigma_t T_{-\infty}} = \frac{1 - \alpha}{\alpha} \quad . \tag{13}$$

Darcy's law (eqn. 4) becomes:

$$\nabla \cdot \underline{N}_g = -N_p \nabla^2 P \quad . \tag{14}$$

To apply AEA to this system of equations, we define $\varepsilon = 1/E_{-\infty} << 1$ as the perturbation parameter, and divide the η space into three zones as shown in Figure 1. For the outer zones I and III, we expand the dependent variables and the basic-state eigenvalue α in terms of ε in the following manner:

$$(T, Y, \rho_s, N_g, \alpha) = (T, Y, \rho_s, N_g, \alpha)_0 + \varepsilon (T, Y, \rho_s, N_g, \alpha)_1 + O(\varepsilon^2) \quad . \tag{15}$$

The inner zone (II) dependent variable expansions are given in the Appendix.

RESULTS

Basic-State Solution

The basic state is characterized by $\partial/\partial t = f = 0$, and a constant upstream gas flux. The basic-state variables are denoted by $(\bar{\ })$. In the outer zones I and III, eqns. (9), (10) and (11) reduce to:

$$\frac{d^2 \bar{T}}{d\eta} = (F_t \bar{N}_g + (1 - F_t)\bar{\rho}_s) \frac{d\bar{T}}{d\eta} \quad ; \tag{16}$$

$$\frac{d\bar{Y}}{d\eta} = \frac{d\bar{\rho}_s}{d\eta} = \frac{d\bar{N}_g}{d\eta} = 0 \quad . \tag{17}$$

Boundary conditions are:

$$\lim_{\eta \to -\infty} \bar{T} = \alpha, \ \bar{Y} = \bar{N}_g = \bar{\rho}_s = 1 \ ; \qquad \eta = 0^{\pm} \ \bar{T} = 1 \ ;$$

$$\lim_{\eta \to +\infty} \bar{T} = 1, \ \bar{Y} = 0, \ \bar{N}_g = 1 + \Delta\nu Y_{-\infty}, \bar{\rho}_s = 1 - \frac{\Delta\nu Y_{-\infty}}{N_3} \quad . \tag{18}$$

Using (18) and the expansions (15) give for zone I to leading order:

$$\overline{T}_0 = \alpha_0 + (1 - \alpha_0)e^{\eta} \quad , \quad \overline{Y}_0 = \overline{N}_{g0} = \overline{\rho}_{s0} = 1 \quad ; \tag{19}$$

and for zone III:

$$\overline{T}_0 = 1 \ , \ \overline{Y}_0 = 0 \ , \ \overline{N}_{g0} = 1 + \Delta\nu Y_{-\infty} \ , \ \overline{\rho}_{s0} = 1 - \frac{\Delta\nu Y_{-\infty}}{N_3} \ . \tag{20}$$

Matching these relations at O(1) with the corresponding relation obtained from solution of the inner zone II equations (see Appendix) gives a formula for calculating the unknown α_0:

$$\frac{D_0\varepsilon}{\alpha_0} = (1 + \frac{\Delta\nu Y_{-\infty}}{2}) \quad . \tag{21}$$

A constant gas flux in zones I and III implies a constant pressure gradient, from (14). The basic-state solution to the momentum equation is then:

$$\text{zone I} \quad \frac{d\overline{P}_0}{d\eta} = -\frac{1}{N_p} \ ; \quad \text{zone III} \quad \frac{d\overline{P}_0}{d\eta} = \frac{(1 + \Delta\nu Y_{-\infty})N_\kappa}{N_p} \ . \tag{22}$$

Stability Analysis

The dependent variables are expanded in terms of β at each order in ε; the superscript (^) denotes the η dependence of these variables in the outer zones I and III:

$$(T,Y,N_g,P) = (\overline{T},\overline{Y},\overline{N}_g,\overline{P}) + \beta f(\hat{T},\hat{Y},\hat{N}_g,\hat{P}) \quad , \tag{23}$$

where $f = e^{(ikz + \sigma t)}$ is a unit-amplitude harmonic disturbance of the planar combustion front, and is also expanded in ε, since we have designated σ as the eigenvalue at $O(\beta)$. We are concerned only with the zeroth-order analysis in the AEA in this study. The solid density is not perturbed.

The analysis begins with the Darcy flow equation (14). In the outer zones I and III, $\nabla\cdot\underline{N_g} = 0$ for all orders in ε and at $O(\beta)$(14) is written:

$$\frac{d^2\hat{P}_0}{d\eta^2} + k^2(\frac{d\overline{P}_0}{d\eta} - \hat{P}_0) = 0 \ . \tag{24}$$

The $O(\beta)$ form of the overall mass balance (11) is written:

$$\frac{\gamma}{F_t}\sigma_0 \frac{d\bar{\rho}_{s0}}{d\eta} = \frac{d\hat{N}_{g0}}{d\eta} \quad . \tag{25}$$

We define $a^{\pm}_{g0}$ as the zeroth-order amplitude of the gas-flux perturbation at $\eta = 0^{\pm}$, and integrate eqn. (25) across the front using (18) to obtain the relationship between a^{-}_{g0} and a^{+}_{g0}:

$$a^{+}_{g0} = a^{-}_{g0} - \frac{\Delta\nu Y_{-\infty}\sigma_0}{1 - F_t} \quad . \tag{26}$$

We now solve eqn. (24) using boundary conditions $\hat{P}(\eta\to\pm\infty)<\infty$ to give:

$$\text{zone I} \quad \hat{P}_0 = \hat{P}^{-}_0 e^{k\eta} - \frac{1}{N_p} \; ; \quad \text{zone III} \quad \hat{P}_0 = P^{+}_0 e^{-k\eta} - \frac{N_\kappa(1 + \Delta\nu Y_{-\infty})}{N_p} \quad . \tag{27}$$

Continuity of pressure across the front demands:

$$\hat{P}^{-}_0 - \frac{1}{N_p} = \hat{P}^{+}_0 - \frac{N_\kappa(1 + \Delta\nu Y_{-\infty})}{N_p} \tag{28}$$

Darcy's law evaluated at $\eta = 0^{\pm}$ provides relations between the pressure and gas-flux perturbations:

$$a^{-}_{g0} = -N_p k\hat{P}^{-}_0 \; ; \quad a^{+}_{g0} = \frac{N_p k\hat{P}^{+}_0}{N_\kappa} \tag{29}$$

Equations (25) through (29) are combined to give an expression for a^{-}_{g0} as a function of the unknowns k and σ_0:

$$a^{-}_{g0} = \frac{k}{N_\kappa + 1}\left(\frac{N_\kappa \Delta\nu Y_{-\infty}\sigma_0}{k(1 - F_t)} + N_\kappa(1 + \Delta\nu Y_{-\infty}) - 1\right) \quad . \tag{30}$$

The $O(\beta)$ forms of the energy and O_2-balance equations are:

$$\frac{d^2\hat{T}_0}{d\eta^2} - \frac{d\hat{T}_0}{d\eta} - (\sigma_0 + k^2)\hat{T}_0 = (F_t\hat{N}_{g0} - \sigma_0 - k^2)\frac{d\overline{T}_0}{d\eta} \quad ; \tag{31}$$

$$\overline{N}_{g0}\frac{d\hat{Y}_0}{d\eta} + \hat{N}_{g0}\frac{d\overline{Y}_0}{d\eta} = 0 \quad . \tag{32}$$

Equation (A.5) (see Appendix) implies $\hat{T}_0^{\pm} = 0$. Using this, boundary conditions $\lim_{\eta\to\pm\infty} \hat{T} = \hat{Y} = 0$ and equations (19), (27) and (29), we arrive at the following:

$$\text{zone I} \quad \hat{Y}_0 = 0 \ ; \tag{33}$$

$$\hat{T}_0 = (1 - \alpha_0)(e^{\eta} + \frac{F_t\overline{a}_{g0}}{k - \sigma_0}e^{(1+k)\eta} - (1 + \frac{F_t\overline{a}_{g0}}{k - \sigma_0})e^{\chi\eta}) \quad ; \tag{34}$$

where,

$$\chi = \frac{1}{2}(1 + \sqrt{1 + 4(k^2 + \sigma_0)}) \ ; \tag{35}$$

$$\text{zone III} \quad \hat{T}_0 = \hat{Y}_0 = 0 \ . \tag{36}$$

Taking the derivative of (34) and matching with the corresponding equation for $d\hat{T}_0^-/d\eta$ obtained from solution of the inner zone equations (see Appendix), and using (30) results in the zeroth-order dispersion relation for σ_0:

$$1 - \chi + \frac{k}{N_\kappa + 1}(\frac{N_\kappa \Delta\nu Y_{-\infty}\sigma_0}{k(1 - F_t)} + N_\kappa(1 + \Delta\nu Y_{-\infty}) - 1) \cdot$$

$$(\frac{F_t(1 + k - \chi)}{k - \sigma_0} - 1) = 0 \quad . \tag{37}$$

DISCUSSION

Basic-State Solution

Figure 2 shows steady front temperature and front speed profiles as functions of the injected air flux, calculated from equations (21) and (13). The reaction $O_2 + C \rightarrow CO_2$ was considered, corresponding to a value of 0.375 for $\Delta\nu$. Values of the apparent activation energy and frequency factor of 145 kJ/mol and 8.28 x 10^{11}/hr respectively were taken from the literature [17], and

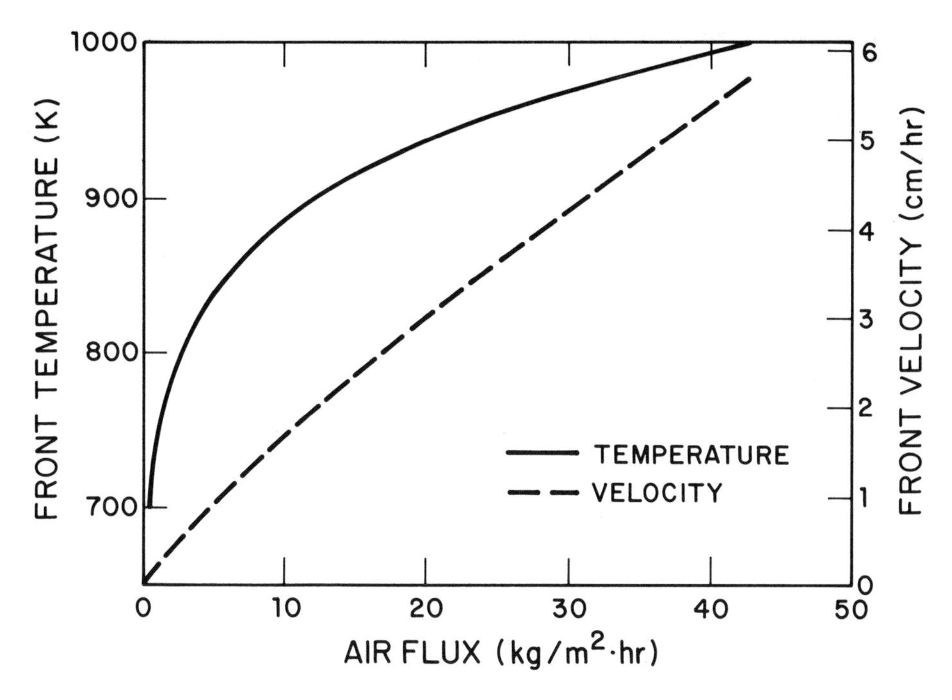

Figure 2. Adiabatic front temperature and front speed as functions of the injected air flux. Kinetic data from ref. [17].

physical property data from ref. [6] were used. No experimental data of RC of char only are available for comparison, to the best of the authors' knowledge.

The simplest kinetic mechanism was used in this study. In reality the order in O_2 for carbon oxidation can vary between 0 and 1 and the observed activation energy can vary over a wide range [18]. These quantities and the frequency factor depend on the relative importance of diffusional limitation of the intrinsic kinetic rate, which makes them functions of the temperature, pressure and the characteristic pore size and/or particle size of the solid. The kinetic parameters are influenced as well by various catalytic impurities possibly present in the medium. Under a wide range of conditions, however, the apparent activation energy is large enough to allow for approximation by AEA. Since the kinetic parameters are subject to such uncertainty, perhaps the best use of this model is as a convenient way of fitting model-dependent rate constants for the combustion. This would give simple equations for predicting the gross effects of process parameters on the front velocity or other unknowns. One must keep in mind, however, the danger of extrapolating the predictions outside the temperature and pressure range of validity of the rate constants determined in such a way.

The present model is valid provided the groups N_1 and N_2 are $\leq O(\varepsilon)$ and $1/Le \leq O(\varepsilon^2)$. The first restrictions are not substantial except at very high absolute pressures, when N_1 and N_2 can become significant. The second restriction is not felt to be significant under conditions of interest. Of the constant physical parameter assumptions, a constant value for λ is the most severe, since the thermal conductivity of coal and char is a very strong function of temperature at typical reaction temperatures.

Stability Analysis

Figures 3 and 4 show the characteristics of the zeroth-order dispersion relation (37). Figure 3 plots neutrally stable wave numbers against a dimensionless gas flux F_t for various values of the permeability ratio N_K. The horizontal axis $k = 0$ is also neutrally stable for all parameter values, and wave numbers inside these envelopes of neutral stability are unstable. Two values of $\Delta\nu$, 0.375 and 0.75, corresponding respectively to CO_2 and CO as the product gases, and an injected gas composition of air were used. As can be seen, the effect of gas generation at the front, of which $\Delta\nu$ is a measure, is negligible, since the curves at each value of N_K are indistinguishable. Of course, in comparing these two reaction schemes, one must remember that for a given air flux, the basic-state conditions characterized by F_t will be substantially different. For the case of $N_K = 0$, all modes are unstable, while for $N_K = 1$, all modes are stable. Both these results require comment. For a zero or finite permeability change ($0 < N_K \leq 1$) no steady state actually exists, since the combustion front speed continually increases, due to the pressure dependence of the kinetics, as the front advances up the pressure gradient forcing the gas flow. What we have done for the basic state is a frozen-time analysis, except for the case of $N_K = 0$, where the pressure gradient in zone III is for practical purposes zero and the combustion occurs at constant pressure. However, the time scale characteristic of such changes in the front velocity is related to d/S_f, where d is the distance over which the pressure field is established, and is very long, except possibly at very high absolute pressures. Thus, the quasi-steady-state assumption is considered very reasonable. Another shortcoming of the model is the assumption of incompressible flow, employed to make the momentum equation analytically tractable in the stability analysis. This assumption precludes the pressure perturbation from influencing the reaction kinetics; the effect of pressure comes in only through the basic state. This pressure perturbation could destabilize the front at values of N_K close to unity, but is not felt to be overly important for the same reasons given above.

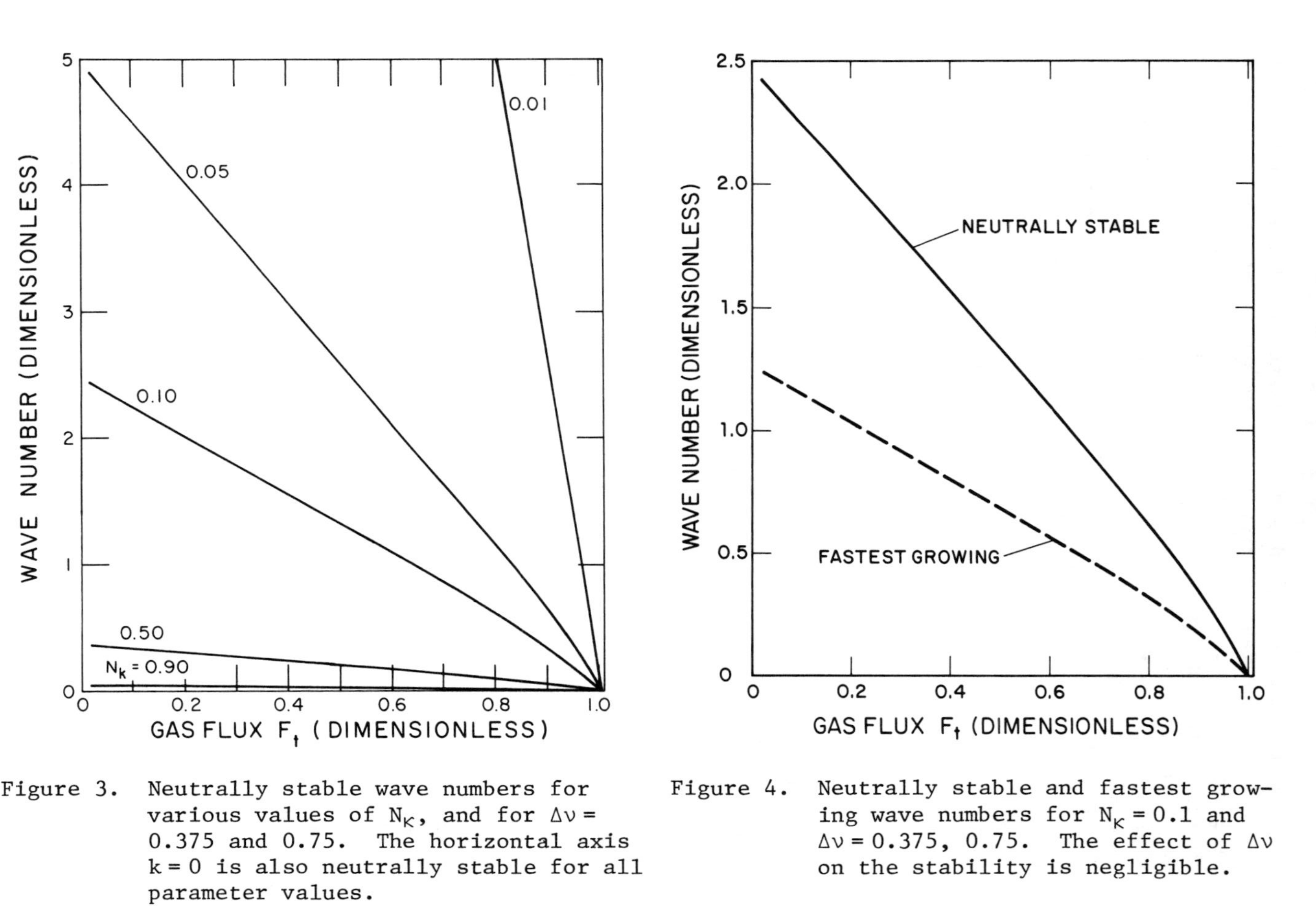

Figure 3. Neutrally stable wave numbers for various values of N_K, and for $\Delta\nu$ = 0.375 and 0.75. The horizontal axis k = 0 is also neutrally stable for all parameter values.

Figure 4. Neutrally stable and fastest growing wave numbers for $N_K = 0.1$ and $\Delta\nu = 0.375$, 0.75. The effect of $\Delta\nu$ on the stability is negligible.

At the other extreme $N_K = 0$, the dispersion relation (37) predicts that the fastest growing disturbances are those of infinite wave number. For finite values of N_K, heat conduction stabilizes the process, but another stabilizing mechanism has been omitted. This is due to the fact that at O(1) in ε, perturbations in the front temperature are not allowed, i.e. all heat generated locally at $O(\beta)$ must propagate the front. Equation (37) with $\Delta\nu = 0$ is identical with the zeroth-order dispersion relation from ref. [5]. The model of [5] is somewhat less complicated and allows for the analytical calculation of the $O(\varepsilon)$ eigenvalues for both basic and perturbed states. At $O(\varepsilon)$, the front temperature is perturbed. This brings in a mechanism which stabilizes large wave number disturbances even for $N_K = 0$. The representation of the pressure gradient as discontinuous at the front is considered reasonable for this model since the solid density and therefore the porosity are discontinuous across the front at zeroth-order.

Figure 4 shows the most highly-amplified wave numbers from (37) using a value of 0.1 for N_K and the two previously mentioned values of $\Delta\nu$, which again has a negligible effect on the stability of the system.

The dispersion relation predicts that an increase in the pressure, with all other parameters held constant, will destabilize the process by decreasing the steady-state value of F_t, thus causing smaller channels to form. This has been experimentally observed [4]. Increasing the O_2 fraction of the injected gas also lowers F_t; this basic-state effect overrides the stabilizing influence of gas generation, resulting in destabilization. An observation of this effect is reported in the Russian literature [19]. An increase in the imposed gas flux mildly stabilizes the process, since F_t is a weakly increasing function of N_g.

APPENDIX

Inner Reactive Zone Structure

In the inner reactive zone, centered at $\eta = 0$, we introduce a stretch variable $\xi = \eta/\varepsilon$ to scale properly the variations of temperature and oxygen concentrations in this region. Recall that to leading order, the temperature is constant in this zone, while the O_2 concentration, gas flux and solid density go through an O(1) change. We therefore expand these variables in the following way:

$$T = 1 + \varepsilon\theta_1 + O(\varepsilon^2) \ ; \quad (Y,N_g,\rho_s) = (y,n_g,r_s)_0$$
$$+ \varepsilon(y,n_g,r_s)_1 + O(\varepsilon^2) \ . \tag{A.1}$$

Introducing ξ and expansions A.1, and the expansion for α (16) into the general balance equations (9),(10) and (11) gives to leading order:

$$\frac{d^2\theta_1}{d\xi^2} = -D_0\varepsilon(1-\alpha)^2 y_0 e^{\alpha_0\theta_1} \ ; \tag{A.2}$$

$$\frac{d(n_{g0}y_0)}{d\xi} = -D_0\varepsilon(1-\alpha)y_0 e^{\alpha_0\theta_1} \ ; \tag{A.3}$$

$$(-N_3 + \beta\frac{\gamma}{F_t}\frac{\partial f}{\partial t})\frac{dr_{s0}}{d\xi} = \frac{dn_{g0}}{d\xi} = \Delta\nu Y_{-\infty}D_0\varepsilon(1-\alpha_0)y_0 e^{\alpha_0\theta_1} \ . \tag{A.4}$$

These equations are valid for both the basic and perturbed states.

The inner and outer expansions for the dependent variables are related by expanding the latter in Maclaurin's series in η, expressing these in terms of the inner variable ξ, and comparing the resulting series with the inner variable expansions at each order in ε. This results in the following:

$$\lim_{\xi\to\pm\infty} \ 1 = T_0^{\pm} \ ; \quad (y,n_g,r_s)_0 = (Y^{\pm},N_g^{\pm},\rho_s^{\pm})_0 \ ; \tag{A.5}$$

$$\theta_1 = \frac{dT_0^{\pm}}{d\eta}\xi + T_1^{\pm} \ ; \quad (y,n_g,r_s)_1 = \frac{d(Y^{\pm},N_g^{\pm},\rho_s^{\pm})_0}{d\eta}\xi$$
$$+ (Y^{\pm},N_g^{\pm},\rho_s^{\pm})_1 \ . \tag{A.6}$$

Equations (A.2) and (A.4) are combined to eliminate the non-linear term and integrated using the basic-state boundary conditions at $\lim_{\xi\to-\infty}$ provided by (A.5),(A.6) and (18) to give an expression for n_{g0} in terms of $d\theta_1/d\xi$:

$$n_{g0} = 1 + \Delta\nu Y_{-\infty} - \frac{\Delta\nu Y_{-\infty}}{1-\alpha_0}\frac{d\theta_1}{d\xi} \ . \tag{A.7}$$

This expression is substituted into (A.3), and (A.2) and (A.3) are combined and integrated similarly to obtain an expression for y_0 in terms of $d\theta_1/d\xi$:

$$y_0 = \frac{\frac{d\theta_1}{d\xi}}{(1-\alpha_0)(1+\Delta\nu Y_{-\infty} - \frac{\Delta\nu Y_{-\infty}}{1-\alpha_0}\frac{d\theta_1}{d\xi})} \quad . \tag{A.8}$$

Equation (A.8) is then used in (A.2) to give an equation in terms of θ_1 and its derivatives:

$$(1 + \Delta\nu Y_{-\infty} - \frac{\Delta\nu Y_{-\infty}}{1-\alpha_0}\frac{d\theta_1}{d\xi})\frac{d^2\theta_1}{d\xi^2} = -D_0\varepsilon(1-\alpha_0)\frac{d\theta_1}{d\xi}e^{\alpha_0\theta_1} \quad . \tag{A.9}$$

The transformation $\psi = d\theta_1/d\xi$ is used and (A.9) is integrated once, using basic-state boundary conditions at $\lim_{\xi\to+\infty}$. The resulting expression is evaluated at $\lim_{\xi\to-\infty}$, noting from (A.6) that at this limit $\theta_1\to-\infty$, giving:

$$(1+\Delta\nu Y_{-\infty} - \frac{\Delta\nu Y_{-\infty}}{2(1-\alpha_0)}\frac{d\bar{T}_0^-}{d\eta})\frac{d\bar{T}_0^-}{d\eta} = \frac{\varepsilon D_0(1-\alpha_0)}{\alpha_0} \quad . \tag{A.10}$$

Use of (19) in (A.10) gives equation (21).

To obtain the zeroth-order dispersion relation for the stability analysis, we need only combine equations (A.2) and (A.3) and integrate using boundary conditions at $\lim_{\xi\to+\infty}$ provided by (A.5), (A.6),(18) and (36), resulting in the following:

$$(1 - \alpha_0)n_{g0}y_0 = \frac{d\theta_1}{d\xi} \quad . \tag{A.11}$$

Evaluation of (A.11) at $\lim_{\xi\to-\infty}$ and at $O(\beta)$, using (A.5),(A.6),(19) and (33) gives:

$$\frac{d\hat{T}_0^-}{d\eta} = (1 - \alpha_0)a_{g0}^- \quad . \tag{A.12}$$

Matching this with the derivative of (34) from the zone I solution and using (30) provides the dispersion relation (37).

ACKNOWLEDGMENTS

Financial support through the Gas Research Institute (Contract no. 5081-260-0570) and the Division of International Programs at NSF (Grant no. INT-8108409) is gratefully acknowledged. The authors thank Professor Norbert Peters of the Rheinisch-Westfälische Technische Hochschule at Aachen, West Germany, for helpful discussions concerning AEA, and also Ms. Ellen Romig for preparing the manuscript.

NOMENCLATURE

A - Frequency factor for reaction (includes surface activity)
a - Amplitude of gas-flux perturbation at the reactive front
C - Heat capacity
D_{eff} - Effective gas diffusivity
E - Activation energy of reaction
f - Harmonic corrugation of planar combustion front in stability analysis
k - Wave number of disturbance in stability analysis
M - Average molecular weight of gas
$\underline{N}_g$ - Outer zone gas flux vector
N_g - Outer zone gas flux in x-direction
n_g - Inner zone gas flux in x-direction
P - Pressure
q - Heat release per unit reaction
R - Universal gas constant
r - Reaction rate
r_s - Inner zone solid density
S_f - Combustion front speed
T - Outer zone temperature
t - Time
x - Coordinate in direction of gas flow
Y - Outer zone O_2 concentration
y - Inner zone O_2 concentration
z - Lateral coordinate

Greek Letters

α - Basic state eigenvalue = $(T_{-\infty}/T_f)$
β - Dimensionless order parameter for stability analysis (β<<1)
γ - = C_g/C_s
$\Delta\nu$ - Stoichiometric coefficient (mass of solid gasified/mass of O_2 reacted)
δ - Characteristic combustion zone thickness
ε - AEA perturbation parameter = $(1/E_{-\infty})$
η - Coordinate referenced to moving combustion front
θ - Inner zone temperature

κ - Permeability of porous medium
λ - Thermal conductivity of coal
μ - Gas viscosity
ξ - Inner zone stretch variable
ρ - Density
Σ_t - Sum of the thermal fluxes (Eqn. 8)
σ - Growth coefficient of disturbance, eigenvalue in stability analysis
ϕ - Solid porosity
χ - Exponential argument defined by Eqn. (38)

Subscripts

f - Evaluated at combustion front
g - Gas
I - Zone I
III - Zone III
s - Solid
$-\infty$ - Evaluated at $\eta \to -\infty$
0 - Zeroth-order in AEA expansion
1 - First-order in AEA expansion

Superscripts

$+$ - Evaluated at $\eta = 0^+$
$-$ - Evaluated at $\eta = 0^-$
$*$ - Dimensional variable
$\bar{}$ - Basic-state variable
$\hat{}$ - Perturbation variable

Dimensionless Groups

D - Damkohler number defined by eqn. (12)

$$F_t = \frac{N_{g-\infty} C_g}{\Sigma_t}$$

$$Le = \frac{\lambda N_{g-\infty}}{D_{eff} \rho_g \Sigma_t}$$

$$N_p = \frac{P_f \rho_{gf} \kappa_I \Sigma_t}{\mu N_{g-\infty} \lambda}$$

$$N_\kappa = \frac{\kappa_I}{\kappa_{III}}$$

$$N_1 = \frac{\phi \rho_g \Sigma_t}{N_{g-\infty} \rho_{s-\infty} C_s}$$

$$N_2 = \frac{S_f \rho_g \phi}{N_{g-\infty}}$$

$$N_3 = \frac{\rho_{s-\infty} S_f}{N_{g-\infty}}$$

REFERENCES

1. Hommert, P. J., S. G. Beard and R. P. Reed, Proc. 3rd Underground Coal Conversion Symp., Lawrence Livermore Lab. Report CONF-770652, p. 268 (1977).
2. Lu, H. S., M.S. Thesis, University of Wyoming, Laramie, WY (1980).
3. Land, C. S., Ph.D. Dissertation, University of Wyoming, Laramie, WY (1974).
4. Breidung, K. P., Ph.D. Dissertation, Rheinische-Westfälische Technische Hochschule, Aachen, West Germany (1982).
5. Britten, J. A. and W. B. Krantz, to be submitted to Combusion and Flame.
6. Kotowsky, M. D. and R. D. Gunn, Laramie Energy Research Center Report LERC/RI-76/4, Laramie, WY (1976).
7. Amr, A., Combustion and Flame 41, p. 301 (1981).
8. Corlett, R. C. and C. F. Brandenburg, 1977 Spring Meeting, Western States Section, Combustion Institute, Seattle, WA (April, 1977).
9. Murty-Kanury, A., *Introduction to Combustion Phenomena*, 2nd Ed., Gordon and Breach Sci. Pub., p. 297 (1977).
10. Gunn, R. D. and W. B. Krantz, Soc. Pet. Eng. J. 20, p. 267 (1980).
11. Mui, K. C. and C. A. Miller, submitted to Soc. Pet. Eng. J.
12. Joulin, G. and P. Clavin, Combustion and Flame 35, p. 139 (1979).
13. Joulin, G. and T. Mitani, Combustion and Flame 40, p. 235 (1981).
14. Clavin, P. and F. A. Williams, J. Fluid Mech. 116, p. 251 (1981).
15. Kassoy, D. R. and P. A. Libby, Combustion and Flame 48, p. 287 (1982).
16. Ludford, G. S. S., J. de Mecanique 16(4), pp. 531 and 553 (1977).
17. Hsia, S. J., T. A. Wellborn and T. F. Edgar, Proc. 4th Underground Coal Conversion Symp., Sandia Laboratory Report SAND-78-0941, Albuquerque, NM (1978).
18. Walker, P. L., F. Rusinko and L. G. Austin, Adv. in Catal. 11, p. 134 (1959).
19. Olness, D. U., Lawrence Livermore Laboratory Translation Report UCRL-53144 (1981).

PROPAGATING FLAMES AND THEIR STABILITY

M. Matalon and B. J. Matkowsky

Dept. of Engineering Sciences and Applied Mathematics
Northwestern University
Evanston, Illinois 60201 USA

Viewed on a hydrodynamical scale, a flame may be considered as a surface of discontinuity, separating burned from unburned gas. Unlike earlier treatments, which ignored the flame structure, the present study accounts for the interaction of the fluid flow with the transport processes and chemical reactions occurring inside the thin flame zone. Thus we derive, rather than prescribe, jump conditions across the flame front and an equation for the flame speed. The model, derived in coordinate invariant form, describes the dynamics of flame fronts including their stability. Particular attention is focused on the stability of curved flames, which reveal some characteristics that do not exist in the corresponding analysis of plane flames. Due to the stabilizing effect of curvature, disturbances of circular flames grow more slowly than those for plane flames. As in the case of plane flames, when the mass diffusivity of the deficient reactant component is sufficiently smaller than thermal diffusivity, curved flames can be stabilized. Finally, in contrast to plane flames, the effect of viscosity on curved flames is comparable to that of diffusion and is destabilizing. This dependence decreases with increasing radius of curvature and disappears entirely for plane flames.

Combustion processes involve the flow of compressible fluids in which exothermic chemical reactions take place. Such processes exhibit a strong coupling, due to thermal expansion, between the chemical reaction rates and the fluid dynamics of the system (including all transport processes). These complex interactions are extremely difficult to analyze mathematically. They involve the Navier-Stokes equations for viscous compressible

G. Nicolis and F. Baras (eds.), Chemical Instabilities, 137–145.

flows coupled with the equations of conservation of species and of energy. Thus the equations involve not only the usual fluid dynamical nonlinearities but also the exponentially nonlinear terms arising from the Arrhenius chemical reaction rates. Therefore, scientists have resorted to approximate treatments by introducing phenomenological models and/or by considering various limiting cases, thus permitting analysis of the resulting equations.

In many situations of practical interest, the burning occurs within a narrow zone which, when viewed on the much larger fluid dynamical scale, may be considered as a moving surface called the "flame front". Across the flame surface, velocity, density, temperature, pressure and chemical compositions undergo sudden jumps or discontinuities. The relations between the values of these variables on either side of the front as well as the equation describing the evolution of the flame surface must be determined by the flame structure, i.e. by the chemical reactions and transport processes occurring inside the flame zone. Early treatments of flames as gasdynamic discontinuities ignored the structure of the flame and therefore did not account for the interactions of chemical reactions and transport processes with the fluid flow. Instead, conservation of mass and momentum was assumed to hold across the flame surface, namely

$$[\rho(\vec{V}\cdot\vec{n} - \nu_n)] = 0 \quad , \tag{1}$$

$$[\vec{V}\times\vec{n}] = 0 \quad , \tag{2}$$

$$[p + \rho(\vec{V}\cdot\vec{n})(\vec{V}\cdot\vec{n} - \nu_n)] = 0 \quad , \tag{3}$$

where $\vec{V}$ is the fluid velocity, $\vec{n}$ a vector normal to the flame front, ν_n the normal velocity of the front and p, ρ the pressure and density of the gas mixture, respectively. The brackets $[\phi]$ denote the jump in the quantity ϕ across the flame. The evolution of the front was determined by prescribing the flame speed

$$S_f \equiv \vec{V}\cdot\vec{n} - \nu_n \quad . \tag{4}$$

Thus, in these models, the fluid dynamical field on either side of the flame is assumed to be inviscid and incompressible, with given density jump $[\rho]$. If in addition S_f is known, a complete description of the problem is given in principle by solving Euler's equations subject to the conditions (1)-(3) and the equation prescribing S_f. For example, Landau [6] and Darrieus [2] assumed the flame speed of a perturbed flame front to be the constant speed S_f^o. Markstein [8] suggested a dependence of S_f on flame curvature via

$$S_f = S_f^o(1 + \mu/\mathcal{R})$$

where $\mathcal{R}$ is the local radius of curvature of the front and μ a phenomenological parameter. These studies and others along similar lines are summarized in Markstein [9].

There have been several recent works [1,14,11,5] which have derived, rather than prescribed, jump conditions for the fluid variables across the flame front and an equation for the flame speed S_f, by considering the flame structure and its interaction with the fluid flow. These analyses share common overall assumptions but differ somewhat in details. In all these works the flame front is described in Cartesian coordinates by the explicit form

$$x - f(y,z,t) = 0 \quad . \tag{5}$$

The flame structure consists of a thin boundary layer where transport processes dominate, in which there is another, much thinner boundary layer where chemical reaction takes place. If ℓ_D represents the characteristic length associated with diffusion and L a typical length of the problem, e.g. the scale of the outer fluid dynamical field, then the flame thickness is $O(\delta)$ where $\delta = \ell_D/L$. The reactive boundary layer is $O(\varepsilon\delta)$ where ε is inversely proportional to a representative activation energy of the chemical reactions. Since many of the reactions occurring in combustion in fact have large activation energies, the reaction rate is strongly temperature dependent, and the chemical reaction is indeed confined to a thin reactive diffusive layer. For example, typical flames have reaction zones ~ 10^{-3} mm and transport zones ~ 10^{-1} mm. On the fluid dynamical scale the flame including the reaction zone embedded in it may be regarded as a moving front. In order to account for the interaction of the chemical reactions and transport processes with the fluid flow, the solution inside each of the boundary layers is determined and matched with the outer flows, by the method of matched asymptotic expansions. This provides the jump conditions for the fluid variables across the flame and the equation for the speed S_f. The results are given as power series in δ, the leading terms of which correspond to the Landau-Darrieus model. In calculating the $O(\delta)$ correction terms, Clavin and Williams [1] and Pelce and Clavin [14] assumed small flame front deformations, i.e. $f = O(\delta)$ and small transverse velocities, i.e. $\vec{V} = u\vec{i} + O(\delta)$. In contrast, Matalon and Matkowsky [11] assumed all deformations to be O(1), thus considering arbitrary flame shapes in general fluid flows. Furthermore, they identified the $O(\delta)$ term in the expression for S_f, i.e. the deviations of the flame speed from its adiabatic value S_f^o, to be proportional to the local flame stretch. They also wrote the jump conditions in the form (1)-(3) with the right hand side replaced by $O(\delta)$ terms indicating that mass and momentum are not necessarily conserved across the flame front. We note that the expression for S_f was also derived by Joulin and Clavin [5]. Finally we point out that unlike the earlier treatments which assumed the flow field on either side of the flame to be inviscid, $O(\delta)$ viscous terms are incorporated in the present models.

The results described above are formulated in Cartesian coordinates and for flames given in the explicit form (5). Clearly it is important to have the jump conditions and the equation for S_f given in coordinate free form, in terms of the general description $F(\vec{X},t) = 0$ of the flame front. In the present paper we express these conditions in invariant form.

The flame speed. An expression for the flame speed was derived by Matalon and Matkowsky [11] as

$$S_f = 1 - \delta\alpha\kappa + o(\delta) \quad , \tag{6}$$

where the stretch factor κ was derived by Matalon and Matkowsky [11] in Cartesian coordinates, and by Matalon [10] in coordinate invariant form. The latter is given by

$$\kappa = \{\nu_n \nabla\cdot\vec{n} - \vec{n}\cdot\nabla\times(\vec{V}\times\vec{n})_{F=0}\}_{F=0} \tag{7}$$

where $\nu_n = -F_t/|\nabla F|$ is the normal velocity of the flame surface and $\vec{n}$ a unit normal vector pointing towards the burned gas as indicated earlier. The parameter α is defined as

$$\alpha = \beta + \frac{\gamma}{2}\ell \tag{8}$$

where ℓ is defined by the Lewis number Le as

$$Le = 1 + \varepsilon\ell \quad , \tag{9}$$

and β and γ depend only on the heat release q, as

$$\beta = \frac{1+q}{q}\ell n(1+q) \quad , \quad \gamma = \int_{-\infty}^{0} n(1+qe^x)dx > 0 \quad . \tag{10}$$

As mentioned above a phenomenological dependence of S_f on flame curvature was anticipated by Markstein [8] and a phenomenological dependence on the nonuniformity of the velocity field was mentioned by Eckhaus [3], both in a linearized form. By contrast the derived equation (6), which is exact to $O(\delta)$, describes non-linear deformations and does not contain any phenomenological parameters. Finally we note that equation (6) reveals an important relation between flame speed and flame stretch. At a point along the flame where the stretch κ is positive, the local flame speed decreases if $\ell > -\ell^*$ with $\ell^* = 2\beta/\gamma > 0$, and increases otherwise. This result should be contrasted with the intuitive argument in Lewis and von Elbe [7] indicating that positive stretch necessarily implies a slower flame speed.

The jump conditions. The conditions relating the fluid variables across the flame front derived in Matalon and Matkowsky [11], become, in coordinate free form:

$$[\rho(\vec{V}\cdot\vec{n} - \nu_n)] = \delta\ell n(1+q)\kappa + o(\delta) \tag{11}$$

$$[\vec{V}\times\vec{n}] = -\delta(Pr+\beta)\{[\nabla\times\vec{V}] + 2q\,\nabla\times\vec{n}\} + o(\delta) \tag{12}$$

$$[p + \rho(\vec{V}\cdot\vec{n})(\vec{V}\cdot\vec{n} - \nu_n)] = \delta\{\beta[\nabla p\cdot\vec{n}] - q(1+\beta)\nabla\cdot\vec{n} + \nu_n \ell n(1+q)\kappa\} + o(\delta) \quad . \tag{13}$$

Note that unlike (1)-(3) mass and momentum are not necessarily conserved across the flame. Equation (11) states that (locally) the net normal mass flux is proportional to the stretch factor κ. The difference between the mass flowing in and out of a volume element of thickness δ, is therefore either accumulated inside the element and/or transferred (tangentially) along the flame surface. Equation (12) states that the tangential fluid velocity vector is not necessarily continuous at the flame front. The $O(\delta)$ jump in $\vec{V}\times\vec{n}$ results from unbalanced tangential stresses. It is proportional to the vorticity $[\nabla\times\vec{V}]$ produced at the flame and to the rotation $\nabla\times\vec{n}$ of a surface element. We note in particular the dependence on both the heat release q, and the Prandtl number Pr. Finally, equation (13) states that a net normal momentum flux across the flame, results from unbalanced pressure gradients, flame curvature and from the momentum associated with the net excess mass given by (11).

The flow field. On either side of the flame, the fluid flow satisfies Euler's equations to leading order, with $O(\delta)$ correction terms accounting for viscous dissipation. These equations are

$$\nabla\cdot\vec{V} = 0 \tag{14}$$

$$\rho\left(\frac{\partial\vec{V}}{\partial t} + (\vec{V}\cdot\nabla)\vec{V}\right) = -\nabla p + \delta Pr\nabla^2 V \quad , \tag{15}$$

where the density function ρ is given by

$$\rho = \begin{cases} 1 & , \quad \text{for} \quad F < 0 \\ (1+q)^{-1} + o(\delta) & , \quad \text{for} \quad F > 0 \end{cases} \quad . \tag{16}$$

We note that up to and including $O(\delta)$ terms, the flow is essentially incompressible on either side of the flame front. In general, a curved flame produces nonuniform temperatures along its front so that the fluid behind is stratified. However these nonuniformities are $O(\varepsilon\delta)$ as discussed later.

The velocity and pressure fields on either side of the flame as well as the instantaneous shape of the flame front can now be found in principle, by solving equations (14) and (15), as power series in δ, subject to (7) and (11)-(13). The problem depends on all the relevant physicochemical parameters, namely the Lewis number Le, the Prandtl number Pr and the heat release q. To leading order, the problem is identical to that formulated in the 1940's and discussed earlier. The $O(\delta)$ terms, included here, account for the flame structure which was ignored in the earlier studies.

<u>The temperature field</u>. As discussed in Matalon and Matkowsky [11], the deviations of the temperature at the front from the adiabatic flame temperature $T_a \equiv 1 + q$ are usually small, and found to be $O(\varepsilon\delta)$. Thus, behind the flame, i.e. for $F > 0$

$$T = T_a + \varepsilon\delta\theta + \cdots \tag{17}$$

where the temperature perturbations θ are determined from

$$\frac{\partial\theta}{\partial t} + \vec{V}\cdot\nabla\theta = 0 \quad , \tag{18}$$

and along the flame front,

$$\theta_f = -\ell\gamma\kappa \quad . \tag{19}$$

Therefore the flame temperature is given by

$$T_f = 1 + q - \varepsilon\delta\ell\gamma\kappa \quad . \tag{20}$$

Thus a positive stretch is associated with a decrease in flame temperature if $\ell > 0$ (Le > 1) and with an increase of flame temperature if $\ell < 0$ (Le < 1). We recall that a positive stretch is associated with a decrease in the local flame speed if $Le > Le^*$ with $Le^* < 1$, and an increase in flame speed otherwise. Thus, at least for some range of Lewis numbers, specifically for $Le^* < Le < 1$, an increase in flame temperature is not associated with an increase in flame speed.

Finally, we remark that having derived expressions for the jumps in the fluid variables across the flame front, we can also derive an expression for the jump in vorticity across the flame (cf. Matalon and Matkowsky [11]). This jump corresponds to the vorticity produced in the flame. Even if the flow in fresh mixture is irrotational, the curved flame will generate vorticity. Our expression quantifies this result.

There are various applications of the simplified model we have derived. For example, there have been attempts to use conditions (1)-(3) together with assumption of constant flame speed, to numerically simulate flame propagation in complicated flow configurations. Our model thus provides more accurate prescriptions of these conditions for such simulations. Another example is the possible determination (in certain circumstances) of turbulent flame speed (cf. Clavin and Williams [1]). The model is particularly well suited for studying the stability properties of flames in various configurations. Particular attention is focused here on the stability of circular flames which reveal some characteristics that do not exist in the corresponding analysis of plane flames.

For plane flames, the linear stability analysis yields a dispersion relation expressing the growth rate σ of a disturbance of wave number k, as [4,14]

$$\sigma = Ak - \ell_D(B_1 + \ell B_2)k^2 + \cdots \quad , \tag{21}$$

where A, B_1 and B_2 depend only on q. For disturbances with large wavelengths, the first term in (21) is dominant; thus thermal expansion destabilizes the flame. However since experiments are performed in a confined environment, disturbances of very large wavelengths can often be ruled out. If under such circumstances the mobility of the deficient reactant is sufficiently low, i.e. $\ell > \ell_c$, plane flames are potentially stable. For $\ell < \ell_c$, an instability occurs which corresponds to the formation of cells (Sivashinsky [15]). We note that (21) does not exhibit the dependence of σ on the viscous effects which are presumably contained in higher order terms. This led to the conclusion that viscosity plays a secondary role compared to diffusion. While this is true for plane flames it is not necessarily true for curved flames as we now show.

For circular flames, stabilized on a point source, the linear stability analysis yields a dispersion relation expressed as power series in 1/R, R being the radius of the steady flame. Thus [12]

$$\sigma = C\frac{1}{R} - \ell_D(E_1 + \ell E_2 - PrE_3)\frac{1}{R^2} + \cdots \qquad (22)$$

where C, E_1, E_2 and E_3 depend on q and on the wavenumber n (an integer). For large circular flames, the first term in (22) is dominant and the flame is shown to be unstable. However circular flames of smaller radius are potentially stable [13] if $\ell > \ell_c$ and $Pr < Pr_c$. Clearly viscosity has a destabilizing effect in this problem and unlike the plane case, its role is comparable to the role of diffusion. We note in particular that equation (22) is derived for $R \gg \ell_D$. Thus even a small curvature is sufficient to amplify the effect of viscosity which plays a more significant role than previously anticipated. Finally, we point out that the dispersion relation (22) depends on the integer n, where n/R is essentially the wavenumber of the disturbance. For large n the flame front behaves locally as if it were plane, and indeed it can be shown that in the limit, $C \to A$, $E_1 \to B_1$, $E_2 \to B_2$ and $E_3 \to 0$ so that we recover (21).

REFERENCES

1. Clavin, P. and Williams, F. A. 1982, "Effects of molecular diffusion and of thermal expansion on the structure and dynamics of premixed flames in turbulent flows of large scales and low intensity, J. Fluid Mech. 116, pp. 251-282.

2. Darrieus, G. 1945, "Propagation d'un front de flamme," presented at Le congres de Mecanique, unpublished.

3. Eckhaus, W. 1961, "Theory of flame front stability," J. Fluid Mech. 10, pp. 80-100.

4. Frankel, M. L. and Sivashinsky, G. I. 1983, "The effect of viscosity on hydrodynamic stability of a plane flame front," Combustion Science and Technology (to appear).

5. Joulin, G. and Clavin, P. 1982, "Note on premixed flames in large scales and high intensity turbulent flow," (submitted).

6. Landau, L. D. 1944, "On the theory of slow combustion," Acta Physicochimica URSS 19, p. 77. (Also collected papers by Landau, L. D., Gordon and Breach, 1967).

7. Lewis, B. and von Elbe, G. 1967, Combustion, Flames, and Explosions of Gases, 2nd Ed., Academic Press.

8. Markstein, G. H. 1951, "Experimental and theoretical studies of flame front stability," J. Aero. Sci. 18, p. 199.

9. Markstein, G. H. 1964, Nonsteady Flame Propagation, Pergamon Press, Oxford.

10. Matalon, M. 1983, "On flame stretch," Combustion Science and Technology 29, pp. 225-238.

11. Matalon, M. and Matkowsky, B. J. 1982, "Flames as gasdynamic discontinuities," J. Fluid Mech. 124, pp. 239-260.

12. Matalon, M. and Matkowsky, B. J. 1983, "The stability of flames: hydrodynamic and diffusional thermal effects," SIAM J. Appl. Math. (to appear).

13. Matkowsky, B. J., Putnick, L. J. and Sivashinsky, G. I. 1980, "A nonlinear theory of cellular flames," SIAM J. Appl. Math. 38, pp. 489-504.

14. Pelce, P. and Clavin, P. 1982, "Influence of hydrodynamics and diffusion upon the stability limits of laminar premixed flames," J. Fluid Mech. 124, pp. 219-238.

15. Sivashinsky, G. I. 1983, "Instabilities, pattern formation, and turbulence in flames," Annual Review of Fluid Mechanics (to appear).

SECONDARY BIFURCATION IN FLAME PROPAGATION

Thomas Erneux and Bernard J. Matkowsky

Dept. of Engineering Sciences and Applied Mathematics
Northwestern University
Evanston, Illinois 60201 USA

1. INTRODUCTION

A great deal of work has been done in attempts to explain the transition from laminar to turbulent fluid flow via successive instabilities, as a critical parameter e.g. the Reynolds number, is varied. The successive instabilities are associated with a sequence of bifurcations, each generating more complex spatial and temporal patterns. Recently, similar attempts have been made to explain the transition from laminar to turbulent flame propagation via a succession of bifurcations and their accompanying instabilities. In this sequence, the steadily propagating planar flame loses stability to successively more complex modes of propagation such as cellular, pulsating or multi-periodic pulsating flames.

In this paper, we present two recent results of our work in this area. First, we consider the transition from periodic pulsating to quasi-periodic pulsating flames. Then, we consider the transition from a stationary cellular flame to a pulsating cellular flame. In each case, we characterize the transition as a secondary (or higher order) bifurcation.

The model we employ in the first study involves a general system of reaction-diffusion equations. This system includes the diffusional-thermal model [1], in which the effect of a given fluid dynamical field is employed in the transport (reaction-diffusion) equations describing the evolution of the temperature field and the concentrations of the chemical species participating in the chemical reactions. Thus the qualitative effects of thermal expansion are considered to be weak, and

G. Nicolis and F. Baras (eds.), Chemical Instabilities, 147–158.

transport effects on the underlying fluid dynamical field are not taken into account. In the second study, such effects are taken into account, but only to a limited extent, so that there is some coupling between the equations of fluid dynamics and the transport equations. Specifically the coupling, due to thermal expansion, enters only in the external force field, though not elsewhere in the fluid dynamical equations, and the result is a Boussinesq type model [1]. The problem we study is a limiting form of this model. For an account of recent work in which thermal expansion is fully taken into account, and the full interaction of the flame with the fluid dynamical field through which it propagates is studied, consult the articles by Clavin and Matalon and Matkowsky in this volume.

In addition, in each of these studies, the Mach number, representing the ratio of the flame speed to a characteristic sound speed is considered to be small, so that the resulting process is essentially isobaric. Finally we also employ the thin flame approximation, which is due to the relatively large activation energies in the chemical reactions considered. Thus, chemical reactions are important only in a thin region surrounding a moving surface called the flame front, where the temperature first approaches its burned value. On the unburned side of this region, chemical reactions are negligible because the temperature is too low, while on the burned side, they are negligible because the reactions have essentially gone to completion.

2. TRAVELLING WAVES ALONG THE FRONT OF A PULSATING FLAME

In this section, we describe both periodic and quasi-periodic waves travelling along the front of a pulsating flame. The model we employ may be written in the form

$$F(u,\psi,\lambda) = 0 \tag{2.1}$$

where F represents the operator describing the nonlinear reaction-diffusion equations, and u represents the vector of concentrations and temperature. The function ψ corresponds to the position of the flame front, and λ is the bifurcation parameter, e.g. the Lewis number Le, defined as the ratio of the thermal conductivity to the molecular diffusivity of the limiting reactant. In the governing equations (2.1), u and ψ are functions of t, x, y and $z' = z - \psi(x,y,t)$ where $-z$ is the direction of propagation of the flame. Thus, we consider a coordinate system moving with the flame.

Equations (2.1) admit a simple solution called the basic state which corresponds to the steady planar flame given by,

$$u = u_0(z',\lambda) \quad , \quad \psi = \psi_0 = - t \quad . \tag{2.2}$$

Its stability can be analyzed by studying the linearized problem

$$F_{u,\psi}(u_0,\psi_0,\lambda)\Phi = 0 \quad . \tag{2.3}$$

Typical results of this analysis in the context of flame propagation are given in Figure 1 (e.g. [2]), which is a plot of the neutral stability curves. Here k is the wave number of a

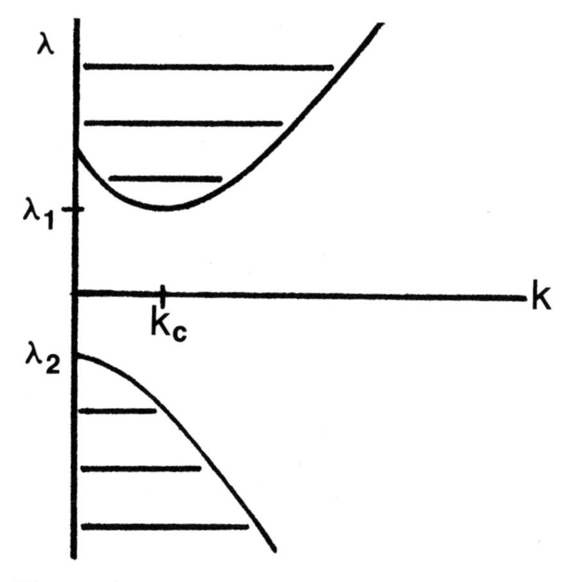

Figure 1

perturbation in the x and y directions. For $\lambda > \lambda_1$, the instability corresponds to the formation of travelling waves along the front of a pulsating flame. For $\lambda < \lambda_2$, the instability corresponds to the formation of stationary cellular flames. Both pulsating and cellular flames have been observed experimentally (cf. [3]).

When λ is slightly above λ_1, we expect a transition from the unstable steady state to travelling waves. The simplest waves correspond to two dimensional (y,z') flames of the form

$$u = u_0(z',\lambda) + [\alpha(\tau,\varepsilon)q(z')e^{i(T+Y)} + \beta(\tau,\varepsilon)q(z')e^{i(T-Y)}$$
$$+ c.c.] + v(z',Y,T,\tau,\varepsilon) \tag{2.4a}$$

$$\psi = \psi_0(T) + \alpha(\tau,\varepsilon)e^{i(T+Y)} + \beta(\tau,\varepsilon)e^{i(T-Y)} + c.c. + w(Y,T,\tau,\varepsilon) \qquad (2.4b)$$

where ε is a small parameter defined by

$$\varepsilon \equiv [(\lambda - \lambda_1)/c]^{1/2} \qquad (2.5)$$

with $c = \pm 1$, and T, Y, τ are given by

$$T \equiv [\sigma_0 + \varepsilon^2\sigma(\varepsilon^2)]t \quad , \quad Y = k_c y \quad , \quad \tau = \varepsilon^2 t \quad . \qquad (2.6)$$

Here c.c. denotes complex conjugate, α and β are two $O(\varepsilon)$, complex amplitudes which are determined from the nonlinear analysis. They multiply the two (complex) eigenfunctions that span the nullspace N of (2.3) with $\lambda = \lambda_1$ ($\varepsilon = 0$). In contrast, v and w are $O(\varepsilon^2)$ quantities and are orthogonal to N. Thus, for small ε, the deviations $u - u_0$ and $\psi - \psi_0$ are, to a first approximation, linear combinations of the modes $\exp(\pm i(T \pm Y))$.

The unknown amplitudes α and β are determined by solving a system of ordinary differential equations, called the bifurcation equations. In [4,5], it is shown that the symmetries present in the governing equations (2.1) and the eigenfunctions of equations (2.3) are reflected in corresponding symmetry properties of the bifurcation equations. Specifically, since the problem is invariant with respect to translations in T and Y, and with respect to reflections in Y, the analysis leads to bifurcation equations of the form

$$\varepsilon^2\alpha_\tau = \alpha f(\varepsilon^2 c, \varepsilon^2\sigma(\varepsilon^2), \alpha\bar{\alpha}, \beta\bar{\beta})$$

$$\varepsilon^2\beta_\tau = \beta f(\varepsilon^2 c, \varepsilon^2\sigma(\varepsilon^2), \beta\bar{\beta}, \alpha\bar{\alpha})$$

$$\alpha(0) = \alpha_0(\varepsilon) \quad , \quad \beta(0) = \beta_0(\varepsilon) \quad . \qquad (2.7)$$

We solve these equations, by expanding α, β and σ as power series in ε in the form

$$\alpha(\tau,\varepsilon) = \varepsilon\alpha_1(\tau) + \varepsilon^2\alpha_2(\tau) \cdots \quad ,$$

$$\beta(\tau,\varepsilon) = \varepsilon\beta_1(\tau) + \varepsilon^2\beta_2(\tau) \cdots \quad ,$$

$$\sigma(\varepsilon^2) = \omega + \varepsilon^2\sigma_2 \cdots \quad . \qquad (2.8)$$

Then, from (2.7), we find that α_1 and β_1 must satisfy:

$$\alpha_{1\tau} = \alpha_1(ac + b\omega + A\alpha_1\overline{\alpha}_1 + B\beta_1\overline{\beta}_1)$$

$$\beta_{1\tau} = \beta_1(ac + b\omega + A\beta_1\overline{\beta}_1 + B\alpha_1\overline{\alpha}_1)$$

$$\alpha_1(0) = \alpha_{10} \quad , \quad \beta_1(0) = \beta_{10} \tag{2.9}$$

The complex coefficients a, b, A and B have been determined by Matkowsky and Olagunju [2], for the specific combustion model derived in [1]. Except for the fact that $Re(a) > 0$, there are in general no restrictions on the values of the coefficients.

The steady state solutions of equations (2.9) correspond to time-periodic solutions of equations (2.1). The linear stability of a specific solution can be determined from an analysis of the Jacobians of equation (2.9) and its complex conjugate evaluated at this solution. The possible steady states and their stability properties are summarized as follows:

(i) the basic state,

$$|\alpha_1| = |\beta_1| = 0 \tag{2.10a}$$

are stable iff

$$c = -1 \tag{2.10b}$$

(ii) the travelling waves,

$$|\alpha_1| = 0 \quad , \quad |\beta_1| \neq 0 \quad (\text{and} \, |\alpha_1| \neq 0 \, , \, |\beta_1| = 0) \tag{2.11a}$$

correspond to "pure" mode solutions. These solutions are stable iff

$$Re(A) < 0 \quad \text{and} \quad Re(B-A) < 0 \tag{2.11b}$$

(iii) the standing waves,

$$|\alpha_1| = |\beta_1| \neq 0 \tag{2.12a}$$

correspond to mixed-mode solutions. These solutions are stable iff

$$Re(A+B) < 0 \quad \text{and} \quad Re(A-B) < 0 \quad . \tag{2.12b}$$

A stability diagram of the possible solutions is shown in Figure 2. We note that standing and travelling waves can never be stable for the same values of the parameters A, B, a and b.

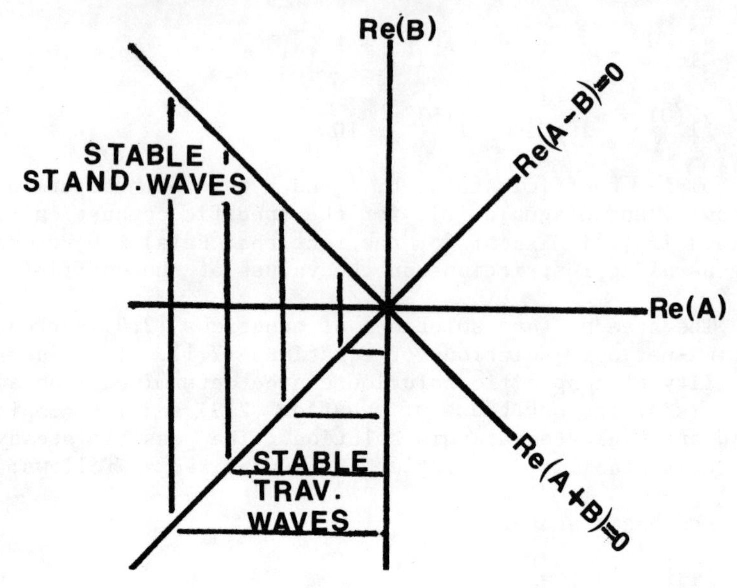

Figure 2

We have determined steady solutions of (2.9). They describe periodic solutions of (2.1). We now wish to describe quasi-periodic solutions of these equations. To do so, we consider equations (2.9) with $\alpha_1 = \rho \exp(i\theta)$, $\beta_1 = \delta \exp(i\phi)$. Then (2.9) becomes

$$\rho_\tau = \rho(P + A_1\rho^2 + B_1\delta^2)$$

$$\delta_\tau = \delta(P + A_1\delta^2 + B_1\rho^2)$$

$$\rho\theta_\tau = \rho(Q + A_2\rho^2 + B_2\delta^2)$$

$$\delta\phi_\tau = \delta(Q + A_2\delta^2 + B_2\rho^2)$$

$$\rho(0) = \rho_0 \quad , \quad \delta(0) = \delta_0 \quad , \quad \theta(0) = \theta_0 \quad , \quad \phi(0) = \phi_0 \qquad (2.13)$$

where

$$P \equiv a_1c + b_1\omega \quad , \quad Q \equiv a_2c + b_2\omega \quad .$$

The subscripts 1 and 2 in a, b, A and B denote the real and imaginary parts of these quantities respectively. When

$$B_1 - A_1 = 0 \tag{2.14}$$

the solution of (2.13) admits the following asymptotic behavior:

$$\begin{pmatrix} \rho^2 \\ \delta^2 \end{pmatrix} \underset{\tau\to\infty}{\longrightarrow} -\begin{pmatrix} \rho_0^2 \\ \delta_0^2 \end{pmatrix} \frac{1}{(\rho_0^2 + \delta_0^2)} \frac{P}{A_1}$$
$$\tag{2.15}$$
$$\begin{pmatrix} \theta \\ \phi \end{pmatrix} \underset{\tau\to\infty}{\longrightarrow} \left[Q - \frac{P}{(\rho_0^2 + \delta_0^2)A_1} \begin{pmatrix} A_2\rho_0^2 + B_2\delta_0^2 \\ A_2\delta_0^2 + B_2\rho_0^2 \end{pmatrix} \right] \tau$$

Thus, we have found that there exists a family of steady state solutions which satisfies the relation:

$$\rho^2 + \delta^2 = -\frac{P}{A_1} > 0 \quad . \tag{2.16}$$

We note that there are solutions with $\rho^2 \neq \delta^2 \neq 0$. Since θ and ϕ evolve differently for $\rho_0^2 \neq \delta_0^2$, they lead to different frequency corrections for the two propagating modes. Thus, the corresponding solution of equations (2.1) is quasi-periodic.

We now consider the more general case of

$$(B_1 - A_1)/e \equiv \eta^2 \ll 1 \tag{2.17}$$

with $e = \pm 1$ and

$$\lambda - \lambda_1 = \eta^2 p + O(\eta^4) \quad , \quad \sigma = \sigma_0 + \eta^2\omega + O(\eta^4) \quad . \tag{2.18}$$

The analysis of the bifurcation equations in this case indicates that in addition to the primary states given by (2.11) and (2.12), there exists a secondary bifurcation to quasi-periodic solutions which satisfies (2.16) with $P \equiv a_1 p + b_1\omega$. The condition for its existence and its location depends on higher order terms in the bifurcation equations. This analysis is carried out by Erneux and Matkowsky in [6]. Figure 3 exhibits a typical bifurcation diagram of the amplitude as a function of λ.

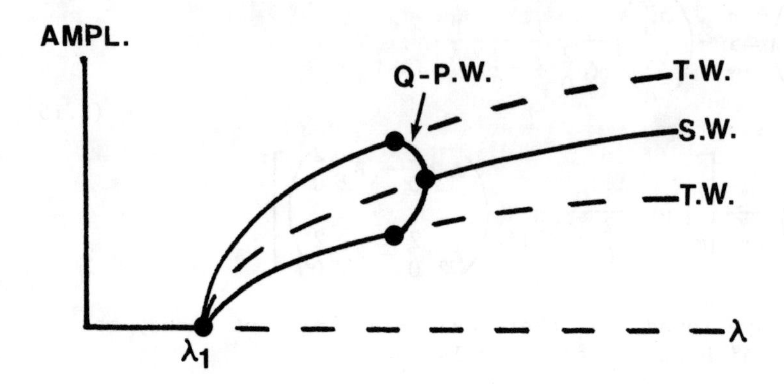

Figure 3

3. PULSATING CELLULAR FLAMES

We have exhibited periodic and quasi-periodic behavior on the upper branch of the neutral stability curves. In this section, we consider the Boussinesq type model [1] near the lower branch of the neutral stability curves and exhibit a tertiary bifurcation to periodic behavior on this branch.

The basic state then becomes unstable with respect to steady non-uniform modes in x and y and we observe a transition to stationary cellular flames. Since the instability is preferential with respect to zero wave number perturbations, the bifurcation problem can be simplified by seeking new evolution equations valid for small wave number perturbations. This analysis leads to a nonlinear equation for the flame shape $\phi(x,y,t) = \psi(x,y,t) - \psi_0$ given by

$$\phi_t + \frac{1}{4} \nabla^4 \phi + \nabla^2 \phi + \frac{1}{2} |\nabla \phi|^2 + G\phi = 0 \tag{3.1}$$

where ϕ is subject to boundary and initial conditions, appropriate to the particular physical situation that is modeled.

In this equation $G > 0$ represents a nondimensional parameter describing the stabilizing effect of e.g. gravity or the curvature of a cylindrical (spherical) flame. These are described in the recent review papers [3,7]. A similar equation has also been obtained for a different class of reaction diffusion equations [8].

We now analyze equation (3.1) in detail. Our purpose is to show that the interaction of steady modes may lead to a bifurcation to time-periodic solutions. The problem is motivated by the observation that polyhedral flames on a Bunsen burner may sometimes rotate about its vertical axis. Similar bifurcation problems were recently discussed in the context of convective instabilities [9]. Here we analyse the simplest case of a two-dimensional (z',y) flame subject to zero-flux boundary conditions in the y-direction:

$$\phi_y = \phi_{yyy} = 0 \quad \text{at} \quad y = 0 \text{ and } L \quad . \tag{3.2}$$

We show that a tertiary bifurcation to stable time-periodic solution can be observed when G is progressively decreased.

From the linear stability analysis of the zero solution $\phi = 0$, we note that when

$$\begin{aligned} L &= L_c = \sqrt{20}\,\pi \\ G &= G_c = 1/25 \quad , \end{aligned} \tag{3.3}$$

the linearized equation admits a double eigenvalue and the eigenfunctions that span its nullspace M are $\cos(\pi Y)$ and $\cos(2\pi Y)$ where $Y \equiv y/L$. Defining a small parameter ν by:

$$\nu \equiv (L - L_c)/d \quad (\text{where } d = \pm 1 \quad \text{when} \quad L - L_c \gtrless 0) \tag{3.4}$$

and assuming that

$$G - G_c = \nu g(\nu) \quad (\text{where} \quad g(\nu) = O(1)) \quad , \tag{3.5}$$

bifurcation theory allows us to seek a solution of (3.1) of the form

$$\phi(Y,t,\tau,\nu) = \alpha_1(\tau,\nu)\cos(\pi Y) + \alpha_2(\tau,\nu)\cos(2\pi Y) + \gamma(Y,t,\tau,\nu) \tag{3.6}$$

where $\tau \equiv \nu t$, α_1 and α_2 are $O(\nu)$ amplitudes and γ is orthogonal to M and is $O(\nu^2)$. The functions α_1 and α_2 satisfy a system of ordinary differential equations. Our analysis considers several cases, the details of which will be presented elsewhere [10]. We found that bifurcation to time-periodic solutions occurs when

$$\mu_2 - \mu_1 > 0 \quad \text{and} \quad G - \mu_2 = O(\nu^2) \tag{3.7}$$

where μ_1 and μ_2 correspond to the primary bifurcation points associated with the modes $\cos \pi Y$ and $\cos 2\pi Y$. They are defined by:

$$\mu_j(L(\nu)) \equiv \frac{j^2\pi^2}{L^2}\left(1 - \frac{4j^2\pi^2}{L^2}\right) \quad (j = 1,2) \tag{3.8}$$

Under the conditions (3.7), $\alpha_1 = O(\nu^{3/2})$, $\alpha_2 = O(\nu)$ and the bifurcation equations are given by

$$\begin{aligned} \nu\alpha_{1\tau} &= \alpha_1[-(\mu_2 - \mu_1) - \tfrac{1}{20}\alpha_2] + O(\nu^{7/2}) \\ \nu\alpha_{2\tau} &= -\alpha_2(G - \mu_2) + \tfrac{1}{80}\alpha_1^2 - \tfrac{1}{180}\alpha_2^3 + O(\nu^4) \end{aligned} \tag{3.9}$$

Equation (3.9) has been derived by a Galerkin procedure. In addition, our results have also been derived by a multi-time perturbation analysis [10].

From a study of equations (3.9), we observe the following sequence of bifurcations (see Figure 4) as G is progressively decreased:

1. When $G = \mu_2$, the basic state, $\alpha_1 = \alpha_2 = 0$ transfers its stability to the primary bifurcated pure-mode, steady state,

$$\alpha_1 = 0 \quad , \quad \alpha_2 = \pm\,[-180(G - \mu_2)]^{1/2} \tag{3.10}$$

2. When $G = \mu_s \equiv \mu_2 - \frac{20}{9}(\mu_2 - \mu_1)^2$, we observe a secondary bifurcation to the steady mixed mode state

$$\begin{aligned} \alpha_1 &= \pm\,[80\,\alpha_{2s}(G - \mu_s)]^{1/2} \\ \alpha_2 &= \alpha_{2s} \equiv -20(\mu_2 - \mu_1) \end{aligned} \tag{3.11}$$

3. When $G = \mu_H \equiv \mu_2 - \frac{20}{3}(\mu_2 - \mu_1)^2$, we observe two tertiary bifurcations to time-periodic solutions. An analysis of

equations (3.1) when $|G - \mu_H| \to 0$ indicates that these time-periodic solutions emerge as stable solutions and that their amplitude grows as $1/\nu$.

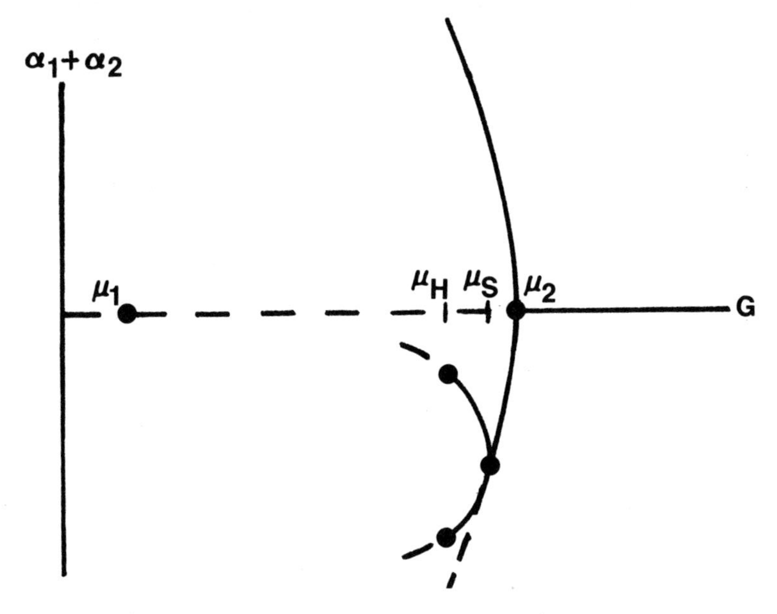

Figure 4

REFERENCES

1. Matkowsky, B. J. and Sivashinsky, G. I. (1979), SIAM J. Appl. Math. 37, pp. 686-700.

2. Matkowsky, B. J. and Olagunju, D. O. (1982), SIAM J. Appl. Math. 42, pp. 486-501.

3. Sivashinsky, G. I. (1983), Ann. Rev. Fluid Mech. 15, pp. 179-199.

4. Sattinger, D. H. (1980), Bull. Am. Math. Soc. 3, pp. 779-819.

5. Erneux, T. (1981), J. Math. Biol. 12, pp. 199-214.

6. Erneux, T. and Matkowsky, B. J. (1983), in preparation.

7. Margolis, S. B. and Matkowsky, B. J. (1983), Comb. Sci. and Tech., in press.

8. Kuramoto, Y. (1980), Prog. of Theor. Physics 63, pp. 1885-1903.

9. Erneux, T. and Reiss, E. L. (1983), SIAM J. Appl. Math. in press.

10. Erneux, T. and Matkowsky, B. J. (1983), in preparation.

MOLECULAR DYNAMICS SIMULATION OF THERMAL EXPLOSION

Dong-Pao Chou[a] and Sidney Yip[b]

[a]Institute of Nuclear Energy Research, Lung-Tan, Taiwan 325. [b]Massachusetts Institute of Technology, Cambridge, Massachusetts 02139 USA

The temperature instability of a two-dimensional reactive fluid of N hard disks bounded by heat conducting walls has been studied by molecular dynamics simulation. The collision of two hard disks is either elastic or inelastic (exothermic reaction), depending on whether the relative kinetic energy at impact exceeds a prescribed activation barrier. Heat removal is accomplished by using a wall boundary condition involving diffuse and specular reflection of the incident particles. Critical conditions for ignition have been obtained and the observations compared with continuum theory results. Other quantities which can be studied include temperature profiles, ignition times, and the effects of local fluctuations.

INTRODUCTION

The fundamental problem in thermal explosion phenomena is the critical behavior of a body of material which is capable of self-heating while being surrounded by a thermal reservoir held at a constant temperature [1-3]. As the system heats up an instability will result if the heat generated within the body exceeds that removed by the reservoir, in which case the body temperature will show a sharp increase in its time evolution. The conditions relating the system size and material characteristics under which such behavior can occur are clearly of paramount interest, along with the induction period (time from initial heating to the onset of ignition) and the final temperature the system can reach.

In the continuum theory description thermal ignition is

G. Nicolis and F. Baras (eds.), Chemical Instabilities, 159–170.

modeled by the time-dependent heat conduction equation with an Arrhenius reaction term [1,3]. Criticality of the system is expressed by a dimensionless parameter δ, the Frank-Kamenetskii parameter, which is a measure of the relative importance of heat generation to heat removal by conduction. Much of the existing analysis has assumed constant reactant concentration, and one finds that not only is there a critical value of δ, but also the ratio of activation energy for reaction to the reservoir temperature must exceed a certain minimum value [4-8]. Because the reaction term introduces exponential nonlinearities, only limited analysis of the time-dependent equation has been made [9]. Correspondingly, results for the induction period are quite incomplete.

Molecular dynamics is a well-established discrete particle simulation technique for calculating structural and dynamical properties of many-body systems [10]. Besides extensive applications in the study of simple fluids [11], a few investigations of chemical reactions have been made [12,13], including reactions with exothermicity [14-16]. In this paper we describe a molecular dynamics study of thermal ignition in a slab of two-dimensional hard sphere fluids bounded by heat conducting walls [14]. We demonstrate the feasibility of simulating thermal explosion at the atomistic level by direct observation of ignition behavior. The critical values of δ derived from the data are found to approach the continuum theory prediction as the system size increases. Because microscopic fluctuations are not averaged out in the basic simulation output, it would appear that one can potentially generate valuable information on nonlinear fluctuations and instabilities in a chemical system.

CONTINUUM THEORY

A complete description of thermal explosion involves the consideration of both the system temperature and the reactant concentration. If one is concerned only with ignition, then it is reasonable to neglect reactant consumption. Moreover, for the analysis of critical conditions it is sufficient to consider only the steady-state solution to the temperature equation. When cast in a dimensionless form, this equation reads

$$\frac{d^2\theta}{dx^2} = -\,\delta\,\exp\left(\frac{\theta}{1+\varepsilon\theta}\right) \qquad (1)$$

where

$$\theta = \frac{E}{RT_o^2}\,(T - T_o) \qquad (2)$$

$$x = r/r_o \tag{3}$$

$$\delta = \frac{ZQEr_o^2}{kRT_o^2} e^{-E/RT_o} \tag{4}$$

$$\varepsilon = RT_o/E \tag{5}$$

In these expressions we have defined activation energy E, wall temperature T_o, heat of reaction Q, collision frequency Z, slab half thickness r_o, thermal conductivity k, and gas constant R. Typically, the boundary conditions are $\theta(x=1) = 0$ and $(d\theta/dx)_{x=0} = 0$.

Eq.(1) is simply a statement of balance between heat conduction and heat generation. The significance of the Frank-Kamenetskii parameter δ is that it is a measure of the relative importance of heat removal by conduction to heat production. We can expect that solutions will exist only for certain values of δ, and hence the condition for criticality. Notice that δ is not the only parameter in the problem, the ratio of wall temperature to activation energy, ε, also appears. Criticality, therefore, will depend on both ε and δ.

Although Eq.(1) has no known analytic solutions, several approximations exist which enable one to examine essentially all the characteristic features of the problem [4-6]. In addition, numerical methods are available for obtaining very accurate results [7,8]. The basic approximation is to simplify the temperature dependence introduced by the Arrhenius factor. Setting the right-hand side of Eq.(1) equal to $\delta f(\theta)$, one can write

$$f(\theta) = \exp\left(\frac{\theta}{1+\varepsilon\theta} \right)$$

$$\approx \exp(\theta) \tag{6}$$

This approximation, valid when activation energy is high compared to ambient temperature, renders Eq.(1) analytically tractable as originally shown by Frank-Kamenetskii [1-3]. The solution can be expressed as

$$\theta(x) = \theta_M - 2 \ln(\cosh\alpha x) \tag{7}$$

where $\alpha = (\delta \exp(\theta_M)/2)^{\frac{1}{2}}$, and $\theta_M = \theta(x=0)$ is the maximum temperature. The expression for θ_M is

$$\theta_M = \ln(2\alpha^2/\delta) \tag{8}$$

where $\cosh\sigma = \sigma(2/\delta)^{1/2}$. From this one sees that for a given δ there will be either two solutions for θ_M or no solutions.

The variation of θ_M is illustrated by the dashed curve in Fig. 1. In the region of doubled-valued solutions the system can exist in a low-temperature or a high-temperature steady-state. As δ increases toward the critical value, the two solutions move toward each other and eventually merge at $\delta_c(\varepsilon=0)$, a value which can be determined by differentiating Eq.(1). (One finds δ_c = 0.878 at which $\theta_M = 1.19$.) For $\delta > \delta_c$, the system will not have a steady-state solution because it is then in a supercritical state.

The Frank-Kamenetskii approximation reduces the analysis to a one-parameter description; it is useful for showing that the competition between heat production and heat removal indeed leads to the existence of criticality. However, this approximation ignores another important aspect of chemical kinetics, that of finite reaction activation energy. The effect of nonzero ε in $f(\theta)$ can be analyzed using a number of different approximations [4-6]. By writing

$$f(\theta) \approx \exp\left(\frac{\theta}{1+\varepsilon\theta_M} \right) \tag{9}$$

one can introduce a change of variables, $\theta' = \theta(1 + \varepsilon\theta_M)$ and $\delta' = \delta(1 + \varepsilon\theta_M)$, and again make use of the Frank-Kamenetskii solution, Eqs. (7) and (8). Eq.(9) is one of several approximations suggested which are all essentially based on the idea of expanding the temperature about the maximum value [4-6].

The finite-ε analyses give a θ_M behavior qualitatively different from the Frank-Kamenetskii description. As shown in Fig. 1, the system now has either a triple-valued or a single-valued solution depending on ε and δ. There exists a special value of ε, which we denote as ε_o, beyond which only single-valued solutions exist. (This value, which varies with geometry, has been estimated to be $\varepsilon_o = 0.245$ in the case of a slab.) For $\varepsilon < \varepsilon_o$, one has triple-valued solutions provided $\delta^c(\varepsilon) < \delta < \delta_c(\varepsilon)$. Notice the appearance of two points of tangency, δ^c and δ_c. The significance of δ_c is similar to the critical δ in the Frank-Kamenetskii description. Suppose $\varepsilon < \varepsilon_o$ and δ is just slightly less than $\delta_c(\varepsilon)$, the system can reach a low-temperature steady-state on curve A. Now, if δ is increased so that it is just slightly greater than $\delta_c(\varepsilon)$, the temperature will go to a much higher steady-state value on curve C. This sudden change in the steady-state temperature is the property that defines criticality; in other words, for a critical system there will be a temperature gap in which no steady-state solution is possible.

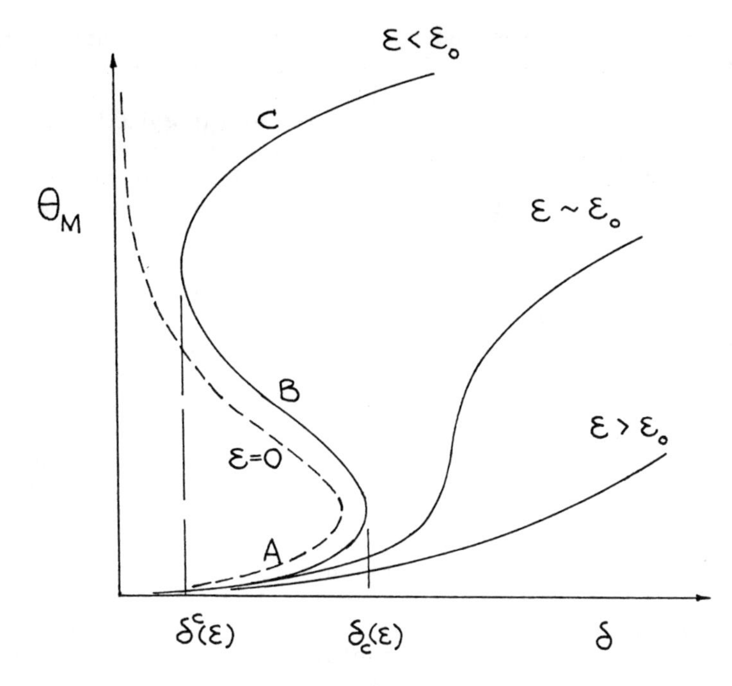

Figure 1. Dependence of θ_M on δ at various ε (schematic), region of multiple steady-state solutions is defined by $\delta^c(\varepsilon) < \delta < \delta_c(\varepsilon)$ with $\varepsilon < \varepsilon_o$. δ_c and δ^c are the critical values for ignition and extinction, respectively. For $\varepsilon > \varepsilon_o$, the solution does not show critical behavior in the sense of a jump condition, large change in θ_M for a small variation in δ.

We will henceforth refer to $\delta_c(\varepsilon)$ as the critical value for ignition. A similar discussion can be given to the high-temperature branch, curve C, to show that $\delta^c(\varepsilon)$ is the critical point for extinction. One also can show that curve B, which lies between $\delta^c(\varepsilon)$ and $\delta_c(\varepsilon)$, is unstable, so that there are still only two steady-state temperatures for the system. For $\varepsilon > \varepsilon_o$, θ_M rises monotonically with δ as any steady-state value can be reached. In this case, there is no criticality and one expects no instabilities.

MOLECULAR DYNAMICS SIMULATION

The initial objective of this work is to demonstrate the feasibility of simulating the temperature instability associated with thermal ignition. To keep the simulation model as simple as possible, we choose to work with a two-dimensional hard sphere system in plane geometry. The use of hard core interaction considerably facilitates the calculation while one expects the essential features of ignition behavior should not be affected by the details of interatomic collisions. For the present study, two modifications of the standard molecular dynamics techniques for hard core systems [17] are needed. The first is that one must introduce reactions into the dynamics, and secondly, one must provide for heat flow across the walls of the simulation system.

A simple way of simulating collision-induced reactions is to calculate the relative kinetic energy E_r of a pair of colliding hard disks at the instant of impact. If this value is greater than a prescribed threshold E, then the collision will be treated as a reaction or inelastic collision, otherwise it will be treated as a normal elastic collision. In the event of a reaction, an amount of energy Q will be released through the outgoing velocities of the colliding disks. The fraction of Q given to each disk is governed by the laws of kinematics as expressed by the conservation of momentum and total energy. One could label the disks and distinguish between those capable of exothermic reaction (excited state particles) and those incapable of reaction (ground state particles). In the present simulation we have assumed no depletion of reactants, so in effect every particle is capable of reaction even if it has reacted in the preceding collision.

In order to maintain close correspondence between simulation and the classical theory of thermal ignition, Eq.(1), it is necessary to verify that our method of simulating reactions does give a reaction rate having an Arrhenius temperature dependence. From the simulation output we have determined the ratio of reaction rate to total collision rate. This ratio should have the behavior of $\exp(-E/RT)$. As shown in Fig. 2, the data indeed follow this expected temperature variation. The same result also was obtained at a lower density $n^* = nd^2 = 0.157$, where n is the number density and d the hard disk diameter.

The modification which allows heat transfer across the system walls consists of treating the boundaries of the square system along one direction (say the x-direction) to have fixed walls, while the boundaries along the other direction are periodic in the usual way. When a particle hits the fixed wall, it is reflected in one of two ways, specular or diffuse. In the case of specular reflection, $v_x' = -v_x$, $v_y' = v_y$, where the prime denotes the reflected velocity component and the unprimed components refer to the incident velocities. In the case of diffuse reflection, the reflected velocity is chosen from a Maxwellian distribution at the wall temperature consistent with detailed balance [14]. The ratio of diffuse to specular reflection is chosen to be unity, since this choice leads to results for the

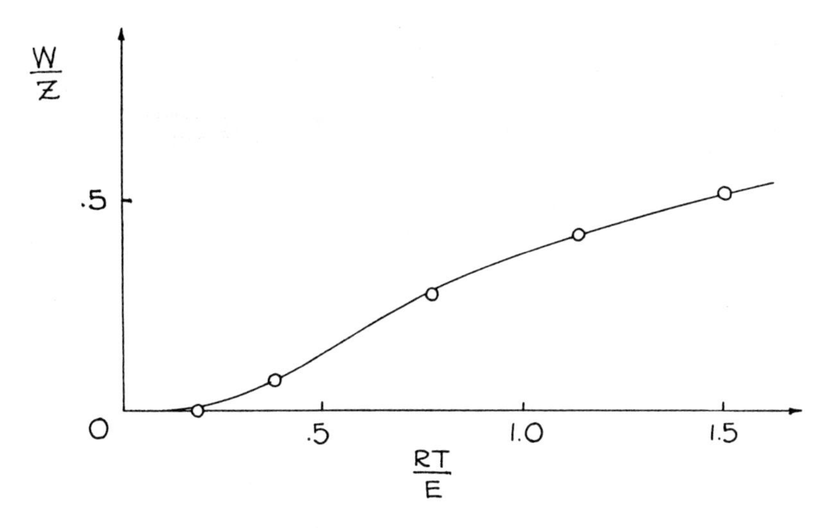

Figure 2. Temperature dependence of the ratio of reaction rate W to collision rate Z, molecular dynamics data (circles) and the Arrhenius dependence exp(-E/RT) (curve).

thermal conductivity, when the boundary condition is tested in a simulation of colliding hard disk fluids without reactions, that is in agreement with the prediction of the Enskog theory for two-dimensional hard sphere fluids.

RESULTS

The critical properties of reacting hard disks bounded by heat conducting walls have been studied with the intention of determining the critical values of δ at a particular value of ε. A system of N disks, with N varying from 72 to 882, was placed in a square area with side dimension of $2r_o$. Each particle was assigned a velocity sampled from a Maxwellian distribution at the temperature of the fixed wall T_o. As the disks began to collide, the system temperature, defined as $RT = \langle K \rangle$, where $\langle K \rangle$ is the average kinetic energy of a particle, was monitored. In addition, the square was divided into ten strips along the x-direction, and a local temperature was calculated for each of the strips.

Two typical results for the temporal evolution of the temperature averaged over the entire system are shown in Fig. 3. One sees that for condition a, which corresponds to a particular set of values, $\varepsilon = 0.175$, $n^* = 0.157$, $r_o = 32.15d$ and Q, the temperature increased only slightly over a time interval of $\sim 8000t_d$, where $t_d \sim 0.3t_c$, t_c being the mean free time between particle

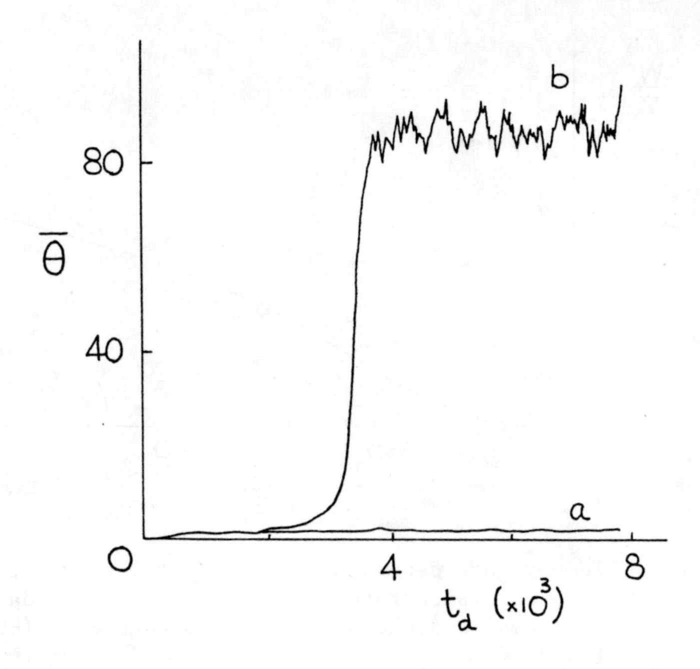

Figure 3. Temporal evolution of the reduced temperature θ at $\varepsilon = 0.175$ averaged over a slab of 648 disks, subcritical condition (a) and supercritical condition (b). One unit of simulation time interval is equivalent $t_d \sim 0.3t_c$, where t_c is the mean time between particle collisions.

collisions. The implication is that at this condition δ is less than δ_c. However, by increasing Q, one can increase δ (cf. Eq. (4)) until it is greater than δ_c. This is the case with condition b. One sees that after a certain amount of incubation time (which we may call ignition time, $\sim 3600t_d$ in this case) the temperature suddenly increases sharply by a factor of ~ 80. Having determined the δ values corresponding to conditions a and b, one can then iterate between subcritical and supercritical conditions until a sufficiently precise value is determined for δ_c. This value is characteristic of the condition specified by n^*, r_o and ε. Results of this type obtained for the same value of ε but different n^* and r_o are summarized in Table 1.

Examination of Table 1 leads one immediately to two observations. The first is that for a given density, δ_c increases with increasing system size and appears to reach an asymptotic value of 1.1. This is just the value predicted by a numerical solution of Eq.(1). Thus, we see that the present molecular dynamics results can make contact with the continuum limit. Secondly, at fixed number of particles N, δ_c decreases with sys-

TABLE 1

Critical Values of Frank-Kamenetskii Parameter for a Self-heating Slab of Hard Disks (ε = .175)

nd^2	N	r_o/d	r_o/ℓ^+	δ_c	B
0.385	72	6.84	13.27	.62	6.7
	128	9.12	17.70	1.04	9.0
0.157	72	10.72	5.84	.48	4.9
	128	14.29	7.79	.68	6.5
	200	17.86	9.74	1.02	8.1
	450	26.79	14.61	1.07	12.2
	648	32.15	17.53	1.12	14.6
0.116	450	31.21	11.84	1.08	10.5
0.077	450	38.23	9.18	.83	8.6
	512	40.78	9.79	.84	9.2
	648	45.87	11.02	.94	10.3
	882	53.52	12.85	1.16	12.0

$\ell^+$is the collision mean free path for hard disks, $\ell = 2\sqrt{2}ndg(d)$, where g(d) is the equilibrium pair distribution at contact.

tem density. This can be understood qualitatively by allowing surface heat transfer to take place at the wall boundary. Operationally, this means replacing the boundary condition $\theta(x=1) = 0$ by $d\theta/dx = -B\theta$, where $B = Hr_o/k$ is the Biot number with H the surface heat transfer coefficient. Now at a fixed ε, δ_c will depend on B. The variation of δ_c with B can be calculated using the approximation Eq.(9) [4,5]. The result is shown in Fig. 4. At fixed N, r_o is inversely proportional to $(n^*)^{1/2}$ and k is $\propto n^{*-1}$. If we assume H is constant, then $B \propto n^{*1/2}$ and Fig. 4 shows δ_c decreasing with decreasing n*. Fig. 4 also shows that at large B, δ_c eventually reaches a value corresponding to the case of no surface heat transfer. At a given n*, H and k are fixed, so B is proportional to r_o, the system size. Thus, Fig. 4 provides a qualitative explanation of the first observation as well.

The above arguments show that the molecular dynamics results can be qualitatively understood on the basis of continuum theory. However, when the results are analyzed quantitatively, there is sufficient discrepancy remaining to justify further study of the

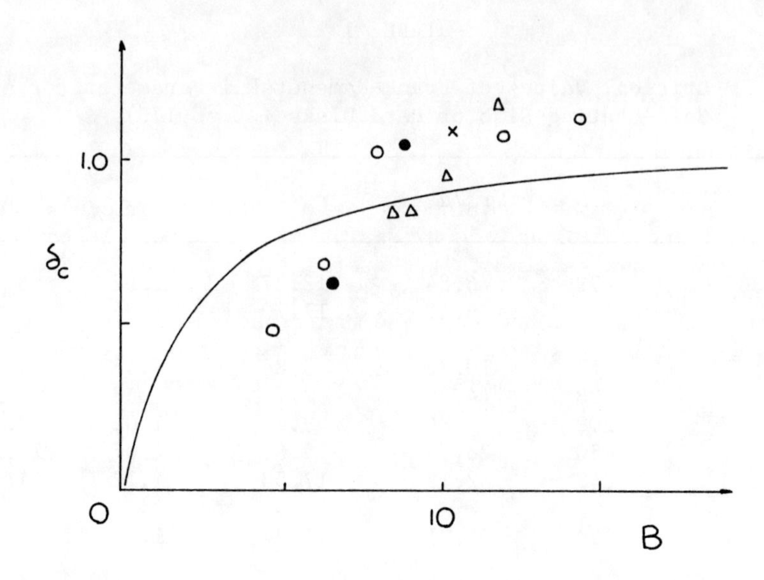

Figure 4. Variation of δ_c with Biot number B at $\varepsilon = 0.175$, molecular dynamics results at several densities ($n^* = 0.770$ (Δ), 0.116 (x), 0.157 (o), 0.385 ($\bullet$)) and continuum theory result using approximation Eq.(9) [14].

observed size dependence of δ_c.

CONCLUDING REMARKS

We have described a molecular dynamics study of thermal ignition and have presented results on the critical behavior of a self-heating slab. Other quantities which have been examined are temperature profiles and the variation of ignition time with δ [14]. Comparison of a steady-state temperature distribution at subcritical condition with Eq.(7) reveals a sharp temperature change at the wall boundary, implying that that other boundary condition [18] besides $\theta(x=1) = 0$ should be used. Ignition times are expensive to simulate because a number of runs are required to average out the statistical fluctuations; they are also difficult to calculate from continuum theory [9,19,20]. Consequently, only a preliminary comparison has been attempted [14]. Further study would be quite worthwhile since induction period is one of the most fundamental ignition properties.

The two basic variables in chemical reactions are species concentration and temperature. In the present work we have decoupled them by assuming constant reactant concentration. Our results are therefore relevant to only the initial phase of the system response. There is no intrinsic difficulty in simulating reactions involving species conversion [13], and it would be of considerable interest to study nonisothermal reactions with composition variation.

We have emphasized the comparison between molecular dynamics results and continuum theory analysis. This is felt to be essential for demonstrating that the effects described by a macroscopic theory can be simulated in terms of particle dynamics. Once it is established that the basic phenomenon of interest can be also properly described by a small system of discrete particles, then one can study the fluctuations in such a system and learn how the fluctuations are related to the macroscopic instabilities. Since fluctuations are beyond the scope of continuum theory, one has then the possibility of obtaining new information from simulation.

ACKNOWLEDGEMENT

This work was supported in part by the U.S. Army Research Office and the National Science Foundation.

REFERENCES

1. P. Gray and P.R. Lee, Combustion and Oxidation Review 2, 1 (1967).
2. A.G. Marzhanov and A.E. Averson, Combust. Flame 16, 89 (1971).
3. D.A. Frank-Kamenetskii, Diffusion and Heat Transfer in Chemical Kinetics (Plenum Press, New York, 1969), 2nd ed.
4. W. Gill, A.R. Shouman and A.B. Donaldson, Combust. Flame 41, 99 (1981).
5. T. Takeno, Combust. Flame 29, 209 (1977).
6. N.W. Bazley and G.C. Wake, Combust. Flame 33, 161 (1978).
7. J.W. Enig, D. Shanks and R.W. Southworth, "The Numerical Solution of the Heat Conduction Equation Occurring in the Theory of Thermal Explosions," NAVORD Rept. 4377, U.S. Ordanance Laboratory, Nov. 7, 1956.
8. A.R. Shouman, A.B. Donaldson and H.Y. Tsao, Combust. Flame 23, 17 (1974).
9. J.W. Bebernes and D.R. Kassoy, SIAM J. Appl. Math 40, 476 (1981).
10. For an extensive bibliography, see W.W. Wood and J.J. Erpenbeck, Ann. Rev. Phys. Chem. 27, 319 (1976).
11. D. Levesque and J.J. Weis, in Monte Carlo Methods in Statistical Mechanics, K. Binder, ed. (Springer Verlag, Berlin, 1979).
12. H.W. Harrison and W.C. Schieve, J. Chem Phys. 58, 3634

(1973); D.L. Jolly, B.C. Freasier, and S. Nordholm, Chem. Phys. 21, 211 (1977); A.J. Stace and J.N. Murrell, Mol. Phys. 33, 1 (1977).
13. See, for example, J. Portnow, Phys. Letters 51A, 370 (1975), P. Ortoleva and S. Yip, J. Chem. Phys. 65, 2045 (1976), and J.S. Turner, J. Phys. Chem. 81, 2379 (1977).
14. D.P. Chou, Ph.D. Thesis, MIT (1981); D.P. Chou and S. Yip, Combust. Flame 47, 215 (1982); D.P. Chou and S. Yip, to be published.
15. F.E. Walker, A.M. Karo, and J.R. Hardy, "Comparison of Molecular Dynamics Calculations with Observed Initiation Phenomena," Lawrence Livermore Laboratory Report UCRL - 85187 (1981).
16. D.H. Tsai and S. Trevino, J. Chem Phys., to be published.
17. W.W. Wood and J.J. Erpenbeck, in Statistical Mechanics, Part B: Time-dependent Processes, B.J. Berne, ed. (Plenum, New York, 1977), chap. 1.
18. W. Gill, A.R. Shouman and A.B. Donaldson, Combust. Flame 41, 99 (1981).
19. D.R. Kassoy and J. Poland, SIAM J. Appl. Math. 39, 412 (1980).
20. A simple approximation based on an extension of a space-independent description, the Semenov model, seems to be reasonably successful, J-C. Lermant, M.S. Thesis, MIT (1983), J-C. Lermant and S. Yip, to be published.

FLUCTUATIONS IN COMBUSTION

G. Nicolis, F. Baras and M. Malek Mansour

Université Libre de Bruxelles,
Faculté des Sciences,
Campus Plaine, Bvd. du Triomphe,
B - 1050 Bruxelles / Belgium.

A stochastic description of explosion phenomena is set up, both for isothermal and for exothermic reaction mechanisms. Numerical simulations and analytic study of the master equation show the appearence of long tail and multiple humps in the probability distribution, which subsist for a certain period of time. During this interval the system displays chaotic behavior, reflecting the random character of the ignition process. An estimate of the onset time of transient bimodality is carried out in terms of the size of the system, the intrinsic parameters, and the initial condition. The implications of the results in combustion are discussed.

INTRODUCTION

Much of this Volume deals with the transition phenomena observed in isothermal or temperature dependent reaction sequences, involving appropriate cooperative interactions like autocatalysis, and functioning far from equilibrium. Classical bifurcation phenomena involving the loss of stability of a uniform steady state and the evolution to a limit cycle or a space pattern, abrupt overshoots associated to ignition and explosion, or transition to chaotic dynamics are some characteristic examples.

In this Chapter we would like to outline an enlarged description of nonlinear systems, which incorporates thermodynamic fluctuations. As well known fluctuations are universal phenomena, generated spontaneously by all physico-chemical systems. In many instances their effect is small because their strength is

G. Nicolis and F. Baras (eds.), Chemical Instabilities, 171–188.

scaled by an inverse power of the size of the system, measured by the number of particles, N. In recent years however, it was realized that in spite of this property an analysis involving fluctuations is necessary in two types of problems : bifurcation of new branches of solutions arising when a reference state loses its stability [1] ; and evolution starting from an initial unstable or marginally stable state [2,3], in a system possessing at least one asymptotically stable attractor.

Now, in many instances rich dynamical behavior is observed in the form of a transient as the system evolves from a certain initial state to the unique final stable state predicted by the phenomenological description. Combustion, the subject of the second part of the present Volume, is one characteristic example. More generally, we want to deal with situations in which an initial induction regime characterized by a very small rate of change of the pertinent variable is suddenly interrupted by a violent explosive behavior, ignited at some characteristic time, t_c, (cf. Fig. 1).

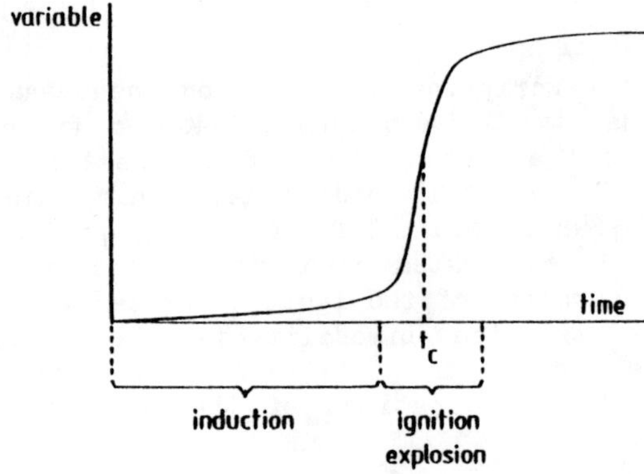

Figure 1. Transient evolution toward a unique stable state exhibiting a long induction period and an abrupt explosion at $t = t_c$.

Our principal goal is to show that in this situation also, it is important to incorporate fluctuations in the description. As it will turn out, the basic reason for this is that the additional smallness parameter related to the "violence" of the phenomenon can in many cases counteract the scaling of the fluctuations by the inverse system size and produce a chaotic regime dominated by stochastic effects.

In the next Section we introduce two simple examples of thermal and chemical explosion. Subsequently we present numerical results establishing the important role of fluctuations and outline an analytic study. The final Section is devoted to the implications of the results.

SIMPLE MODELS OF THERMAL AND CHEMICAL EXPLOSION

Consider a thermal explosion that is to say, a process in which the rate of removal of the heat released by an exothermic reaction is insufficient to maintain a relatively low level of temperature in the system. Let us idealize as much as possible and treat the process as adiabatic (all the reaction heat is disposed in the heating of the mixture). It is true that this condition is never strictly realized in real world situations. However, it should still provide a reasonable description of the ignition period in an open system, whereby an abrupt transient is observed during the evolution toward the final steady state.

The simplest nontrivial case is that of a single irreversible exothermic reaction

$$X \xrightarrow{k(\bar{T})} A \tag{1}$$

where $\bar{T}$ is the temperature and $k(\bar{T})$ the temperature-dependent rate constant. Let $\bar{x}$ be a suitable scaled intensive variable describing chemical composition, r_V and C_V respectively the heat of reaction and the specific heat at constant volume. Assume furthermore that the system is well stirred, so that one can discard transport phenomena. The mass and energy balance equations then read

$$\begin{aligned} \frac{d\bar{x}}{dt} &= - k(\bar{T})\,\bar{x} \\ C_V \frac{d\bar{T}}{dt} &= - r_V \frac{d\bar{x}}{dt} = r_V\, k(\bar{T})\,\bar{x} \end{aligned} \tag{2}$$

Each of these equations is closed thanks to the conservation condition

$$C_V T_0 + r_V x_0 = C_V \bar{T} + r_V \bar{x} = \text{const} \equiv C_V T_{max} \tag{3}$$

in which (T_0, x_0) are the initial values of $(\bar{T}, \bar{x})$ and $(T_{max}, \bar{x} = 0)$ the final ones, after the reaction has been completed and X has been exhausted. Fom eq. (2) and (3) we obtain

$$\frac{d\bar{x}}{dt} = - k(T_{max} - \frac{r_V}{C_V}\bar{x})\,\bar{x} \tag{4a}$$

or

$$\frac{d\bar{T}}{dt} = k(\bar{T})\,(T_{max} - \bar{T}) \tag{4b}$$

where $k(\bar{T})$ is given by the Arrhenius law

$$k(\bar{T}) = k_0 \exp(- \frac{U_0}{R\bar{T}}) \tag{5}$$

U_0 being the activation energy.

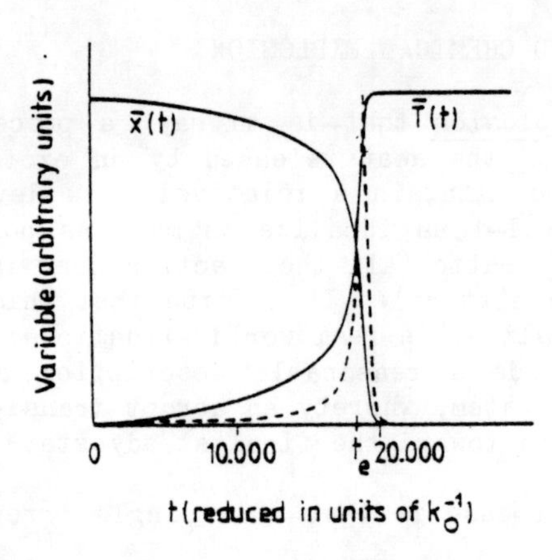

Figure 2.

Solution of eqs. (4). Parameter values : U_0/R = 10000, T_{max} = 2000, r_V/C_V = 1200.

Fig. 2 describes the solution of eqs. (4). It is seen that the reaction rate reaches abruptly its maximum value at a time t_e which will be referred to as the "explosion time" [4].

Consider next a system in which explosion occurs primarly because of chemical kinetics. Real world systems of this type involve multiple steps and competition between various pathways, many of which contain autocatalytic or inhibitory effects associated with the appearance of chain reactions and free radicals [4]. Instead of developping the analysis of such a system however, we present hereafter a mathematical model which captures the essence of the phenomenon while still allowing a fairly complete mathematical treatment. The specific example we choose is the autocatalytic mechanism suggested by Schlögl [6]. Further comments on the role of autocatalysis are to be found in the Chapters by P. Gray and S.K. Scott and by I. Epstein.

The model consists of two reversible reactions.

$$A + 2X \rightleftharpoons 3X$$
$$X \rightleftharpoons B \quad (6)$$

The concentration of A and B is supposed to remain constant. We are therefore dealing with an open system which, depending on the ratio of concentrations of A and B, will function close to or far from composition equilibrium. Using by now standard notation [5] we can write the rate equation for the suitably scaled concentration $\bar{x}$ of X in a form exhibiting two control parameters δ and δ'

$$\frac{d\bar{x}}{d\tau} = -\bar{x}^3 + 3\bar{x}^2 - (3+\delta)\bar{x} + 1 + \delta' \quad (7)$$

where τ is a scaled time. For $\delta = \delta' = 0$ one has a bifurcation point, corresponding to the triple root $\bar{x} = 1$ of eq. (7) at the steady state. For δ, $\delta' \neq 0$ the usual analysis leads to Fig. 3,

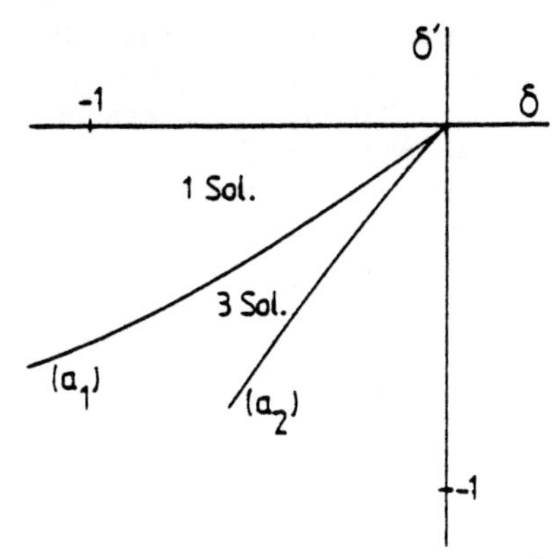

Figure 3. State diagram for the Schlögl model.

in which the regime of three real solutions is contained between lines (a_1) and (a_2) of parameter space. On the lines (a_1) and (a_2) themselves the system exhibits one marginally stable steady state, arising from a limit point bifurcation, and one asymptotically stable steady state. Moreover, outside but close to (a_1) and (a_2), the system is under the influence of a slow mode, the remnant of the limit point bifurcation. As a result it exhibits a long induction period followed by a quick evolution to the unique stable attractor. It is this regime of chemical explosion that interests us primarly here.

It is useful to visualize the evolution of both systems (1) and (6) in terms of a kinetic potential, the integral of the right hand side of eqs. (4) or (7) over $\bar{x}$. The two potentials are depicted in Figs. 4a and 4b.

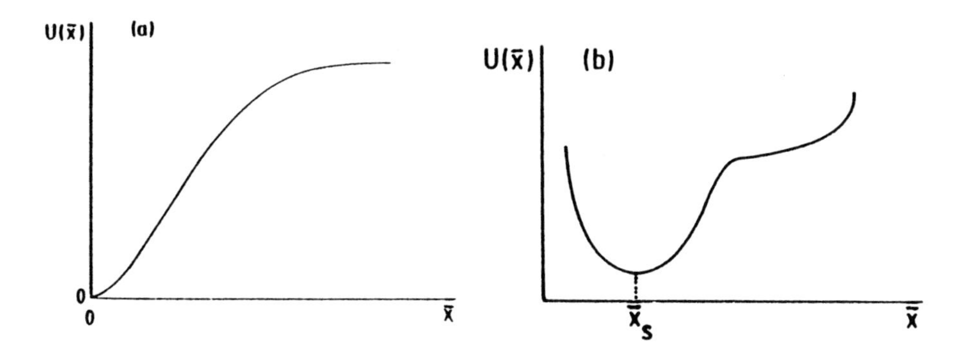

Figure 4. Kinetic potential $U(\bar{x})$ versus concentration $\bar{x}$ for thermal explosion model (a), and for Schlögl model (b).

They both have a single minimum, an inflexion point for values of $\bar{x}$ attained at the explosion time, and no further extremum in the region of high values of $\bar{x}$. The only difference between them is that in the thermal case the minimum is at $\bar{x} = 0$ (closed system, complete combustion) whereas in the chemical case it is located at a finite value of $\bar{x}$ (open system) corresponding to the unique stable state of the rate equation in the region outside the curves (a_1) and (a_2) in Fig. 3. In a typical situation the system is started in the region of $\bar{x}$ values for which the potential is flat. A slow evolution, reflected by this flatness, first take place ; but when the vicinity of the inflexion point is reached, $\bar{x}$ becomes quickly depleted and finally evolves to the unique final state.

STOCHASTIC FORMULATION

We now enlarge our description to include fluctuations, motivated by the qualitative arguments advanced in the Introduction. The central quantity to evaluate becomes thus the probability P(X,t) of having, at time t, a number of particles equal to X within the reaction vessel.

Clearly, in the thermal explosion case [7] the underlying stochastic process is a pure death process, since the concentration of X can only decrease. We can therefore write the following Master equation for the probability P(X,t) :

$$\frac{dP(X,t)}{dt} = \mu(X+1)\, P(X+1,t) - \mu(X)\, P(X,t) \tag{8a}$$

where the death rate is given by (cf. eqs. (3-5))

$$\mu(X) = k_o\, X \exp\left(- \frac{U_o}{k_B \left(T_{max} - \frac{r_v X}{C_v N}\right)} \right) \tag{8b}$$

On the other hand in the chemical explosion case [8] we deal with a birth and death process for which the Master equation reads

$$\begin{aligned} \frac{dP(X,t)}{dt} = {} & \lambda(X-1)\, P(X-1,t) - \lambda(X)\, P(X,t) \\ & + \mu(X+1)\, P(X+1,t) - \mu(X)\, P(X,t) \end{aligned} \tag{9a}$$

where the birth and death rates are, respectively,

$$\begin{aligned} \lambda(X) &= \frac{3X(X-1)}{N} + N\,(1 + \delta') \\ \mu(X) &= \frac{X(X-1)(X-2)}{N^2} + X\,(3 + \delta) \end{aligned} \tag{9b}$$

N is a size parameter (typically $N \gg 1$). In writing (8a) as well as (9a), we incorporated the requirement that in a thermodynamic system the transition rates have to be extensive.

We have solved numerically the above Master equations and, in addition, we simulated the underlying stochastic processes using a Monte-Carlo type method. A first result is shown in Fig. 5 for the thermal explosion case.

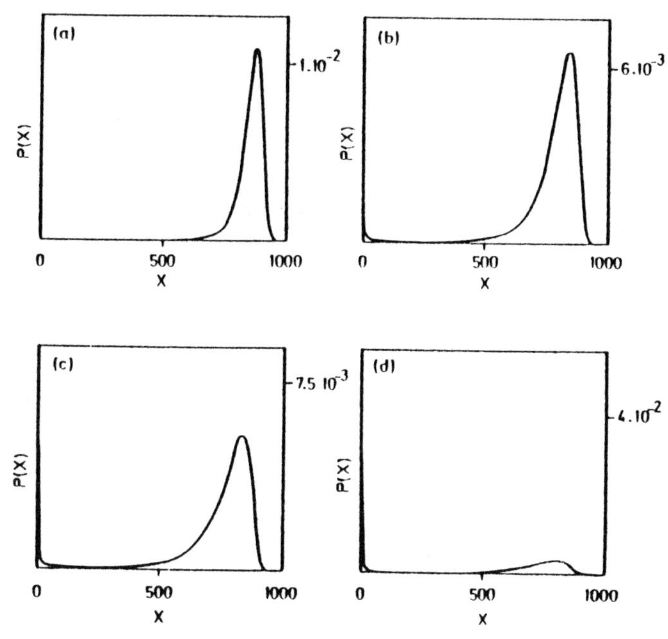

Figure 5. Succesive steps of the evolution of the probability distribution for the thermal explosion model leading from a unimodal regime (part (a)) to a long tail (part (b)) and a bimodal regime (parts (c) and (d)). Parameter values : $U_0/R = 10000$, $T_{max} = 2000$, $r_V/C_V = 1200$.

At t = 0 we start with exactly N = 1000 particles in the region in which the kinetic potential, as function of $\overline{x}$ (Fig. 4a), is very flat. Shortly after this initial condition the probability distribution develops a width, while its maximum moves only slightly to the region of low values of X. A pronounced flattening of the distribution then takes place, followed by the appearence of a second peak located at a value of X close to zero. Meanwhile the first peak is still centered at values of X well above the value characteristic of explosion. The long tail and the two peaks subsist for some time, but eventually the system collapses to zero, which is an absorbing state attained with probability one [7].

An interesting way to summarize the above described stages of the evolution is to plot X_{max}, the most probable value of X, as a function of time. This is done in Fig. 6.

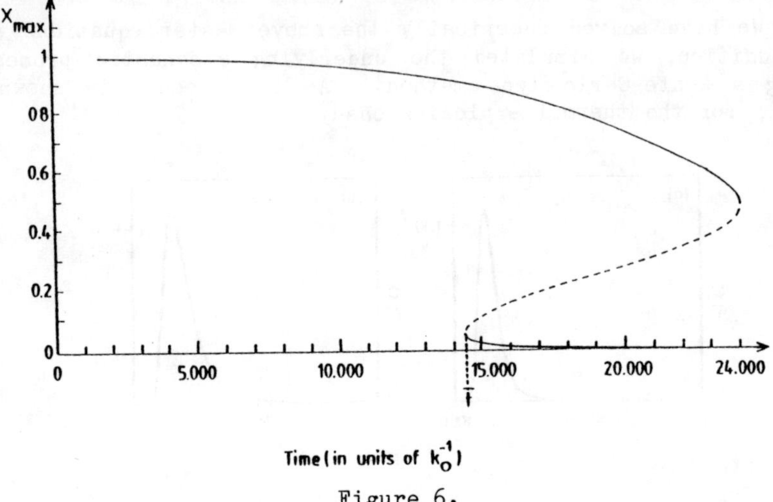

Figure 6.

We see that for short times we obtain a unique solution. But there exists a critical time, $\bar{t}$, beyond which new branches of solution appear, reflecting the formation of a second peak of the probability distribution. Eventually the upper branch dissapears, and the system evolves to extinction as combustion is completed.

The situation described in Fig. 6 is strongly reminiscent of the phenomenon of bifurcation, whereby new branches of solutions come into play when some suitable control parameters are varied. The difference is of course that in the present case the appearence of new branches can only be a transient. We coined the term "bifurcation unfolding in time" [7] to this phenomenon in order to capture both the similarities and the differences with its more familiar "static" counterpart.

From the standpoint of combustion our result means that temporarily the population of molecules will split into a part for which combustion has not yet taken place, and a part for which combustion is pratically terminated. In other words, ignition time becomes a random variable, whose variance is directly related to the coexistence time of the two probability peaks. Moreover, in a spatially distributed system the population of molecules involved in the process will tend to separate in space, thereby producing a "nucleus" of combustion which eventually will propagate in the system. Such a phenomenon will be the precursor of hot spots or flames that have been widely referred to in the

Chapter by P. Clavin.

Similar results have been obtained for the stochastic behavior in the chemical explosion case [8]. In Fig. 7 we propose a useful visualization of the evolution, by plotting the "probability surface" P(X,t) in (X,t) space.

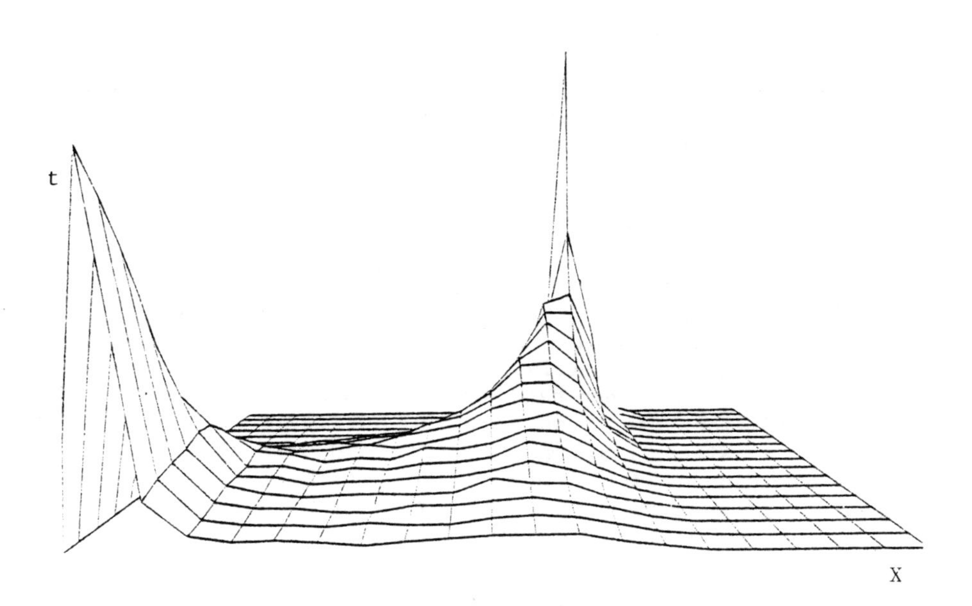

Figure 7. Transient bimodality for the chemical explosion model, visualized by plotting the probability as a function of both the number of particles X and the time t. Parameter values : $\delta = -0.38$, $\delta' = -0.5$.

A section by a line parallel to the X axis gives the probability profile for a given t, while a section by a line parallel to the t axis gives the evolution of P for given X. From this picture one can easily determine the way the most probable values evolves in time. The existence of a two-hump distribution will be reflected by a curve exhibiting limit points and hysteresis. Again however, contrary to ordinary situations in which hysteresis behavior appears when a parameter is varied, in the present case both the new branches of most probable values and the hysteresis region will be observed as time follows its course. We can refer to this as <u>transient bimodality</u>.

The results just described are not universal. In particular, the occurrence of transient bimodality depends on the size of the system, on its intrinsic parameters (U_o, r_v ; δ , δ'), and on the initial conditions. Generally speaking, for very large systems and for fixed parameters and initial conditions bimodality is bound to disappear, but the dependence of this trend on size is a weak one. Denoting by Δt the time interval of bimodality, one finds that

$$\Delta t \approx \psi N^{-1/2}$$

in which ψ is a large factor related to the violence of explosion. Estimations using reasonable parameter values [7] easily yield $\psi \approx 10^3$, so that for $N \approx 10^{12}$ Δt is of the order of milliseconds. In as much as 10^{12} is the order of magnitude of the number of particles contained in a volume of several hundred cubic microns, we conclude that bimodality should be observable -and important- in real world situations.

Coming to the role of the intrinsic parameters, the general trend is that bimodality is enhanced for the parameter range for which the deterministic evolution displays two widely separated time scales. On the other hand -and this leads us to the role of initial conditions- to "probe" such a time scale difference the system has to start from a state located sufficiently before the inflexion point of the deterministic potential (cf. Fig. 4). Otherwise it undergoes a rapid relaxation to the final state following essentially the deterministic path.

QUALITATIVE INTERPRETATION AND ANALYTIC RESULTS

Many of the results presented in the preceding Section reflect the influence of elements which are somewhat unconventional in a typical problem of stochastic theory. Indeed, in addition to the intrinsic parameters built in the phenomenological equations of evolution, we saw that the size and the initial conditions played an equally important role in determining the qualitative behavior. In the present Section we would like to advance a qualitative explanation of these findings and to set the basis for an analytic description.

Let us refer to Fig. 8, in which the deterministic potential is once again drawn as a function of the composition variable X in the thermal explosion problem. Suppose that we start with exactly N particles at t = 0, and that N is well on the right of the position of the inflexion point of U(X), in a region in which the potential is rather flat. As mentioned already in the previous Section, for t > 0 the probability function will develop a width, and its peak will begin to travel to the left slowly,

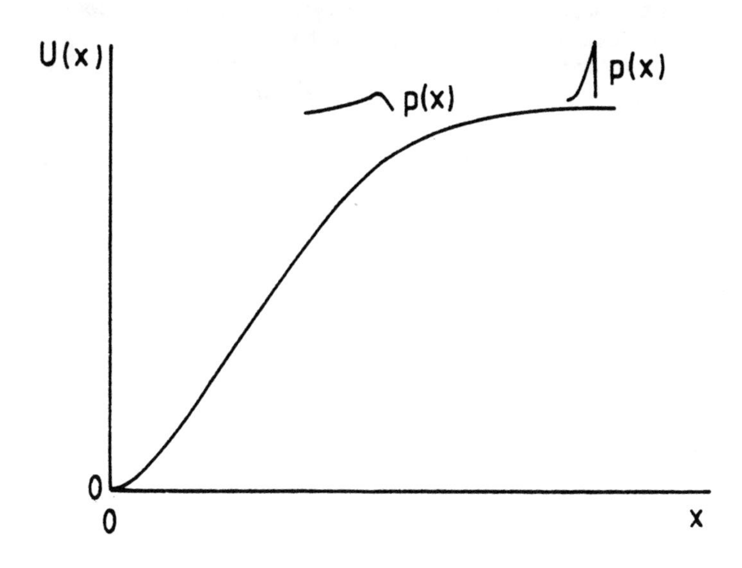

Figure 8. Gradual development of long tail and two humps as the probability mass migrates towards the ignition region.

owing to the smallness of $U'(X)$. Meanwhile the width around the peak will broaden because of the fluctuations, and at some moment a substantial probability mass will have reached the vicinity of the inflexion point. As $U'(X)$ is much larger on the left of this latter point, that part of the probability mass will evolve according to a much faster time scale. As a result the imbalance on the two sides of the inflexion point will become further amplified. This should mark the beginning of long tail and transient bimodal behavior described in the preceding Section.

A fully quantitative treatment of the above intuitive ideas is difficult at the present time, for two reasons : in the thermal case the death rate depends exponentially on the state variable ; and in the chemical case one deals with a birth and death process with highly nonlinear transition probabilities whose time-dependent behavior remains poorly known, despite recent significant progress [2,3]. In a preceding paper [7] we circumvented this difficulty in the thermal case by adopting an idealized piecewise linear representation of the transition rates, which captures their essential features while allowing a rather exhaustive analytic treatment. Here we present an alternative description using the full form of the transition rates, and the more limited aim we fix to ourselves is to determine the critical time beyond which transient bimodality is expected to occur.

We first switch from the Master equation description, eq. (8), to a Fokker-Planck equation for the probability distribution. As the system has a single asymptotically stable state this passage is legitimate [9] provided that a systematic expansion in inverse system size $\varepsilon = N^{-1}$ is performed in which the fluctuations around the deterministic path $\bar{T}$,

$$\theta = T - \bar{T}(t) \tag{10}$$

are scaled by the adequate power of ε. We obtain in this way (from now on it is understood that both θ and T are normalized by the initial temperature T_o):

$$\frac{\partial P(\theta,t)}{\partial t} = - \frac{\partial}{\partial \theta} [f(\bar{T}+\theta) - f(\bar{T})] P(\theta,t) + \frac{\varepsilon}{2} q f(\bar{T}) \frac{\partial^2 P}{\partial \theta^2} \tag{11a}$$

where

$$f(\bar{T}) = (T_\infty - \bar{T}) \exp(- \frac{U_o}{k_B \bar{T}})$$

$$q = \frac{r_V x_o}{C_V T_o} \tag{11b}$$

We seek for solutions of (11a) of the form

$$P = \exp(\varepsilon^{-1} \phi)$$

$$\phi = \sum_{n=1}^{\infty} a_n \theta^n \tag{12}$$

Moreover we introduce a new time scale τ measured by the progress of the temperature variable $\bar{T}$ as given by the phenomenological balance equation [10]

$$\frac{d\tau}{dt} = f(\tau) \tag{13}$$

Substituting into eq. (11a) and equating to zero the various powers of θ we obtain $a_1 = 0$, and

$$\frac{da_2}{d\tau} = - 2 d_1(\tau,\varepsilon') a_2 + 2 q a_2^2 + [- 3 d_3(\tau,\varepsilon') + 6 q a_4] \tag{14}$$

in which ε' is a second small parameter related to the violence of the explosion,

$$\varepsilon' = \frac{k_B T_o}{U_o} \tag{15}$$

and

$$d_1(\tau,\varepsilon') = \frac{-1}{\tau_\infty - \tau} + \frac{1}{\varepsilon' \tau^2} \qquad (16)$$

$$d_3(\tau,\varepsilon') = \frac{1}{2} \left\{ \left(\frac{6}{\varepsilon' \tau^4} \quad \frac{6}{\varepsilon'^2 \tau^3} + \frac{1}{\varepsilon'^3 \tau^6}\right) + \frac{6}{\varepsilon' (\tau_\infty - \tau) \tau^3} - \frac{3}{\varepsilon'^2 (\tau_\infty - \tau) \tau^4} \right\} \qquad (17)$$

with $\tau_\infty = T_{max}/ T_o$.

If the truncated form of eq. (14), in which all ε-dependent terms are set equal to zero, predicted a finite negative value of a_2 for all values of τ, then the probability distribution would essentially be Gaussian, with a width given by a_2^{-1}. Additional terms in the expansion (12) would be superfluous, as they would give corrections to the moments of P vanishing as a power of ε. The evolution of the system would thus be essentially deterministic, since the most probable value would remain uniquely defined and would evolve according to the deterministic rate equation.

Conversely, a necessary condition for transient bimodality to occur is that a_2 reaches values which are of the order of some power of ε , and subsequently changes sign from negative to positive values. In such a case one would have to push the expansion of ϕ at least up to fourth order terms. For suitable values of the coefficients the function $\{- \phi(\theta)\}$ would represent a "stochastic potential" having two minima and a maximum. In that sense the evolution of our system could be viewed as the motion of a "particle" in a time-dependent potential, which is similar to the deterministic one (Fig. 4) for the initial and final stages but is qualitatively different from it for intermediate times.

Let us analyze the conditions under which a_2 can indeed vanish, limiting ourselves to the determination of the corresponding critical time of bimodality onset, starting from the unimodal regime. It turns out that this allows to neglect the coupling with the fourth moment, and reduce eq. (14) to the form [7]

$$\frac{da_2}{d\tau} = - 2\, d_1(\tau,\varepsilon')\, a_2 + 2\, q\, a_2^2 - 3\, \varepsilon\, d_3(\tau,\varepsilon') \qquad (18)$$

$$\varepsilon,\ \varepsilon' \ll 1$$

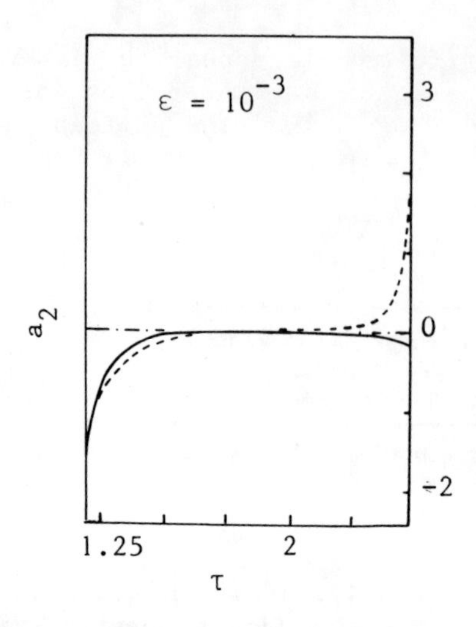

Figure 9.
Time evolution of the coefficient a_2 obtained by numerical integration of eq. (18) for $\varepsilon = 0$ (full line) and for $\varepsilon \neq 0$ (dashed line).
Parameter values : $U_o/R = 10000$, $T_{max} = 2000$, $r_V/C_V = 1200$.

Fig. 9 describes the results of numerical integration of this equation. The full line gives the evolution of a_2 in the limit in which ε is set equal to zero. We see that a_2 never changes sign. However, it does become very small for some interval of time during which, in view of eq. (12), the probability distribution will be very flat. Both the smallness of a_2 and the length of this interval become more and more pronounced as the parameter ε' becomes very small. On the other hand for $\varepsilon \neq 0$ the sign of a_2 can change at some critical value of time in the explosion region (Fig. 9, dotted lines), which is in good agreement with the value obtained earlier from the numerical simulation of the full Master equation reported. For fixed parameter values this transition to bimodality occurs only as long as ε is not less than some threshold value.

A fully analytic approach to this phenomenon of long tail and bimodality, based on a perturbative analysis of eq. (18), is reported in ref. [11]. Here we outline some important elements allowing a preliminary interpretation of the numerical results.

We first observe from eq. (16) that the coefficient d_1 vanishes at the deterministic explosion time τ_e, for which the rate of change of $\bar{T}$ calculated from eq. (4) presents a maximum. Setting $d_1(\tau_e, \varepsilon') = 0$ we obtain a quadratic in τ_e, from which we find the approximate solution

$$\tau_e = \tau_\infty - \varepsilon' T_\infty^2 + \ldots \tag{19a}$$

Moreover, the following inequalities hold

$$\begin{aligned} d_1 > 0 \quad &\text{for} \quad \tau < \tau_e \\ d_1 < 0 \quad &\text{for} \quad \tau > \tau_e \end{aligned} \tag{19b}$$

The behavior of the coefficient d_3 appearing in eq. (18) in the vicinity of $\tau = \tau_e$ can also be determined from the expli- expression, eq. (17). In view of eq. (19a) the dominant contribution in this range is given by

$$d_3 \approx \frac{1}{2} \left(\frac{1}{\varepsilon'^3 \tau^6} - \frac{3}{\varepsilon'^2 (\tau_\infty - \tau) \tau^4} \right) \tag{20}$$

The following properties follow then straightforwardly

$$\begin{aligned} d_3 &= 0 \quad \text{for} \quad \tau_0 = \tau_e - 2 \varepsilon' T_\infty^2 \\ &\qquad\qquad\quad \cong \tau_\infty - 3 \varepsilon' T_\infty^2 \\ d_3 &> 0 \quad \text{for} \quad \tau < \tau_0 \\ d_3 &< 0 \quad \text{for} \quad \tau > \tau_0 \end{aligned} \tag{21}$$

Let us now follow from eq. (18) the evolution of a_2 starting from a negative value $a_2(0)$ representing a sharply peaked initial distribution. It is convenient to decompose the evolution into successive stages.

(i) Initial stage

Since a_2 is finite initially ($\tau = 1$) the terms in ε can be neglected in eq. (18). The resulting equation can be solved exactly, and the result is

$$a_2(\tau) = \frac{1}{(\tau_\infty - \tau)^2} e^{\frac{2}{\varepsilon'}\left(\frac{1}{\tau} - 1\right)} \cdot \left\{ \frac{1}{(\tau_\infty - 1)^{-2} a_2^{-1}(0) - 2q \int_1^\tau d\tau' \, e^{\frac{2}{\varepsilon'}\left(\frac{1}{\tau'} - 1\right)} \frac{1}{(\tau_\infty - \tau)^2}} \right\} \tag{22}$$

This expression remains negative for all value $\tau \leqslant \tau_\infty$. It first decreases exponentially as long as τ is in an "initial layer" $\tau - 1 = O(\varepsilon')$; it subsequently reaches in a non-exponential fashion a value close to zero which is smaller the smaller the ε' ; and finally, after a lapse of time whose magnitude increases when ε' decreases it diverges as $\tau \to \tau_\infty$. This is in agreement with the trend shown in the full curve of Fig. 9. Note that beyond the initial layer a_2 depends on ε' in a non-analytic way. Moreover, since in this range $|a_2|$ becomes very small the ε-dependent terms can no longer be neglected in eq. (18).

(ii) Intermediate stage

Once $|a_2|$ has become small, the quadratic terms can be neglected. Eq. (18) reduces to

$$\frac{da_2}{d\tau} = - 2\, d_1(\tau, \varepsilon')\, a_2 - 3\, \varepsilon\, d_3(\tau, \varepsilon') \quad ; \quad \tau > \tau^* \tag{23}$$

τ^* being a suitable "matching time" with the initial regime. The solution of this equation is

$$a_2(\tau) = \frac{1}{(\tau_\infty - \tau)^2}\; e^{\frac{2}{\varepsilon'}\left(\frac{1}{\tau} - \frac{1}{\tau^*}\right)}$$

$$\cdot \left\{ (\tau_\infty - \tau^*)^2\, a_2(\tau^*) - 3\, \varepsilon\, e^{\frac{2}{\varepsilon'}\frac{1}{\tau^*}} \int_{\tau^*}^{\tau} d\tau'\; e^{-\frac{2}{\varepsilon'}\frac{1}{\tau'}}\; (\tau_\infty - \tau')^2\, d_3(\tau', \varepsilon') \right\} \tag{24}$$

Since $a_2(\tau^*)$ is negative, it is clear from eq. (24) that $a_2(\tau)$ cannot change sign unless τ is larger than the time τ_0, eq. (21), at which the coefficient d_3 in the integrand switches from positive to negative values. Moreover, if this change is to take place, it will have to be before the time at which d_1 vanishes ($\tau = \tau_e$, see eq. (19b)), otherwise the trend of increase of a_2 will be reversed. We are therefore led to the estimate that the critical time of bimodality onset, $\bar{\tau}$ must lie between τ_0 and τ_e, or

$$\tau_\infty - 3\, \varepsilon'\, \tau_\infty^2 < \bar{\tau} < \tau_\infty - \varepsilon'\, \tau_\infty^2 \tag{25}$$

This is in complete agreement with the results of the numerical simulations.

DISCUSSION

We have seen that the evolution of the probability distribution of a chemical system showing explosive behavior can be decomposed into three stages : an initial one, characterized by a Gaussian-like distribution whose maximum travels with a speed close to the one predicted by the deterministic equations ; the explosion proper, characterized by a flattening of the distribution, and eventually the appearence of a second hump for some period of time ; and finally, a regime of relaxation toward the unique stable attractor, characterized by the gradual disappearance of the peak generated by the initial condition.

Our results suggest that the above dynamics can be viewed as an evolution in a stochastic potential whose qualitative aspect depends on time : at the beginning it is similar to the deterministic potential, but subsequently it deforms (the deformation depending on the volume and initial conditions) and develops a second minimum. This minimum is responsible for the transient "stabilization" of the maximum of P(X,t) before the inflexion point. As the tunneling towards the other minimum on the stable attractor goes on, the first minimum disapears and the asymptotic form of the stochastic potential, determining the stationary properties of P(X,t), reduces again to the deterministic one. This phenomenon of "phase transition in time" is somewhat reminiscent of spinodal decomposition.

Needless to say, real world chemical explosions are multistep phenomena involving the competition between various pathways, many of which contain autocatalytic of inhibitory effects associated with the appearance of free radicals and chain reactions. We expect that in such a complex dynamics the role of fluctuations will be even more important than in the simple models studied in the present Chapter. More generally, it seems to us that chain reactions and explosive behavior should be characteristic examples of a <u>fluctuation chemistry</u> [1] , in which probabilistic elements are built into the system and confer to the process an essentially statistical character.

Finally, it is expected that in addition to thermal and chemical explosion similar behavior will arise in a host of other problems involving transient evolution and multiple time scales. The formation and propagation of fractures in a variety of materials [12] and the switching phenomena in lasers [13] are two tempting examples.

ACKNOWLEDGMENT

We are grateful to Professor I. Prigogine, Dr. J.W. Turner and Dr. M. Frankowicz for suggestions and fruitful discussions. The research reported in this paper is supported, in part, by the U.S. Department of Energy under contract number DE-AS05-81ER10947 and by the Belgian Government : A.R.C., Convention n° 76 / 81.II.3.

REFERENCES

1. Nicolis, G. and Prigogine, I., 1977, "Self-Organisation in Nonequilibrium Systems", Wiley, New York
2. Suzuki, M., 1981, in "Proc. XVIIth Solvay Conf. Phys.", Wiley, New York
3. Caroli, B., Caroli C. and Roulet, B., 1979, J. Stat. Phys., 21, 415
   Caroli, B., Caroli, C. and Roulet, B., 1980, Physica 101A, 581
4. Kondratiev, V.N. and Nikitin, E.E., 1981, "Gas-Phase Reactions", Springer-Verlag, Berlin Heidelberg New York
5. Nicolis, G. and Turner, J.W., 1977, Physica, A 89, 245
6. Schlögl, F., 1972, Z. Physik, 253, 147
7. Baras, F., Nicolis, G., Malek Mansour, M. and Turner, J.W., 1983, J. Stat. Phys., 32, n°1, 1
8. Frankowicz , M. and Nicolis, G., J. Stat. Phys., in press
9. Van Kampen, N.G., 1976, Adv. Chem. Phys., 34, 245
10. Kassoy, D.R., 1975, Comb. Sc. Tech., 10, 27
    Kassoy, D.R., 1977, Q. Jl. Mech. Appl. Math., 30, 71
    Kassoy, D.R. and Linan, A., 1978, Q. Jl. Mech. Appl. Math., 31, 99
11. Baras, F., 1984, Ph. D. dissertation
12. see for instance, Bottani, C. and Gaclioti, G., 1982, Physica Scripta, T1, 65
13. see for instance, Ciftan, M. and Robi, H. and Bowden, C., eds., 1980, "Optical bistability", Plenum Press, New York

STUDIES OF THERMAL FLUCTUATIONS IN NONEQUILIBRIUM SYSTEMS BY MONTE CARLO COMPUTER SIMULATIONS

Alejandro L. Garcia, Jack S. Turner

Center for Studies in Statistical Mechanics and Thermodynamics, The Univerisity of Texas, at Austin, Austin, TX

In recent years, there has been much interest in the nature of the fluctuations in nonequilibrium systems [1]. Most of the work in this field has consisted of studying the composition fluctuations for a given system through the Master Equation, the Fokker-Planck Equation or a Stochastic Differential Equation. Recently, these methods have been applied to the study of thermal systems by Nicolis, Baras, and Malek Mansour [2]. In this paper, we review their analysis of the two reservoir model. We discuss a computer simulation which has been developed to study this system and present a confirmation of their thermal fluctuation predictions.

1. ANALYSIS

In this section, we essentially review the analysis for the two reservoir model presented in Nicolis, Baras, and Malek Mansour [2]. The two reservoir model consists of a system connected to two reserviors by Knudsen apertures. We assume that the system interacts with the two reservoirs, exchanging both particles and energy. We assume that the system is spatially homogeneous and that the time scale for thermalization is small compared to the time scale for the flow so that the system is almost always in a state of thermal equilibrium. With this condition we may define a temperature analogous to the equilibrium temperature and may assume that the velocity distribution is Maxwell-Boltzmann. If we assume Knudsen flow through a hole of cross-section, σ, then we

G. Nicolis and F. Baras (eds.), Chemical Instabilities, 189–195.

may write the macroscopic equations for the number density n, and the energy density e,

$$\frac{dn}{dt} = \frac{\sigma}{V}\sqrt{\frac{k}{2\pi m}}\left[n_1T_1^{1/2} + n_2T_2^{1/2} - 2n\bar{T}^{1/2}\right]$$
$$\frac{de}{dt} = \frac{2\sigma}{V}\sqrt{\frac{k^3}{2\pi m}}\left[n_1T_1^{3/2} + n_2T_2^{3/2} - 2n\bar{T}^{3/2}\right] \tag{1}$$

where $\bar{T}$ is the temperature, m the molecular mass, V the system volume, and k is Boltzmann's constant. We solve Eq. (1) for the steady-state number density and temperature, n_s and $\bar{T}_s$,

$$\bar{T}_s = \frac{n_1T_1^{3/2} + n_2T_2^{3/2}}{n_1T_1^{1/2} + n_2T_2^{1/2}} \qquad n_s = \frac{n_1T_1^{1/2} + n_2T_2^{1/2}}{2\bar{T}_s^{1/2}} \tag{2}$$

In order to write the Master Equation for the two reservoir system we need the transition rate between the state (particle number N, total energy E) and the state (N+r, E+ ε). We know that the rate at which particles reach the hole is proportional to their velocity, and that we have a Maxwell-Boltzmann velocity distribution. From this we may write that the transition rate, W(N,E,N+r,E+ ε), is,

$$W(N,E,N+r,E+\varepsilon) = \frac{4\pi\sigma}{m^2V}\left(\frac{m}{2\pi k}\right)^{3/2}|\varepsilon|\begin{cases} N_1T_1^{-3/2}\exp(-\varepsilon/kT_1) & r=1,\ \varepsilon>0 \\ + N_2T_2^{-3/2}\exp(-\varepsilon/kT_2) & \\ 2\,N\,T^{-3/2}\exp(\varepsilon/kT) & r=-1,\ \varepsilon<0 \end{cases} \tag{3}$$

where T is the fluctuating temperature and so we may write the Master Equation as,

$$\frac{d}{dt}P(N,E) = \sum_{r=-1}^{1}\int_{-\infty}^{\infty}d\varepsilon \;\; W(N-r,E-\varepsilon,N,E)\;P(N-r,E-\varepsilon) - W(N,E,N+r,E+\varepsilon)\;P(N,E) \tag{4}$$

We now wish to take the thermodynamic limit of our Master Equation. In this limit the realizations for our Markov process tend to continuous paths. If the first two truncated differential moments exist then the Markov process is said to be a diffusion process. The general definition for the (1,k)th differential moment is,

$$A_{\ell,k}(n,e) \equiv \sum_{r=-1}^{1} \int_{-\infty}^{\infty} d\varepsilon \left(\frac{r}{V}\right)^{\ell} \left(\frac{\varepsilon}{V}\right)^{k} \quad W(N,E,N+r,E+\varepsilon) \tag{5}$$

We define for convenience f and G^2 as,

$$f \equiv \begin{pmatrix} A_{1,0} \\ A_{0,1} \end{pmatrix} \qquad G^2 \equiv V \begin{pmatrix} A_{2,0} & A_{1,1} \\ A_{1,1} & A_{0,2} \end{pmatrix} \tag{6}$$

It can be demonstrated that under some weak conditions of f and G^2 the transition probability density obeys a Fokker-Planck Equation of the form,

$$\partial_t p(n,e) = -(\partial_n \partial_e)\ (fp) + \frac{1}{2V}(\partial_n \partial_e)\left[(\partial_n \partial_e)\ (G^2 p)\right]^T \tag{7}$$

Horsthemke and Brenig [3] stress a very important point about this analysis; since a continuous Markov process is completely characterized by its first two differential moments it is unnecessary to consider the asymptotic behavior of the higher order differential moments.

At this point, we assert that from physical considerations we can assume that the state-dependent diffusion term, G(n,e), can be replaced by the state-independent diffusion, $G(n_s, e_s)$, since the system, in the thermodynamic limit, will almost always be near the steady state. Malek Mansour, et.al. [4] demonstrated that this replacement was valid for a large class of chemical systems. If we switch from the independent variables (n,e) to (n,T), integrate the Fokker-Planck Equation, expanding f about the steady state, we obtain,

$$\langle(\delta n)^2\rangle = n_s\left[1 - \frac{\Delta}{5}\right], \quad \langle\delta n \delta T\rangle = \frac{2}{5}\Delta, \quad \langle(\delta T)^2\rangle = \frac{\bar{T}_s^2}{3n_s/2}\left[1 - 14\Delta/5\right] \tag{8}$$

where

$$\Delta \equiv \frac{(\bar{T}_s - T_1)\ (\bar{T}_s - T_2)}{2\ \bar{T}_s^2} \tag{9}$$

We notice that the departure from the equilibrium values appears quadratically and that it is most prominent for the thermal fluctuations.

II. SIMULATIONS

In developing the simulation for the two reservoir system we tried to meet two important criteria. First, the simulation had to be fast. Since we were interested in measuring fluctuations in a large system we required very accurate statistics, at least millions of events. Secondly, we tried to keep an eye towards developing simulations for the next generation of thermal fluctuation problems, those with exothermic chemistry and spatial extent.

In deciding what type of code to use, we had several frameworks to chose from. Almost immediately, we rejected using a molecular dynamics code [5] because of our first criterion; also a molecular dynamics code would contain much more detail than we were really interested in. Our second choice was to use a collisionless Monte Carlo code [6]. In this context, by a collisionless code we mean one in which the collisions in the system were not explicitly calculated but rather were assumed to always keep the system in a Maxwell-Boltzmann distribution. This would certainly meet our first criterion but we were not certain whether it would provide enough microscopic detail for more complex systems. For the two reservoir system though it was certainly adequate. A collisional Monte Carlo algorythm was our third choice [7]. It seemed to best meet our criteria, however since most such codes are concerned with flow problems and not with the careful modeling of thermal fluctuations or of sensitive chemical reactions, we had to make some modifications. In conclusion, we wrote three codes, one collisionless and two collisional Monte Carlo codes. Only the collisionless code will be discussed in any detail in this paper.

In the collisionless code, we monitor only the total number of particles and the total energy. We assume that the particles are always thermalized to a Maxwell-Boltzmann distribution by collisions in the system. In the code, three fundamental events may occur. A particle may leave the system, a particle may enter the system from reservoir 1, or one may enter from reservoir 2. We compute these rates, W, from the macroscopic equation, Eq. (1), so,

$$W_0 = 2KnT^{1/2}, \quad W_{1,2} = Kn_{1,2}\, T_{1,2}^{1/2}$$
$$K \equiv \frac{\sigma}{V}\sqrt{\frac{k}{2\pi m}}, \quad W_s = W_0 + W_1 + W_2 \qquad (10)$$

The time till the next event, τ, is chosen in the appropriate Monte Carlo way as,

$$\tau = -\ln(R)/W_s \tag{11}$$

and R is a random number in (0,1). The type of the next event, ℓ, is chosen as,

$$\sum_{i=0}^{\ell-1} W_i < RW_s < \sum_{i=0}^{\ell} W_i \qquad \ell = 0,1,2 \tag{12}$$

All that is left is to determine the energy of the particle which enters or leaves. From the transition rate of the Master Equation, Eq. (3), we can write that the probability of a particle of energy ε leaving the system is,

$$P(\varepsilon)d\varepsilon = \frac{\varepsilon}{(kT)^2} \exp(-\varepsilon/kT)d\varepsilon \tag{13}$$

Then we can make the appropriate Monte Carlo choice as to the particle's energy ε as,

$$\varepsilon = -kT \ln(R_1 R_2) \tag{14}$$

A similar analysis can be done for particles entering the system. We have now entirely specified the dynamics of the process and the code simply has to successively choose an event and update the system given this event.

The collisionless Monte Carlo code has been very successful for the two reservoir system. It is small and can do ten million events in about one hour on a VAX 11/780 computer. For systems of some 500 and 1000 particles, we have made runs at several temperatures. Figure 1 summarizes our results so far for the temperature fluctuations and we are pleased at the agreement with the values predicted by Eq. (8). Though statistics were taken for the other fluctuations, the number fluctuations and the number-temperature correlations will not differ substantially from the equilibrium values and so are not conclusive.

The only results we have for the collisional Monte Carlo codes are timing comparisons with the collisionless code. In the collisional codes, the state of the system is fixed by the velocity distribution of the system. A collisional code which uses the average energy per particle to approximate the collision frequency is virtually as fast as the collisionless code when considering the number of events processed, however, the collisionless code is actually considerably faster since most of the events in the

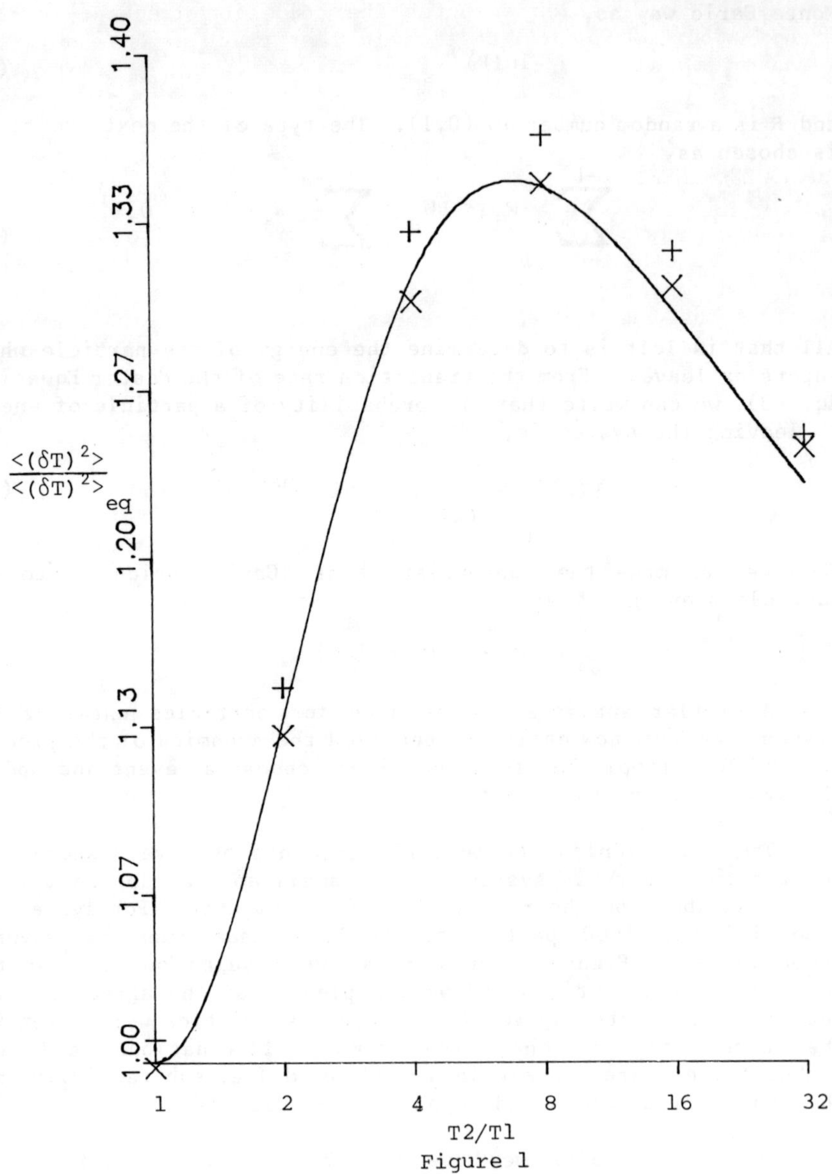

Figure 1
Thermal Fluctuations, Normalized at Equilibrium

Nl = N2; Tl = 200; NOE = Number of Events
+ Nl = 500, NOE = 20 Million
× Nl = 1000, NOE = 40 Million

collisional code are elastic thermalizing collisions. A collisional code which computes the collision frequency exactly spends about 90-95% of its time in that calculation and is thus considerably slower than the other two codes.

ACKNOWLEDGEMENTS

This work was supported in part by ONR contract N00014-80-C-0490. We would like to thank Werner Horsthemke for helpful discussions on stochastic processes and the computational staff of the Acoustical Measurements Division of Applied Research Laboratories, The University of Texas at Austin, Austin, Texas, for computational support.

REFERENCES

1. G. Nicolis and I. Prigogine, 1977, "Self-Organization in Nonequilibrium Systems," Wiley, New York.
2. G. Nicolis, F. Baras, and M. Malek Mansour, 1981, "Nonlinear Phenomena in Chemical Dynamics," ed. C. Vidal, A. Pacault, Springer-Verlag, Berlin, pp. 104-144.
3. W. Horsthemke and L. Brenig, 1977, Z. Physik B27, pp. 341-348.
4. M. Malek Mansour, C. Van Den Broeck, G. Nicolis and J. W. Turner, 1981, Ann. Phys. 131, pp. 283-313.
5. J. S. Turner, 1977, J. Phys. Chem. 81, pp. 2379-2408.
6. Gillespie, 1976, J. Comp. Phys. 22, pp. 403-434.
7. G. A. Bird, 1976, "Molecular Gas Dynamic," Claredon Press, Oxford.

PART III

INTERFACES

MECHANICAL INSTABILITY AND DISSIPATIVE STRUCTURES AT LIQUID INTERFACES

A. Sanfeld and A. Steinchen

Université Libre de Bruxelles, Chimie-Physique II,
C.P. 231, Campus Plaine, Bd. du Triomphe,
B - 1050 Bruxelles (Belgique).

The authors review the theoretical analysis of the hydrodynamic stability of fluid interfaces under nonequilibrium conditions performed by themselves and their coworkers during the last ten years. They give the basic equations they use as well as the associate boundary conditions and the constraints considered. For a single interface (planar or spherical) these constraints are a Fickean diffusion of a surface-active solute on either side of the interface with a linear or an erfian profile of concentration, sorption processes at the interface, surface chemical reactions and electrical or electrochemical constraints for charged interfaces. General stability criteria are given for each case considered and the predictions obtained are compared with experimental data. The last section is devoted to the stability of thin liquid films (aqueous or lipidic films).

1. INTRODUCTION

It is well known that surface motion and deformation may be induced by chemical and physical constraints. These instabilities may lead to ordered behavior with spatial and temporal patterns [1-3].

There is a fundamental difference between these last phenomena and the usual volume instabilities. Indeed the non linear character of the phenomena responsible for the mechanical instability is due to the coupling between mechanical, electrical, chemical or thermal processes through the physicochemical local properties of the interface (boundary conditions).

G. Nicolis and F. Baras (eds.), Chemical Instabilities, 199–221.

The non autonomous character of the interface is responsible for the propagation in the adjacent bulk phases of the dynamical instability generated in the surface. The interface between two immiscible fluids is a transition region in which the chemical composition and the related physical properties abruptly change [4]. It is usually described by a geometrical surface model with singular surface properties [5,6] : surface mass density Γ, surface tension σ, surface charge etc... Out of mechanical equilibrium, the dynamical properties may also be described in terms of singular surface quantities [7-9] balancing the discontinuity of momentum fluxes from the neighbouring bulk phases.

Our purpose is to determine the constraints and the conditions responsible for the onset of surface mechanical instabilities and their influence on the adjacent phases for single interfaces and for two interacting interfaces (aqueous - and dielectric films). The constraints acting on a system at rest hold the reference state removed from thermodynamic equilibrium, we consider here :

1) transfers of matter by diffusion.
2) transfers of matter by sorption
3) surface chemical reactions
4) electrical field distributions
5) non equilibrium interactions between interfaces in films.

In order to obtain analytical predictions, we restrict our study to a linear stability analysis of a reference state at mechanical rest with constant and uniform temperature.

2. BASIC EQUATIONS FOR SINGLE INTERFACES.

Motion at a single fluid interface can be modelized by considering the transversal T and longitudinal L (compression) waves coupled by a coupling term C. The stability is then ruled by a general dispersion equation [10] :

$$L \times T - C = 0 \qquad (1)$$

The contribution of the transversal (longitudinal) displacement to the coupling term C, together with the longitudinal L (transversal T) mode is the tangential (normal) stress.

Tangential stress or Marangoni effect and longitudinal stress or Laplace-Kelvin generalized laws are the fundamental boundary conditions acting on the interfacial layer.

Other boundary conditions are the surface mass and charge balances and electrical continuity or discontinuity conditions.

Moreover a surface state equation is required for the closure of the system of equations.

According to the conditions ascribed, various dynamical behaviors are observed in the interface and in its vicinity :

longitudinal or transveral deformations,
dissipative spatio-temporal structures,
motion in toto,
interfacial turbulence,
spontaneous emulsification.

The interfacial dissipative structures are examples of self-organization predicted by the theories of stability of non equilibrium system of Prigogine, Glansdorff and Nicolis [11] [12] extended to capillary systems by Steinchen and Sanfeld [13]. We will now give the synthetic framework of the basic equations for charged and polarized surfaces in a pure electrostatic approximation [14].

In the incompressible volume phases ($\bar{\nabla}.\bar{v} = 0$), we write :

1. Momentum balance

$$\rho \dot{\bar{v}} + \bar{\nabla}.\bar{\bar{P}} = \bar{F} \tag{2}$$

with the pressure tensor

$$(\bar{\bar{P}})^i_j = p\delta^i_j - \mu\,(v^i_{,j} + v^j_{,i}) \tag{3}$$

and the body forces $\bar{F}$

$$\bar{F} = \rho\bar{g} + \bar{\nabla}.\bar{\bar{T}} \tag{4}$$

with the Maxwell stress tensor defined by

$$(\bar{\bar{T}})^i_j = \frac{\varepsilon}{4\pi} E^i E_j - \frac{1}{8\pi}\sum_k E^2_k \delta^i_j \tag{5}$$

In this description, p is the Kelvin pressure.

2. Maxwell equation

$$\bar{\nabla}\cdot(\varepsilon\,\bar{E}) = 4\pi z\rho \tag{6}$$

where ρz is the charge density.

3. Mass balance

$$\partial_t \rho_\gamma = -\bar{\nabla}\cdot(\rho_\gamma \bar{v}) - \bar{\nabla}\cdot\bar{J}_\gamma + R_\gamma \tag{7}$$

where $\bar{J}_\gamma$ are the diffusion-migration fluxes and R_γ the chemical sources.

In the absence of chemical reactions, the Fick-Nernst laws read, for molar concentrations

$$\partial_t C_\gamma = - \bar{\nabla}C_\gamma \cdot \bar{v} + D_\gamma \bar{\nabla}\cdot (\bar{\nabla}C_\gamma + \frac{z_\gamma C_\gamma}{\mathcal{R}T} \bar{\nabla}\psi) \tag{8}$$

To describe the dynamics of moving charged and polarized interfaces, we have to take into account an electrical double layer composed of [15]

i) a thin region of molecular dimensions (compact layer) containing adsorbed ions

ii) an external continuous region (diffuse layer) in which adsorption forces are negligible

The boundary conditions are :

1. Gauss equation

$$\Delta_s (\varepsilon \bar{E}).\bar{n} = 4\pi z\Gamma \tag{9}$$

The surface charge density $z\rho$ is related to the surface concentrations of the ions adsorbed in the compact layer Γ_γ

$$z\Gamma = \sum_\gamma z_\gamma \Gamma_\gamma \tag{10}$$

2. Jump of electrical potential ψ

$$\Delta_s \psi = \bar{P}^s \cdot \bar{n} \tag{11}$$

where the surface dipole density $\bar{P}^s$ includes both the contribution from the oriented dipole moments of the adsorbed molecules and from the potential drop through the compact layer

3. Continuity of velocities in each phase β

$$\bar{v}^\beta |_s = \bar{v}^s \tag{12}$$

4. Surface momentum balance

$$\Gamma \dot{\bar{v}}^s = \bar{\nabla}_s (\bar{\bar{\pi}}) + \bar{F}^s + \Delta_s (-\bar{\bar{P}} + \bar{\bar{T}}) . \bar{n} \qquad (13)$$

where Γ is the total surface mass density, $\bar{\bar{\pi}}$ is the intrinsic surface stress tensor, $\bar{F}^s$ is the total surface intrinsic body force.

In the horizontal plane, Eq. (13) is the Marangoni condition while along the normal coordinate z, the same relation is the generalized Laplace-Kelvin condition. Along the horizontal coordinates x,y, we assume for a two-dimensional Newtonian system [16]

$$(\overline{\nabla \bar{\bar{\pi}}})_{\{x \atop y\}} = (\bar{\nabla}\sigma)_{\{x \atop y\}} - \eta_{dil} (v^s_{z,z}) + \eta_{sh} \nabla^2_s (\bar{v}_s)_{\{x \atop y\}} \qquad (14)$$

where the phenomenological coefficients η_{dil} and η_{sh} are the intrinsic surface dilational and shear viscosities, and σ is the interfacial tension. An analogous equation may be written in the curvilinear coordinates [16]. Assuming the superposition of all contributions to the total force, we get [17]

$$\bar{F}^s = \Gamma\bar{g} + \bar{F}^s_E + \bar{F}^s_m \qquad (15)$$

where $\Gamma\bar{g}$ is the surface weight, $\bar{F}^s$ is the excess chemical force due to very short range interactions [15].

The surface tension σ is thermodynamically defined by a mechanical contribution σ_M due to the surface composition and an electrical contribution σ_E due to the influence of the double layer [18].

$$\sigma_T = \sigma_M - \sigma_E \qquad (16)$$

where

$$\sigma_E = \frac{1}{4\pi} \int \varepsilon E^2 \, (\sim\sqrt{g^*})_{,1} \, \delta x^1 \qquad (17)$$

With g^* the determinant of the matrix of the space fundamental tensor and 1 is the coordinate curve of the field lines in the general curvilinear orthogonal coordinates.

5. Surface mass balance

$$\dot{\Gamma}_\gamma = - \Gamma_\gamma (\bar{\nabla}.\bar{v}^s + a^*) - \bar{\nabla}_s.\bar{J}^s_\gamma - \Delta_s \{\bar{J}^\beta_\gamma\}.\bar{n} + R^s_\gamma \qquad (18)$$

where $\bar{\nabla}_s.\bar{v}^s$ is the surface divergence of the surface velocity $\bar{v}^s$, a^* is the change of the surface metric, R^s_γ is the source of surface chemical reactions, $\Delta_s \bar{J}^\beta_\gamma.\bar{n}$ accounts for the interchange of mass between the adjacent bulk phases and the surface, $\bar{J}^s_\gamma$ is the singular diffusion-migration flux on the surface

$$\bar{\nabla}_s.\bar{J}^s_\gamma = D^s_\gamma \left(\bar{\nabla}_s \Gamma_\gamma + z_\gamma \Gamma_\gamma \bar{\nabla}_s \psi \right) \tag{19}$$

with D^s_γ the surface diffusion coefficient, Γ_γ the surface concentrations of the adsorbed ions. The sorption fluxes $\bar{J}^\beta_\gamma$ are related to the difference of electrochemical potentials between the surface and the sublayers.

6. Change of interfacial tension

$$\delta\sigma = -\varepsilon^s_d(\omega,k) \frac{\delta}{\omega} v^s_{z,z} + k\,\Psi(\omega,k) \frac{\delta}{\omega} v^s_z \tag{20}$$

where ε^s_d is the dynamical surface elasticity [10,19] related to the longitudinal displacement $\mathcal{D}v^s_z$, Ψ is a phenomenological dynamical quantity related to the normal displacement, ω and k are the frequency and the wavenumber of the perturbations. The longitudinal displacement is connected to the local variations of surface area A

$$\frac{\delta}{\omega} v^s_{z,z} = -\delta \ln A = \frac{1}{\omega} \mathcal{D}v^s_z \tag{21}$$

The phenomenological coefficients ε^s_d and Ψ are related to all the relaxation processes due to mass exchanges, chemical reactions and electrical effects.
Restricting our analysis to plane and spherical interfaces, we solved Eqs. (2)-(5) in terms of velocities along the normal or radial coordinates. Taking into account the boundary conditions (Eqs. (6)-(20)), we obtain the general characteristic equations Eq. (1). Let us now analyze various situations related to the different constraints imposed to the reference state.

3. RESULTS

3.1. Fickean Diffusion for plane interfaces

The transfer of a solute by diffusion is due to the difference of the chemical potential in the two phases I and II. Let us first consider an unperturbed plane interface. For the linear profile

$$c^{\alpha} = c_s - \beta z \qquad (22)$$

$\beta > 0$ (transfer I →II)

$\beta < 0$ (transfer II→ I)

Assuming that there is no accumulation of matter at the interface and for local equilibrium, we get

$$\beta^{I} D^{I} = \beta^{II} D^{II} \qquad (23)$$

$$\Gamma \sim c|_s \qquad (24)$$

For longitudinal waves ($v_z^s = 0$) and for

$\lambda > \sqrt{D/\omega}$ (diffusion penetration depth)

$\lambda > \sqrt{\nu/\omega}$ (impulsion penetration depth)

we analyze the aperiodic and periodic regimes [9] [20]

i) Aperiodic regime

Transfer I → II leads to unstable steady states when

$$\frac{D^{I}}{D^{II}} \frac{\nu^{II}}{\nu^{I}} < 1 \qquad (25)$$

A necessary and sufficient condition for the onset of surface motion is that diffusion occurs from the liquid with the smallest value of D towards the fluid with the largest value of D . This result was previously obtained by Sternling and Scriven [21] who neglected the transversal mode, the accumulation of matter at the surface in the perturbed state, the surface viscosity, the surface acceleration and the surface diffusion. Moreover, the mode of maximum instability is characterized by a wavelength λ_m.

We found $\lambda_m \simeq 10^{-1}$ cm for the transfer of acetic acid from ethylene glycol to ethyl acetate, in good agreement with experiments performed by Orell and Westwater [2]. Our theory also predicts that the surface viscosity has a damping effect and that the critical constraint $\beta^{I} - \beta^{II}$ increases with viscosity, with surface elasticity and with diffusion coefficient.

ii) Periodic regime

For large λ, transfer ($I \rightarrow II$) leads to unstable states when

$$\frac{D^{II}}{D^{I}} \frac{\nu^{I}}{\nu^{II}} < 1 \qquad (26)$$

This situation is in agreement with the experimental result obtained by Linde and Künkle [1] for propionic acid transfered from water to hexane.

In conclusion, Marangoni effect leads to unstable periodic and aperiodic regimes, according to the values of the experimental quantities ν and D. For the erfian profile (non steady state)

$$C(z,t) = C|_{s} + \{ \sim \mathrm{erf}(z / \sqrt{Dt}) \} \qquad (27)$$

We assume that the characteristic time of perturbation is smaller than the time of evolution of diffusion. We recover the same results as for the linear profile. However the instability criteria are nomore sufficient conditions because the penetration depth of diffusion increases with time. The critical time for the onset of cells at the marginal state is calculated by means of the relation

$$\beta = (Dt)^{-1/2} \qquad (28)$$

For the transfer of acetic acid from ethylene glycol to ethyl acetate the critical time observed [2] is a few second. Our theory predicts [19]

$$0.1 < t_c < 9 \text{ sec}$$

according to the value assumed for the surface viscosity η. Indeed t_c increases with η.

Remark : The characteristic equation for two diffusing species has recently been obtained [22]. It reveals new possibilities of instabilities.

3.2. Fickean diffusion for spherical interfaces.

We consider a radial stationary distribution of matter induced by a source (or a sink), for the unperturbed spherical interface. Solving the characteristic equation the conditions for the deformation modes (spherical harmonics $l \geqslant 2$) and for the mode of translation in toto ($l = 1$) are [23]

i) Aperiodic regime

For $l \geqslant 2$ unstable states are obtained for the mode of deformation when the transfer is from phase I to phase II for

$$\frac{D^{I}}{D^{II}} < \frac{(2l-1)\ (2l-3)}{(2l+3)\ (2l+5)} = \left(\frac{D^{I}}{D^{II}}\right)_{cr} \tag{29}$$

and when the transfer is from II to I, for

$$\frac{D^{I}}{D^{II}} > \frac{(2l-1)\ (2l-3)}{(2l+3)\ (2l+5)} \tag{30}$$

The characteristic equations shows a stabilizing effect of surface tension. The above result is in agreement with the experimental observations of Linde [24] for the transfer of 0.04% n butanol from a drop of water to diethylether. For $l = 1$ the characteristic equation shows the incompatibility of non-oscillatory unstable states.

ii) Periodic regime

For $l = 1$, there is a translation in toto of the sphere. It is only the surface of the drop which is uniformly translated.

Streaming inside and outside the drop is non-uniform. This mode is relevant for chemiotaxis (for example motion of protozoas in a gradient of concentration, but it cannot be the mechanism of chemiotaxis for bacteria with rigid cell walls).

The axisymmetric mode $l = 2$ shows new possibilities of oscillatory instabilities. It is interesting from the point of view of cell division and non-equilibrium stability of emulsions.

iii) Kicking drop

A drop of liquid I is hanging in a continuous liquid II. For $l \geqslant 1$ there is a mixing of all the modes corresponding to local deformations together with translation. Matter transfer from one phase to the other may induce the kicking of the drop as well as local deformations and eddies around the drop. This instability may be interpreted in terms of relaxation oscillations due to the coupling between Marangoni effect, viscous drag in adjacent phases, homogeneization due to convection and restoring of the diffusion profile. In a restricted range of the parameters governing the transfer process, both non oscillatory deformations ($l \geqslant 2$) and oscillatory translation ($l = 1$) can occur simultaneously. Our criteria Eqs. (29) and (30) may then be compared with the experiments of Davies [3]. This author indeed observes mechanical instabilities during the transfer of acetic acid or of acetone from a continuous phase II (benzene or toluene) to a hanging drop of water (I). For these two examples $D^{I}/D^{II} < 1$ and ineq. (30) is valid.

3.3. Sorption kinetics for plane interfaces

Pure diffusion kinetics in the bulk phase doesn't always account for onset of motion in the interface and relaxation mechanisms have to be considered. For example non-equilibrium sorption processes may occur between sublayers and surface due to orientation of polar head groups (bolaform molecules) [25]-[27]. Assuming no accumulation of matter in the sublayers in the reference state as well as in the perturbed state and no accumulation of matter in the surface in the reference state we write for the perturbation of the adsorption-desorption flux

$$\delta J^{a-d} = a\delta C - b\delta\Gamma = - D \frac{d\delta C}{dz} \qquad (31)$$

with a and b the adsorption and desorption constants. From the analysis of the dispersion relation we obtain the following conlusions [28] [29].

i) Aperiodic regime

Transfer I $\rightarrow$ II by diffusion + sorption leads also to unstable states for

$$D^{I} < D^{II} \qquad (32)$$

Davies [3] mentions several examples illustrating this instability : transfer of propanol or butanol from water to benzene and transfer of acetic acid from water to CCl_4.

Transfer from I $\rightarrow$ II with $D^{I} < D^{II}$ new possibilities of instabilities arise. Up to now we have no experimental evidence for onset of convection in this last case.

ii) Periodic regime

For large wavelength and for an adsorption-desorption controlled transfer, the stability seems to be always guaranteed whatever be the sorption kinetics (perfect gas, Langmuir, B.E.T.)

For diffusion and sorption acting together in both phases and assuming the usual relation

$$\frac{a^{I}}{a^{II}} = \left(\frac{D^{I}}{D^{II}} \right)^{1/2} \qquad (33)$$

where a^I and a^{II} are the adsorption constants from phase I and II, we predict unstable steady states if

$$\frac{D^{I}}{\nu^{I}} > \frac{D^{II}}{\nu^{II}} \tag{34}$$

and

$$J^{I} > J^{II} \tag{35}$$

For diffusion controlled transfer in phase I and sorption controlled transfer in phase II, we obtain for transfer I $\rightarrow$ II periodic instability (without conditions on ν) and for transfer II $\rightarrow$ I unstable domain between two stable zones.

These results are in agreement with experimental observations of Brian on the transfer of butylamine from water to air [30] and with the extended analysis given by Smith [31].

In conclusion a potential barrier due to controlled sorption kinetics has a stabilizing effect in the oscillatory regime.
Remark. Recent preliminary experiments [32] on transfer of molecules with two polar heads such as propanediol, butanediol and pentanediol between organic solvent (toluene, benzene or hexane) and a drop of water seem to give no instability while the monoalcool of the same length does. However the poor solubility of the diols doesn't allow to perform the experiments in a wide range of concentrations as for the monoalcools.

3.4. Surface chemical reactions at plane interfaces

From Eq. (18), we may define the kinetic matrix element $C_{\gamma\gamma'}$, (surface chemical reactions and sorption steps)

$$C_{\gamma\gamma'} = \frac{\partial R_{\gamma}}{\partial \Gamma_{\gamma}} + \frac{\partial J_{\gamma}}{\partial \Gamma_{\gamma'}} \qquad (\gamma =, \neq \gamma') \tag{36}$$

For one fluctuating species the condition of marginal stability reads [33]

i) Aperiodic regime
for one normal mode with wavenumber k

$$C_{cr} - k^2 D^s = \frac{\varepsilon^s k}{k\eta + 2\Sigma\mu^{\beta}} \tag{37}$$

with ε^s the surface Gibbs elasticity [5].
The domain of unstable wavelengths decreases with increasing viscosities.

ii) Periodic regime

$$C_{cr} - k^2D^s = \frac{k^3\eta + 2\ \Sigma\mu^\beta k^2}{k\ \Gamma + \frac{1}{2}\ \Sigma\rho^\beta} \qquad (38)$$

where $\eta = \eta_{sh} + \eta_{dil}$ cf. Eq.(14).

Oscillatory instabilities only occur in a restricted domain of wavelengths.

Conclusions for the two regimes.

The marginal states are reached for balancing chemical C and surface diffusional-contributions k^2D^s, and mechanical properties (densities ρ, Γ, viscosities η and μ and Gibbs-elasticity ε^s). The onset of mechanical instability (longitudinal motion) thus requires $C > 0$ which means autocatalytic reaction or cooperativity [34].

For several fluctuating species this approach has been recently extended by Dalle-Vedove and Sanfeld [35]. The marginal conditions then involve summations over C_γ, D^s_γ. New possibilities of instabilities are predicted even for intrinsically stable chemical mechanisms. The general conditions for mechanochemical surface instabilities may be summarized as follows :

- Equilibrium surface chemical reactions never induce mechanical instability.
- For only one fluctuating species, the chemical reaction in itself has to be unstable to obtain the onset of surface motion.
- For two (or more) species, the conditions are not so drastic : an intrinsically stable chemical reaction coupled with the hydrodynamic process may induce mechanical instability.
-A stable chemical reaction may be destabilized by mechanical constraints.

3.5. Surface chemical reactions at spherical interfaces.

We obtain qualitatively the same results as for the plane interface, indeed for one fluctuating species the marginal non oscillatory conditions read [35-36] for $l > 1$:

$$C_{cr} = l(l+1)\ [D/R^2 + \varepsilon^s/\delta'] \qquad (39)$$

with R the radius of the drop and δ' a function of the radius, of the viscosities and of the spherical harmonics. It is to note that for $R \to \infty$ Eq. (39) is compatible with Eq. (37) for l = 1

$$C_{cr} = \frac{2D^s}{R^2} + \frac{\varepsilon^s}{\delta^\dagger} \tag{40}$$

where $\delta^\dagger$ is a function of the radius and of the viscosities.

The analysis of Eqs. (39) and (40) as well as Eq. (37) shows that the unstable domain increases with decreasing viscosities. Moreover the minimum value of C_{cr} is obtained for l = 1 which means that translation occurs for lower constraints than deformation. Translation in toto of the drop is then observed before deformation [23] for systems with low crispation number [37 - 39].

- For two fluctuating species new varieties of instabilities may occur, indeed even intrinsically stable chemical reactions may induce mechanical instabilities by coupling with hydrodynamics [23].

Remark : Transfer of matter combined with surface chemical reaction for one fluctuating species has been studied by Hennenberg [19]. It is shown that only marginal overstability can be expected.

Examples of surface motion induced by chemical reactions at planar or spherical interfaces have been observed by Dupeyrat and Nakache [40] at the nitobenzene water interface for quaternary ammonium salts with a long carbon chain (C_{12}-C_{16}-C_{18}) reacting with picric acid in the neighbourhood of the interface. Motion, emulsification and kicking of a drop have been also observed [41] during the extraction of Ni^{++} ions at the liquid-liquid interface. The schemes of reactions proposed seem to be very simple and do not give rise to chemical instabilities per se, the coupling with hydrodynamics through the surface boundary conditions gives however rise to mechanical instabilities.

Another example is the extraction of uranyl nitrate from an aqueous to an organic phase by means of an extractant (TBP) : turbulent motion is observed [42]. The importance of such phenomena in biology could be fundamental to explain membrane deformation or cell motion observed in many biological events (phago-or pinocytosis, chemiotaxis, cell fusion, cellular division etc...). Interesting models were recently suggested by Marquez et al. [22] by Sørensen et al [43], by Velarde et al. [44] and by Sanfeld et al. [34].

The relatively simple schemes of reactions able to induce surface motion and deformation lead us to think that in biological systems also simple biochemical processes such as for example Michaelis-Menten enzymatic reactions at the cell membrane could induce motion and deformation of cells.

3.6. Electrical an electrochemical constraints at plane interfaces.

i) Continuous model (dilute systems)

The system considered consists of a charged interface including a compact layer between two immiscible ionic solutions with dielectric constants ε^{I} and ε^{II}. The surface charge density is compensated by the integral charges of two electrical diffuse layers of thickness $\kappa^{-1} = (\varepsilon RT/8\pi z^2 C_\infty)^{1/2}$ (with C_∞ the concentration in bulk phase $\bar{E} = 0$) extending in the neighbouring phases. The interfacial tension of such systems involves a mechanical term due to short range interactions and an electrical term due to the free energy of the electrical double layers [18]

$$\sigma = \sigma_M - \sigma_E \tag{41}$$

with $$\sigma_E = \int_{-\infty}^{\infty} \frac{\varepsilon E^2}{4\pi} dz \tag{42}$$

Let us consider two situations :

- Restored Boltzmann macroscopic distribution in the diffuse layers.

The constraint is the jump of electrochemical potential between both phases which remains uniform in each phase. From the analysis of the characteristic equation it is shown that only the negative contribution σ_E to the total surface tension σ may be responsible for the onset of surface motion [45].

The marginal stability condition is then

$$\sigma = 0 \quad \text{or} \quad |\sigma_M| = |\sigma_E| \tag{43}$$

Moreover the viscosity increases the wavelength of the fastest rate of growth and reduces the fastest rate of growth ; it thus has a stabilizing contribution.

The condition for instability $|\sigma_E| > |\sigma_M|$ was already predicted by Miller and Scriven [46] with a pure thermostatic analysis.

- Non restored Boltzmann distribution in the diffuse layers.

In this case, the relaxation of diffuse layers ($\kappa D / \lambda$) is of the order of magnitude of the characteristic time of the perturbation (ω^{-1}). The constraint is also the discontinuity of the electrochemical potential but in the perturbed state, this quantity doesn't remain uniform in each phase. We restricted our analysis to ideally polarized systems (no net fluxes through the interface) and to uni-univalent electrolytes. Instabilities are obtained even for non vanishing total surface tension σ .

For low surface charge the general rules for the stability are [47]

$$\rho^I \mu^I > \rho^{II} \mu^{II} \quad \begin{cases} (\varepsilon^{II} C^{II})^{1/2} D^{II} > (\varepsilon^I C^I)^{1/2} D^I & \text{STABLE} \\ (\varepsilon^{II} C^{II})^{1/2} D^{II} < (\varepsilon^I C^I)^{1/2} D^I & \text{UNSTABLE} \end{cases} \qquad (44)$$

Criterion [44] is in agreement with the experiments of Watanabe et al. [48] on electrical emulsification of the system water + KCl in contact with a solution of sodium dodecylsulfate in methylisobutylketone.

For small potential drop or large surface charge the general rules are [47]

$$\rho^I \mu^I > \rho^{II} \mu^{II} \quad \begin{cases} \psi^{Is} - \psi^I > \psi^{IIs} - \psi^{II} & \text{STABLE} \\ \psi^{Is} - \psi^I < \psi^{IIs} - \psi^{II} & \text{UNSTABLE} \end{cases} \qquad (45)$$

The system thus becomes unstable if the phase where the potential drop is the largest is also the the phase with the largest $\rho\mu$. These effects could also partially explain the mechanical instabilities observed by Nakache and Dupeyrat [40] described above.

ii) Discrete electrical and chemical interactions

For large surface charges (for example for ionized monolayers spread at an oil(O)/water(w) interface) discreteness of charge effect have to be considered [49]. The counterions are then located in an outer Helmholtz plane near the plane of primary charges (the inner Helmholtz plane). These two planes are separated by a layer of strongly oriented water molecules. The constraint is due to the absence of exchange between the inner Helmholtz plane and the solution. When dipoles are spread at an interface, discrete interactions also exist between dipoles. The surface tension for these two types of discrete systems (charged or dipolar layers) also consists of a mechanical part σ_M and an electrical part σ_E due to the electrical interactions between charges or dipoles [50]. Mechanical instabilities are predicted in such systems even for $|\sigma_E|<|\sigma_M|$.

The general rules are

$$\varepsilon^w > \varepsilon^o \quad \begin{cases} \rho^w\mu^w > \rho^o\mu^o & \text{STABLE} \\ \rho^w\mu^w = \rho^o\mu^o & \text{MARGINALLY STABLE} \\ \rho^w\mu^w < \rho^o\mu^o & \text{UNSTABLE} \end{cases} \tag{46}$$

The theoretical background of the influence density-viscosity and dielectric constant on the onset of surface motion is discussed elsewhere [51].

Excellent agreement with criteria (Eq. (46)) is obtained for experiments performed on medicinal paraffin - water with cholesterol and sodium dodecylsulfate [52]. Ineqs (46) are also the conditions for emulsifications and demulsification.

3.7. Instabilities in dielectric or in aqueous thin films.

When a liquid phase becomes very thin, both faces of this film interact. The nature of the interactions may be electrical or of shorter range of interaction [53] (attractive van der Waals forces) or even of very short range of interaction (steric repulsive forces between hydrocarbon chains in lipid bilayers or repulsive hydration forces between oriented water molecules around polar heads of molecules merging in aqueous films between two lipid drop or in soap films). When two faces of such films approach one another, repulsive and attractive forces are unbalanced, giving rise to a constraint, corresponding to a non-equilibrium value of the thickness of the liquid film h. Hydrodynamic instabilities of planar films (dielectric or aqueous) have been widely investigated in the last ten years [54] [55] [59]. They are usually characterized by two modes of deformation : stretching or bending mode corresponding to the in phase motion of the two surface boundaries of the film and squeezing mode corresponding to the 180° out of phase motion of these two faces. However these two modes are not always uncoupled and the general dispersion relation may be written :

$$ST \times SQ - C = 0 \tag{47}$$

For symmetrical films (with same surface tension on both faces, same surface Gibbs elasticities, same surface charge, same surface potential, same mass density), the stretching and squeezing modes are uncoupled and may then be analyzed separately. The instability of the stretching mode may account for the deformation observed in the formation of pseudopodes or microvillies on living cells or of the spiculation of erythrocytes. The instability of the squeezing mode may be responsible for the rupture of lipidic films or of soap films and could explain the fusion of cells or of liposomes by the rupture of the aqueous film between them. The stability of symmetric dielectric films charged or not has been investigated by Bisch [10], Gallez et al. [57,58,59] and by Maldarelli [60]. In all the situations investigated the dispersion relation is very complicated. Asymptotic analytic solutions have been obtained [57-59]. The conditions for instability obtained can be compared with those obtained for a single interface by introducing a renormalized surface tension for the film

$$\sigma_F = 2\,(\sigma + \frac{3}{2}\,V(h)) \tag{48}$$

including a renormalization term due to the van der Waals potential V(h) between both faces of the film. A renormalized film elasticity is also defined by

$$\varepsilon_F = 2\,(\varepsilon + h^2 \frac{d^2 V(h)}{dh^2}) \tag{49}$$

The stability of the stretching mode is ruled, as the transversal mode for a single interface, by the positive value of σ_F.

In addition to the roots corresponding to the transversal and longitudinal modes for a single interface, there exists in the limit of long wavelengths, a new root typical to the film dynamics, for both stretching and squeezing modes.

- For charged dielectric films, the additional root for the stretching mode is always negative, while for the squeezing mode the root typical for the film dynamics may become positive for wavelengths larger than

$$\lambda_{cr} = 2\pi \left(\frac{2}{2\sigma + 2\sigma_E}d.1 \cdot \frac{d^2V}{dh^2}\right)^{1/2} \tag{50}$$

The film thus becomes unstable for all peturbations with a wavelength larger than λ_{cr}.

- For a dielectric film sandwiched between two identical ionic solutions undergoing an electrochemical constraint (positive affinity of transfer of ions) the linear stability analysis shows that for both the stretching and the squeezing mode, in the limit of the long wavelentghs, the electrical potential drop through the film tends to destabilize the system [10]. The renormalized surface tension defined by Eq. (48) contains now an electrical destabilizing contribution given by Eq. (42). All the wavelengths larger than

$$\lambda_{cr} = 2\pi \left\{ \frac{(\sigma-\varepsilon E)/8\pi\kappa}{-2\,\frac{d^2V}{dh^2}} \right\}^{1/2} \tag{51}$$

are unstable with regard to squeezing perturbations.

The typical root for the film in the squeezing mode shows a wavelength corresponding to a maximum value of ω_R (mode of maximum instability). This wavelength shifts towards the short wavelengths with increasing the "transmembrane" potential difference with a marked increase of the value of the amplification coefficient ω_R. This prediction is in a perfect agreement with experimental observations on lipid bilayers [61] [62]. For a Newtonian as well as a Maxwellian [63] rheological behavior of the dielectric films, the viscosity has a damping contribution, the amplification coefficient decreases with increasing viscosity without changing the wavelength for maximum instability.

- For lipid films with $h \leqslant 50\text{Å}$ new repulsive forces due to the steric repulsion of the hydrocarbon tails of the lipids have been taken into account by D. Gallez [64]. The squeezing mode is stabilized by a new disjoining pressure due to the steric repulsion. The "black films" are stable with regard to stretching and squeezing perturbations whatever the applied potential, in agreement with experimental observations.

- The stability of thin aqueous films between two lipid phases has been investigated by M. Prevost et al. [65]. The results of

Felderhoff [54] were recovered and an extension to large surface potentials was performed. Moreover for very thin films, an attempt to introduce the repulsive hydration force gives interesting preliminary results corroborating the ideas of Exerova [67] on the high stability of the second Newton black films.

- For both the dielectric and the aqueous films, the role of asymmetry (either mechanical or electrical) has been investigated [64,65]. In both systems, the asymmetry increases the unstable domain. For thin lipid films [64] agreement is obtained with experimental observations on asymmetric black lipid membranes [68].

4. NOMENCLATURE

a, b - Adsorption-desorption constants
a^* - Change of surface metric
C_γ - Mole concentration of component γ
$C^{\gamma\gamma'}$ - Kinetic matrix element
$\mathcal{D}$ - Differentiation operator along z coordinate normal to the interface
$\underline{D}_\gamma^\alpha$ - Diffusion coefficient of component γ in phase α
E - Electric field
$\underline{E}_z$ - Component of electric field along z
$\underline{F}$ - Body force
$\overline{g}$ - Gravity vector
g^* - Determinant of the matrix of space fundamental tensor
h - Thickness of a film
$\overline{J}_\gamma$ - Diffusion-migration flux of component γ
k - Wavenumber of a Fourier mode
l - Spherical harmonics
$\overline{n}$ - Normal vector to the interface
$\overline{\overline{P}}$ - Pressure tensor
$\overline{P}_s$ - Surface dipole density vector
$\mathcal{R}$ - Universal gas constant
R - Radius of curvature
R_γ - Chemical source of component γ
$\underline{T}$ - Temperature
$\overline{\overline{T}}$ - Maxwell stress tensor
t_c - Critical time of onset of instability
$\overline{v}$ - Barycentric velocity
$V(h)$ - Van der Waals potential between two faces of a film of thickness h
x, y - Coordinates in the plane interface
z - Coordinate normal to the plane interface
z_γ - Charge of component γ
$z\Gamma$ - Total surface density of charges
$z_\gamma \Gamma_\gamma$ - Surface density of charges of component γ

Greek letters

β - Concentration gradient
Δ_s - Discontinuity of a property across the surface
δ', δ^{+} - Functions of radius and viscosity
ε - Dielectric constant
ε^s - Static surface Gibbs elasticity
ε^F - Renormalized film elasticity
ε^s_d - Dynamical surface elasticity
η_{dil} - Surface dilational viscosity coefficient
η_{sh} - Surface shear viscosity coefficient
κ - Reciprocal Debye length of the double layer
λ - Wavelength of the perturbation
λ_m - Wavelength for the maximum instability
λ_{cr} - Critical wavelength for the onset of instability
μ - Shear viscosity coefficient
ν - Kinematic viscosity coefficient
$\overline{\overline{\pi}}$ - Intrisic surface stress tensor
ρ - Mass density
ρ_γ - Mass concentration of component γ
$z\rho$ - Charge density
σ - Surface tension
σ_M - "Mechanical" contribution to the surface tension
σ_E - "Electrical" contribution to the surface tension
$\sigma_E^{d.l}$ - "Electrical" contribution of a double layer to the surface tension
σ_F - Renormalized surface tension of a film
ψ - Electrical potential
Ψ - Phenomenological quantity related to normal displacement
ω - Angular frequency of a Fourier mode
ω_R - Real part of the angular frequency of perturbation in normal mode analysis

Subscripts

s - Surface index
γ - Component index
cr - Critical

Superscripts

α - Phase index, α = I, II or s

4. REFERENCES

1. Linde, H. and Kunkel, E. 1969, Warme und Stoffübertragung, 2, pp. 60.
2. Orell, A. and Westwater, J.W. 1962, A.I.Ch.E.J., 8, pp. 350.
3. Davies, J.T. 1972, "Turbulence Phenomena", Ac. Press, New YorK.
4. Steinchen, A., Defay, R. and Sanfeld, A. 1971, J. Chim. Phys., 68, pp. 835 ; ibid., 68 pp. 1241
5. Gibbs, J.W. 1961, "The Scientific Papers", Vol. I, Dover, New York.
6. Defay, R., Prigogine, I., Bellemans, A. and Everett, D.H. 1966 "Surface Tension and Adsorption", Longmans.
7. Bedeaux, D., Albano, R.M. and Mazur, P. 1976, Physica, 81A, pp. 430.
8. Napolitano, L.G. 1982, Acata Astronautica, 9, pp. 199-215.
9. Hennenberg, M., Sørensen, T.S. and Sanfeld, A. 1977, J. Chem. Soc. Farad. Trans II, 73, pp 48.
10. Bisch, P.M. 1980, Ph. D. Dissertation, Brussels University
11. Glansdorff, P. and Prigogine, I. 1971, "Thermodynamic Theory of Structure Stability and Fluctuation", Wiley-Interscience, New York.
12. Nicolis, G. and Prigogine, I. 1977, "Selforganization in Nonequilibrium Systems", J. Wiley, New York.
13. Steinchen, A. and Sanfeld, A. 1980, in "Modern Theory of Capillarity" Ed. Goodrich, F.C. and Russanov, A.I., Akad. Verlag, Berlin.
14. Sanfeld, A., Steinchen, A., Hennenberg, M., Bisch, P.M., van Lamsweerde-Gallez, D. and Dalle Vedove, W. 1979, "Lecture notes in Physics", Ed. Sorensen, T.S. 105, pp. 229.
15. Steinchen, A. 1970, Ph. D. Dissertation, Brussels University
16. Aris, R. 1962, "Vectors, Tensors and The Basic Equations of Fluid Mechanics", Prentice Hall, N.J.
17. Gallez, D., Sanfeld, A. and Bisch, P.M. 1982, P.C.H, 3 pp.1
18. Sanfeld, A. 1968, "Introduction to the Thermodynamics of Charged and Polarized Layers", Wiley-Interscience, London.
19. Hennenberg, M. 1980, Ph. D. Dissertation, Brussels University
20. Sorensen, T.S. and Hennenberg, M. 1979, "Lecture notes in Physics" Ed. Sorensen T.S., 105, pp. 276.
21. Sternling, C.V. and Scriven, E.K. 1959, A.I.Ch.E.J., 5, pp. 514.
22. Marquez, A.R., Dalle-Vedove W. and Sanfeld A. 1981, J. Chem. Soc. Far. Trans. II. 77, pp. 2303.
23. Sorensen, T.S., Hennenberg, M., Steinchen, A. and Sanfeld, A. 1976, J. Coll. Interf. Sci., 56, pp. 191.
24. Linde, H. and Schwartz, E. 1967, Monatsheft d.D. Akad. Wiss. Berlin 6, pp. 330.
25. Defay, R. and Pétré, G. 1971 in "Surface and Colloid Science" (Ed. E. Matijevic) 3, pp. 27 (Wiley-Interscience), New York.

26. Joos P., Bleys, G., and Pétré G. 1982, J. Chim. Phys., 79, pp. 387.
27. Sanfeld, A., Steinchen, A. and Defay, R. 1969, J. Phys. Chem. 73, pp. 4047-4055.
28. Hennenberg, M., Sanfeld, A. and Bisch P.M. 1981, A.I.Ch.E.J., 27, pp. 1002.
29. Hennenberg, M., Bisch, P.M., Vignes-Adler, M. and Sanfeld A., 1979, J. Coll. Interf. Sci. 69, pp. 128 ; 1980 ibid 74 pp.495
30. Brian P.L.T. 1971, A.I.Ch.E.J., 17, pp. 765.
31. Brian P.L.T. and Smith, K.A. 1972, A.I.Ch.E.J., 18, pp. 582.
32. Lin, M. and Steinchen, A. (Marseille) Private communication.
33. Sorensen, T.S., Hennenberg, M., Steinchen, A. and Sanfeld, A 1976, Prog. Coll. Polym. Sci. 61, pp. 6470.
34. Sanfeld, A. and Steinchen, A. 1975, Biophys. Chem., 3 pp. 99 ibid., 1973, Chem. Phys., 1, pp. 156.
35. Dalle-Vedove, W. and Sanfeld, A. 1981, J. Coll. Interf. Sci., 84, 318.
36. Dalle-Vedove, W. and Sanfeld, A. 1981, J. Coll. Interf. Sci., 84, 328.
37. Feuillebois, A. 1983, E.S.A., Madrid, Proceedings of "4th Eur. Symp. Materials Sci. under Microgravity".
38. Velarde, M.G. and Castillo, J.L. 1982 in "Convective Transport and Instability Phenomena" Ed. Zierep., J. and Oertel, H. pp. 235, Braun Karlsruhe.
39. Velarde, M.G. 1982, J. Fluid. Mech., 125 pp 463-474.
40. Dupeyrat, M. and Nakache, E. 1977 in "Electrical Phenomena in the Biological membrane" (Ed. E. Roux), Elsevier, Amsterdam.
41. Durani, K., Hanson, C. and Hughes, C. 1977, Metall. Trans. B. 8B, pp. 169-174
42. Masson, H. 1977, Report C.E.A. Fr.
43. Sorensen, T.S. and Castillo, J.L. 1980, J. Coll. Interf. Sci. 76, pp. 399.
44. Ibanez, J.L. and Velarde M.G. 1977, J. Mathem. Phys., 38, pp. 1479-1483.
45. Prévost, M, Bisch, P.M. and Sanfeld A., 1982, J. Coll. Interf. Sci. 88, pp. 353-371.
46. Miller, C.A. and Scriven, L.E. 1970, J. Coll. Interf. Sci., 33, pp. 360.
47. Bisch, P.M., Steinchen, A. and Sanfeld, A. 1983, submitted to J. Coll. Interf. Sci.
48. Watanabe, A., Higashitsuji, K. and Nishizawa K., (1983) J. Coll. Interf. Sci. 64, pp. 378.
49. Bisch, P.M., Van Lamsweerde-Gallez, D. and Sanfeld, A. (1978) J. Coll. Interf. Sci. 64 pp. 378.
50. Van Lamsweerde-Gallez, D.,Bisch, P.M. and Sanfeld, A. 1978, Bioelectrochem. and Bioenergetics, 5 pp. 401; 1979, J. Coll.Interf. Sci. 71, pp. 513.
51. Sanfeld, A., Lin, M., Bois, A., Panaïotov, I.,and Baret, J.F. 1983, Adv. Coll. and Interf. Sci. under press.

52. Sanfeld, A., Lin, M., Bois, A., Panaïotov, I.,and Baret J.F., 1983, C.R. Ac. Sci., Paris, 296, pp. 609.
53. Verwey, E.J., and Overbeek, J.Th.G., 1948 "Theory of Stability of Lyophobic Colloids", Elsevier, Amsterdam.
54. Felderhoff, B.U., 1968, J. Chem. Phys. 48, pp. 1178; Ibid 49, pp. 44.
55. Ivanov, I.B., and Jain, R.K., 1979, "Lecture notes in Physics" Ed. Sorensen, 105, pp. 120.
56. Ivanov, I.B., and Dimitrov D.S., 1974, Coll. and Polymer Sci.,252, pp. 982.
57. Wendel, H., Gallez, D., and Bisch, P.M.,1981, J. Coll. Interf. Sci., 84, pp. 1.
58. Bisch, P.M., Wendel, H., and Gallez, D., 1983, J. Coll. Interf. Sci., 92, pp. 105.
59. Gallez, D., Bisch, P.M., and Wendel, H., 1983, J. Coll. Interf. Sci., 92, pp. 121.
60. Maldarelli, C., 1981, Ph. D. Dissertation, University of Columbia, New York
61. Chizmadzev, Y.A., Abidor, I.G., Pastushev, V.F., and Arakelya, V.B., 1979, Bioelectrochem. and Bioenergetics, 6, pp.37,.
62. Andrews, D.M., Manev, E.D., and Haydon, D.A., 1970, Spec. Disc. Farad. Soc., 1, pp. 46.
63. Steinchen, A., Gallez, D., and Sanfeld, A., 1982, J. Coll. Interf. Sci., 85, pp. 5.
64. Gallez, D., 1983, Biophys. Chem., under press.
65. Prévost, M., Gallez, D., and Sanfeld, A., 1983, J. Chem. Soc. Trans. Farad. Soc. II, 79.
66. Prévost, M., Gallez, D., 1983, submitted to J. Chem. Soc. Trans. Farad. Soc. II.
67. Exerova, D., Christov, Ch., Penev, I.,1976, in "Foams" (Ed. R.J. Akers) Ac. Press. London.
68. Papahadjopoulos, D., and Okhi, S., 1969, Science, 164, pp. 1075.

INTERFACIAL INSTABILITY IN BINARY MIXTURES : THE ROLE OF THE INTERFACE AND ITS DEFORMATION

Manuel G. VELARDE

U.N.E.D. - Fisica Fundamental
Apartado 50.487, Madrid (Spain)

The role of interfacial deformation is considered in the stability analysis of fluid layers heated from below or above when there is an open interface to ambient air, and double diffusive transport of heat and solute thus leading to variations of interfacial tension that compete or cooperate with buoyancy phenomena. The onset of both oscillatory convection and steady patterns is described.

1. HEURISTIC ARGUMENTS, PARAMETERS AND EQUATIONS [1-5] .

Salt-fingering is the kind of instability that develops at the interface of a warm, salty water overlying a layer of cold and fresh fluid. It is originated by the large separation between two diffusion scales : heat diffusivity ($\kappa \sim 0.001\ cm^2 s^{-1}$) is about two orders of magnitude different from mass diffusivity ($D \sim 0.00001\ cm^2 s^{-1}$) for standard liquid mixtures.

When the fresh and cold layer rather sits on top of a salty warmer water there is no tendency to fingering but to the appearance of oscillations in the thermal and hydrodynamic variables and once more a simple explanation originates from the same large separation of diffusion scales. Thus knowledge of the actual value in an experiment of the Lewis number (or inverse Lewis number, according to author), $Le = D/\kappa$, is crucial for the understanding of the evolution of the fluid layer under varying thermal constraints.

Other relevant parameters in a Rayleigh-Bénard geometry are the thermal Rayleigh number and the solutal Rayleigh number

G. Nicolis and F. Baras (eds.), Chemical Instabilities, 223–232.

$$Ra = \alpha g d^3 \Delta T / \nu\kappa \quad \text{and} \quad Rs = \gamma g d^3 \Delta N / \nu D \quad ,$$

respectively, where α and γ are the thermal and solutal expansion coefficients, g is the gravitational acceleration, $\Delta T / d$ is the thermal gradient across the layer of thickness d, and ν is the kinematic viscosity of the mixture. $\Delta N / d$ denotes a concentra-gradient.

With an open interface several other parameters enter the problem. Surface tension tractions must be considered if there is variation of surface tension with either temperature or solute (an impurity). This is accounted with the inclusion of the thermal and solutal Marangoni numbers (the latter is usually called the Elasticity),

$$M = - (\partial\sigma/\partial T)\, d\Delta T / \mu\kappa \quad \text{and} \quad E = - (\partial\sigma/\partial N_1)\, d\Delta N_1 / \mu D \;,$$

respectively. Here σ is the surface tension (liquid-air, say) and N_1 is the mass-fraction of the impurity, which for convenience is considered the heavier component of the binary mixture. Consideration of surface tension tractions does not necessarily forces the consideration of the deformation of the interface for they may be operating even if the interface remains level. If, however, the deformation is to be considered at least two more dimensionless groups appear, the Bond number (Bo or G) and the capillary (or crispation) number :

$$Bo = \rho g d^2 / \sigma_o \quad \text{and} \quad C = \mu\kappa / \sigma_o d \quad ,$$

respectively, where ρ accounts for the density of the fluid mixture and σ_o is some mean value of the surface tension σ. $\mu = \nu\rho$ is the dynamic viscosity. The Bond number estimates the strength of the gravitational forces with respect to the surface tension and thus large values of G correspond to rather flat, and level interfaces whereas low values of the Bond number exist when interfaces tend to be spherically shaped (at thermodynamic equilibrium this means minimization of free energy). Rather low Bond numbers appear in experiments aboard spacecrafts where gravity might decrease in four or six orders of magnitude the value on Earth. Then capillary lenghts may reach the order of the meter and for this reason there is no need of a container for the handling of liquids at very low Bond numbers. The capillary number compares dissipation to surface tension forces. To a first approximation dissipation tends to damp out all inhomogeneities whereas, as before, surface tension tends to bend interfaces. For standard fluids and standard modes of operation C varies between 10^{-2} and 10^{-7}. Large values of C appear, however, when the operation takes place near a critical point (in temperature for a single component liquid layer, the consultal/demixtion point,...) where the interfacial tension goes to zero or when we handle extremely thin films.

It is of some interest to read the definition of the capillary number from another perspective. Consider an interface where a capillary wave may develop. Let ξ be an estimate of its lenght. The actual value could eventually be d, the cell gap in a Rayleigh-Bénard (Marangoni) experiment. Then a quantity, the (mechanical) time constant of such a disturbance upon the interface (the interface may be likened to a stretched membrane) can be defined through the relation

$$\tau_\xi^2 = \rho\xi^3 / \sigma \qquad \text{from now on } \xi = d.$$

With heat and mass diffusion the other two time constants are

$$\tau_V = d^2 / \nu \qquad \text{and} \qquad \tau_T = d^2 / \kappa .$$

We have

$$C = \tau_d^2 / \tau_V \, \tau_T$$

and thus for large values of the capillary number correspond to the case of disturbances that decay so fast on the thermal and momentum dissipation scales that this happens before the mechanical disturbance, the "wave" decays, i.e., before the interface returns to the level position. Restoring mechanical forces (potential energy) make the interface overshoot the level position thus leading to interfacial oscillations. It is this potential energy that provides in a fluid layer the possibility of reversing an initially given fluid motion. Note that the above given argument can be extended to the case of an interface where some chemical reaction takes place. It suffices to replace the heat diffusivity by the appropriate chemical constant ans abnormally large separation in the different time constants of the problem may result in oscillatory motions of the interface.

Still two more parameters are needed in the problem considered here : the Prandtl number, $P = \nu/\kappa$ and the Schmidt number ν/D. Low Prandtl number fluids also tend to show oscillatory instabilities as there inertial terms tend to dominate dissipation. Finally the combination

$$A = (Ra + Le\ Rs)\ C / G$$

is a quantity that estimates the validity of the Boussinesquian approximation (for a single component $A = \alpha\Delta T$). In the following we shall consider values of A and C small with emphasis however, on the role of C when the Rayleigh and Marangoni numbers compete or cooporate for the onset of instability. All results will show the dependence on P, Le and G and, for illustration, cases with vanishing Rayleigh numbers are also discussed.

We use the following conventions and notations (Figure 1.) : d is the mean distance between two infinitely extended surfaces; the lower is a rigid, heat conducting plate held at constantly controlled temperature. The upper surface is the one open to the ambient air. For simplicity is considered adiabatic (poor heat conductor). The fluid enclosed between these two surfaces is an incompressible binary mixture and we restrict consideration to a two-dimensional problem. Thus x and z denote, respectively, the horizontal and vertical coordinates. The ambient air is assumed to have negligible density and *dynamic* viscosity.

For universality in the description the following units (scales) are introduced : d for lenght, d^2/κ for time, κ/d for velocity, ΔT and ΔN_1 for temperature and mass fraction of the components, respectively. $\mu\kappa/d^2$ and σ_o for pressure and surface tension, respectively. The open surface, S(t), is located at

$$z = 1 + \eta(x,t) \ .$$

$\vec{n}$ denotes the outward unit normal vector to S,

$$\vec{n} = (-\frac{\partial\eta}{\partial x}, 1) / N$$

whereas $\vec{t}$ is the unit tangent vector

$$\vec{t} = (1, \frac{\partial\eta}{\partial x}) / N$$

and the curvature is K(η),

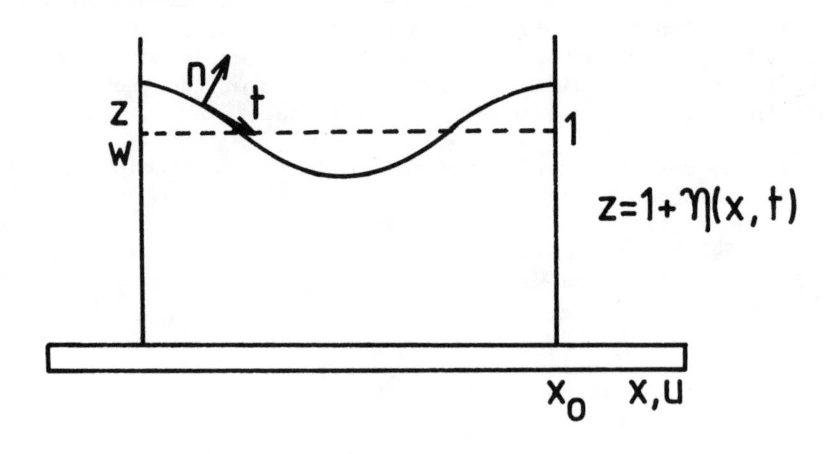

Figure 1. An exaggerated view of the open interface with bottom plate at z = 0.

$K(\eta) = (\partial^2\eta/\partial x^2)/N^3$ with $N = \{1 + (\partial\eta/\partial x)^2\}^{\frac{1}{2}}$.
The stress balance at the deformable interface is

$$\tau_{ij} n_j = -(G/C)(\eta + A\eta^2/2) n_i + (K/C)\{1 - MC(\theta - \eta) - ELe(\Gamma - \eta)\} n_i + t_i(\vec{t}.\vec{\nabla})\{M(\theta - \eta) + ELe(\Gamma - \eta)\} \quad i,j = 1,2 \tag{1}$$

where θ, Γ and $\vec{v}$ account for disturbances upon the temperature, mass fraction and velocity of the initially steady rest state.
The stress tensor is

$$\tau_{ij} = -p\delta_{ij} + \varepsilon_{ij} \quad \text{with} \quad \varepsilon_{ij} = \partial v_i/\partial x_j + \partial v_j/\partial x_i \tag{2}$$

δ_{ij} is the Kronecker delta and the summation convention over repeated indices is assumed.
The kinematic boundary condition at the interface is

$$\partial\eta/\partial_t = N v_i n_i \quad \text{on} \quad z = 1 + \eta \tag{3}$$

The convention that the heat flux is prescribed at the open interface leads to the condition

$$(\vec{n}.\vec{\nabla})\theta = (1 - N)/N \quad \text{on} \quad z = 1 + \eta \tag{4}$$

For the impurity we also prescribe at the interface its flux. This leads to a simpler analysis.
For a fluid layer bounded by a copper plate at the bottom we take there a heat conducting plate, impervious to matter transfer and mechanically rigid . Thus we have

$$v_i = \theta = \Gamma = 0 \quad \text{on} \quad z = 0. \tag{5}$$

We assume that originally the fluid layer is in motionless state with steady linear distributions of temperature and solute. Thus the evolution equations for disturbances upon the motionless steady state are the Navier-Stokes, Fick, and Fourier equations for the region $0 \leqq z \leqq 1 + \eta(x,t)$, $-\infty < x < +\infty$

$$P^{-1}(\partial v_i/\partial t + \vec{v}.\vec{\nabla} v_i) = \partial\tau_{ij}/\partial x_j + Ra\,\theta k_i + LeRs\Gamma k_i \tag{6}$$

$$\partial\theta/\partial t + \vec{v}.\vec{\nabla}\theta = \nabla^2\theta + w \tag{7}$$

$$\partial\Gamma/\partial t + \vec{v}.\vec{\nabla}\Gamma = Le\,\nabla^2\Gamma + w \tag{8}$$

$$\vec{\nabla}.\vec{v} = 0 \tag{9}$$

where $k_i = (0,1)_i$. Note that we have not included the Soret effect in Fick's mass transport equation as, for simplicity, we shall focus on the competition of the two possible gradients of temperature and solute .

2. ANALYSIS OF ARBITRARILY FINITE DISTURBANCES [6,7].

We define the integral over the free interface for a quantity f as

$$\int_{S(t)} f\, ds = \int_0^{S_o(t)} f\, ds = \int_0^{x_o} f(z=1+\eta)\, N dx \qquad (10)$$

where ds is an element of arc length along S(t), and $S_o(t)$ is the length in one period along x. A two-dimensional volume integral of f over a period is

$$< f > = \int_0^{x_o} \int_0^{1+\eta} f(x,z,t)\, dz\, dx \qquad (11)$$

Then we define the energy

$$E = P^{-1} < v^2/2 > + \lambda < \theta^2/2 > + \Lambda Le < \Gamma^2/2 > + (G/C) \int_{S(t)} ds \frac{1}{2} (\eta^2 + A\eta^3/3)/N \qquad (12)$$

where $|\eta| < 3/A$ (Boussinesquian approximation). λ and Λ are the linking parameters whose choice is dictated by the convenience in obtaining the largest parameter region of stability of the initialy motionless steady state of the fluid layer. Thus a variational condition is introduced $\delta(dE/dt) = 0$ where δ accounts for an arbitrary variation subjected, however, to the conditions earlier indicated (mass conservation, boundary conditions and all that) . We have

$$\delta(dE/dt + < 2p \vec{\nabla} . \vec{v} > + \beta \int_x \eta N\, dx) = 0 \qquad (13)$$

The consideration of arbitrary values of the capillary or crispation number produces a formidable problem and a reasonable approach is to consider its contribution, i.e., the role of the surface deformation to a first order approximation . Thus we set

$$\eta = \eta^{(0)} + \eta^{(1)} C + 0(C^2) \ ; \quad \eta^{(0)} = 0 \qquad (14)$$

(note that first it moves and then it gets deformed which in turn affects the motion) together with similar expansions for the remaining quantities. Then the evolution problem for disturbances upon the initial state is reduced to the following Euler-Lagrange equations

$$2\tau_{ij}^{(0)} + (Ra+\lambda)\theta^{(0)} k_i + (Rs+\Lambda)Le\Gamma^{(0)} k_i = 0 \tag{15}$$

$$(Ra+\lambda) w^{(0)} + 2\lambda\nabla^2\theta^{(0)} = 0 \tag{16}$$

$$(Rs+\Lambda) w^{(0)} + 2\,\Lambda\, Le\, \nabla^2\Gamma^{(0)} = 0 \tag{17}$$

$$\vec{\nabla}\,.\,\vec{v}^{(0)} = 0 \tag{18}$$

together with the conditions on $z = 0$:

$$v_i^{(0)} = \theta^{(0)} = \Gamma^{(0)} = 0 \tag{19}$$

On $z = 1$ we have

$$w^{(0)} = 0 \tag{20}$$

$$2\tau_{ij}^{(0)}\, n_j^{(0)}\, t_i^{(0)} + M\,\partial\theta^{(0)}/\partial x + ELe\,\partial\Gamma^{(0)}/\partial x = 0 \tag{21}$$

$$2\lambda\partial\,\theta^{(0)}/\partial x - M\,\partial u^{(0)}/\partial x = 0 \quad \text{and} \tag{22}$$

$$2\,\Lambda Le\,\partial\Gamma^{(0)}/\partial x - E\,\partial u^{(0)}/\partial x = 0. \tag{23}$$

Solutions of the above posed problem can be sought in the form of exponentials $\exp(iax)$ where $\underline{a}$ denotes a Fourier decomposition mode. We can set

$$\theta^{(0)} = \{ \sum_{i=1}^{6} \Omega_i \exp(q_i z) + \Omega_7 (\exp(az)-\exp(-az))\} \exp(iax) \tag{24}$$

where the q_i are the roots of the polynomial equation

$$(q_i^2 - a^2) + (\lambda + \Lambda)\, a /4 = 0 \tag{25}$$

There are similar expressions for the remaining quantities. To the lowest order approximation, necessary conditions for the onset of instability are obtained in the form of Taylor expansion in the capillary number. For instance, using the Marangoni number the motionless steady state of the fluid layer is stable provided that its value remains below

$$M^{(0)} + C\, M^{(1)} \quad , \text{ where}$$

$$M^{(1)} = - 2a^{-2} \; (M + Le\, E) \int_x dx \;\; \tau_{ij}^{(0)} \; n_j^{(0)} \; n_i^{(0)} \; \left(\frac{\partial w}{\partial z}\right)^{(0)}$$

$$\left\{ \int_x dx \;\; \theta^{(0)} \; \left(\frac{\partial w}{\partial z}\right)^{(0)} \right\}^{-1} \tag{26}$$

Here the superscript (0) accounts for values at vanishing crispation number. Similar expressions have been found for the remaining control parameters (Rayleigh and Marangoni numbers). When, however, C vanishes there is no gravity and both Marangoni numbers are positive, energy theory yields $M + E = 56.7$ for a critical wavenumber $a_c = 2.22$. Below this line fluid layer is absolutely stable (Figure 2.).

3. THE CASE OF INFINITESIMAL DISTURBANCES [8] .

Sufficient conditions for instability of the motionless steady initial state of the layer can be obtained by means of a linear stability analysis using normal modes. As the results of this and the preceding energy analysis do not coincide there appear possibilities of subcritical modes of instability (finite amplitude steady states or oscillations) and transient oscillations. Figure 2. gives an illustration of some of the results obtained for the case of vanishing gravitational acceleration ($g = 0$). For a given value of the temperature gradient the role of the impurity is rather clear. In accordance with the sign of E (solutal Marangoni or elasticity number) there is a dramatic lowering of the threshold for thermoconvective instability or the possibility of overstable modes (oscillations). When all buoyancy phenomena in the bulk are negligible ($Ra = Rs = 0$) transition to steady convection is expected above the line

$$M+E = 8a\,\frac{(a^2+G)\{\beta^3-\beta^2(4a-1)-\beta(4a+1)-1\}+4a^3(\beta^2+\beta)ECS}{(a^2+G)\{\beta^3-\beta^2(4a^3+3)+\beta(3-4a^3)-1\}+32a^5C(\beta^2+\beta)} \qquad (27)$$

where S=1-Le, $\beta=\exp(2a)$ and a is the wavenumber. Thus in the limit of vanishing capillary number (or deformation, C=0) we have M+E=79.6 with $a_c=1.99$. When the Bond number is negligible, G=0, the correction induced by the deformation is

$$M^{(1)} = -1.11\ 10^3\{M^{(0)} + ELe\} \qquad (28)$$

Thus, to a first-order approximation with positive Marangoni numbers the deformation of the interface reduces the region of finite amplitude instability.

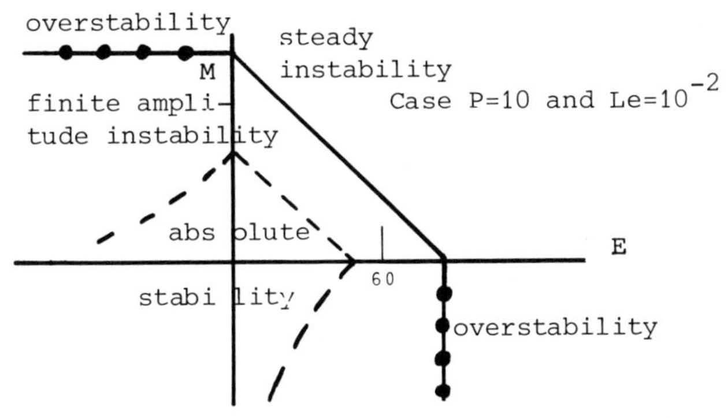

Figure 2(a). Zero-gravity Bénard-Marangoni convection in a binary liquid mixture. Solid and broken lines refer to linear and energy analyses, respectively. Dotted parts correspond to oscillations whose onset depends on both P and Le.

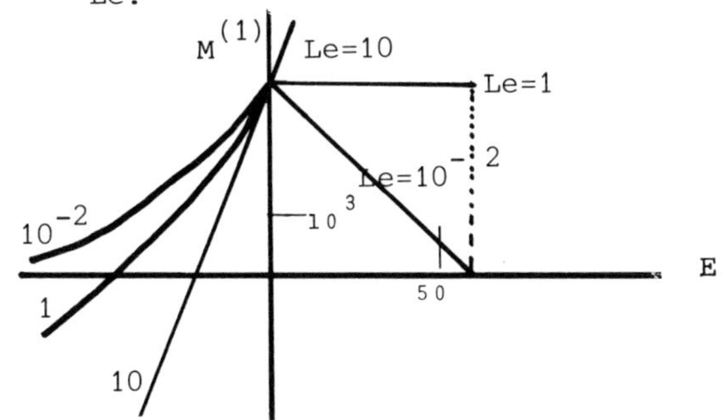

Figure 2(b). Zero-gravity Bénard-Marangoni convection. First-order corrections in the thermal Marangoni numbers produced by the deformation of the interface. To compute the actual Marangoni number one uses Equation (28).

ACKNOWLEDGMENTS

The present report summarizes work done in collaboration with J.L. Castillo .It has been sponsored by the Comisión Asesora de Investigación Científica y Técnica(Spain).

REFERENCES

1.Ostrach,S.,1980,Physicochem.Hydrodyn.,1,p.4
2.Velarde,M.G. and Normand,Ch.,1980,Sci.Am.,243,p.92
3.Velarde,M.G. and Castillo,J.L.,in *Convective transport and instability phenomena*,Zierep,J. and Oertel,H.,editors,Braun-Verlag, Karlsruhe,1982,pp.235-264
4.Velarde,M.G.,1982,in *Nonlinear phenomena at phase transitions and instabilities*,Riste,T.,editor,Plenum Press,N.Y.,pp.205-48
5.Velarde,M.G.,1982,in *Evolution of Order and Chaos in Physics,Chemistry and Biology*,Haken,H.,editor,Springer-Verlag,Berlin,pp.132-45
6.Davis,S.H. and Homsy,G.H.,1980,J.Fluid Mech.,98,p.527
7.Castillo,J.L. and Velarde,M.G.,1982,J.Fluid Mech.,125,p.463.
8.Castillo,J.L. and Velarde,M.G.,paper in preparation with details concerning the linear stability analysis to be published elsewhere.

CHEMICALLY DRIVEN INTERFACIAL INSTABILITIES

M.Dupeyrat[o], E.Nakache[o], M.Vignes-Adler[x]

[o]Lab.Chimie Phys.Univ.Paris6 ;11 rue P.et M.Curie Paris 05
[x]Lab.Aérothermique CNRS 4ter route des Gardes 92 Meudon

Large scale motions have been observed at the interface between an aqueous solution of a long chain alkyltrimethyl ammonium halogenide and a nitrobenzene solution of picric acid in proportions far removed from the equilibrium partition state. These motions differ from the usual Marangoni effect because the desorption of the surface active material, required for sustained movements, depends on a chemical interfacial reaction. Such a reaction coupled to the transfer processes permits an instability to occur without any chemical reactions with complex kinetics.

INTRODUCTION

This paper is devoted to the study of a large scale interfacial turbulence chemically driven at a liquid-liquid interface due to the simultaneous transfer in opposite directions of two solutes one of them being surface active.

These motions of the interface were observed by Dupeyrat et al.in 1971 [1] at a plane interface between an aqueous solution of a long chain alkyltrimethyl ammonium halogenide and a solution of picric acid or potassium iodide in nitrobenzene (or nitroethane). If the system is contained in a polyethylene beaker a pseudo periodic motion is observed in the plane of the interface without noticeable perpendicular deformation. It appears as contractions and expansions in the interfacial plane, observable by means of a spontaneous emulsion, visible a moment after the contact of the phases. When the solutions are poured into a pyrex beaker a strong normal deformation of the interface is observed. The

G. Nicolis and F. Baras (eds.), Chemical Instabilities, 233–245.

motion appears as waves propagating with an amplitude 1 cm along the glass wall, followed by a total disturbance of the interface. These motions must be attributed to the Marangoni effect whereby an interfacial flow is driven by gradients in interfacial tension usually due to local concentration variations resulting from mass transfer across the interface [2,3,4,5,6,7,8] of from convection currents accompanying temperature variations [9,10]. Criteria for the onset of this type of interfacial instability were derived theoretically by Pearson [11] and Sternling and Scriven [12] on the basis of a linearized stability analysis. Considerable improvement in the criteria, based upon a similar approach adapted to the experimental situation studied by the authors, has been achieved [13,14]. The purpose of this work is to understand from experimental data the mechanism of the observed phenomena particularly the influence of the interface on both the physicochemical effects and the hydrodynamic effects.

EXPERIMENTAL

The system studied in some detail consisted of an aqueous solution of hexadecyl trimethylammonium chloride ($C_{16}Cl$)-a gift from a private laboratory - purified three times by crystallization in methanol. This solution was carefully poured into a paraffin coated beaker containing a solution of Picric acid (HPi) - supplied by Merck - dissolved in nitrobenzene-chromatography grade supplied by UCB. The solvents were previously saturated by each other. The experiments were performed at 24°C ±1. In this way, the only movements observed are those in the plane of the interface. Indeed the wettability of the wall involved in the three - dimen-

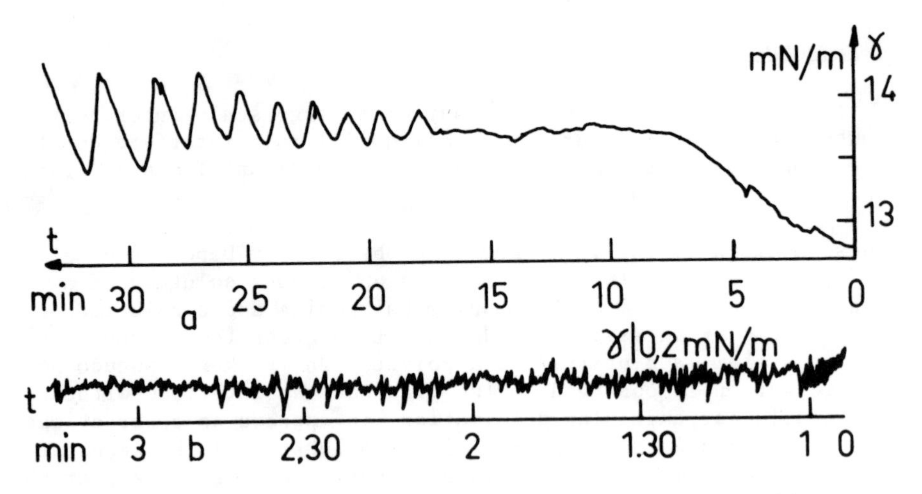

Figure 1. Variation of γ versus time. Interface (nitrobenzene) HPi $3.5\ 10^{-4}$ mol.1^{-1}/(water) $C_{16}Cl$: a) 4.10^{-4} ; b) 2.10^{-4} mol.1^{-1}

sion motion complicates the problem."Expansions"were correlated with a decrease in the interfacial tension. Therefore, the motion was analyzed from the variations in the interfacial tension γ as a function of time, measured by a detachment method using a "stirrup" instead of a ring so that corrective terms were excluded [15,16] .

RESULTS

No motion is observed unless there are two chemical compounds and one is surface active. They were dissolved one in each phase and each one in the phase where it is less soluble . Thus $C_{16}Cl$, a compound lipophilic, fully dissociated in both phases, is dissolved in water and HPi, a compound only slightly dissociated in nitrobenzene, but much more in water so that it is hydrophilic when it is dissociated, is dissolved in nitrobenzene.Thus the system is far from equilibrium.Since the solvents are saturated by each other,the mass transfer essentially deals with these solutes, the concentrations of which are much lower (fifty times at least) than those used by Linde [6] or Davies and Haydon [17]. Typical record of γ versus time are given in figures 1 and 2.

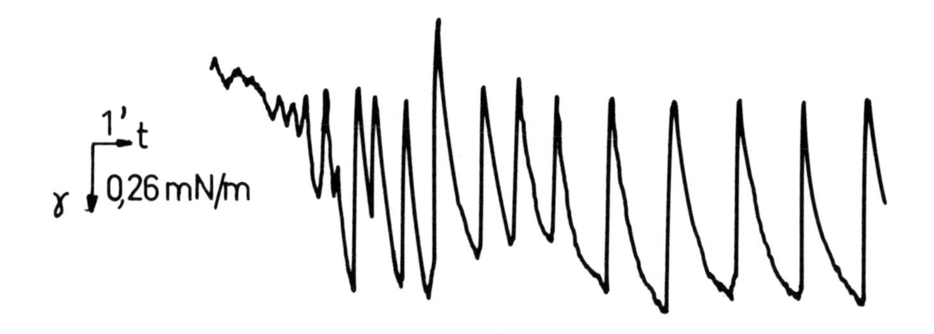

Figure 2. Variation of γ versus time, interface (nitroethane) HPi 3.5 10^{-4} mol.1^{-1} / (water) $C_{12}Br$ 5.10^{-3} mol.1^{-1}

After a more or less long "induction"period,oscillations arise, the period and amplitude of which depend on the concentration of $C_{16}Cl$ and HPi. The oscillations are not really periodic. Indeed the system is closed. However the two bulks can be considered as reservoirs large enough to allow the observation of several oscillations more or less damped. On the figure 2, one can notice the progressive increase then the abrupt decrease in the interfacial tension during the oscillation.

DISCUSSION

The solute transfer from one bulk solution to the other during the establishment of equilibrium involves diffusion and convection in the bulk and adsorption - desorption processes across the interface. Both hydrodynamical and physicochemical effects are simultaneously involved. We will analyze them successively, separating them rather arbitrarily, then examine how they are coupled.

1 - Hydrodynamical analysis

The transfer of surface active molecules generates a special interfacial convection described for example by Davies [18]. Due to the local convective currents a local increase of surface molecules near the interface produces at some point an increase of surface pressure $\Delta\pi$. The monolayer tends to spread further over the surface dragging some adjacent liquid with it. If conditions of viscosity, diffusivity, concentration are fullfilled, according to Sternling and Scriven for example, an eddy of fresh solution would occur to this point amplifying the movement which appears as an "expansion"(figure 3a). Then due to the momentum transfer,

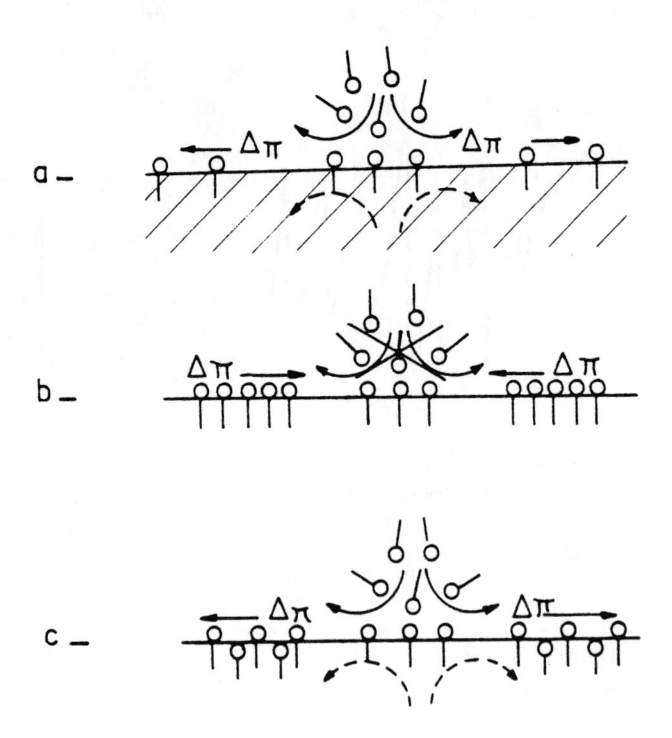

Figure 3. Hydrodynamical effects related to the density of the adsorbed layer.

eddies would occur on the other side of the interface too. If the density of surface active material becomes such as the adsorbed layer is sufficiently condensed, the surface pressure of this monolayer resists to the eddy (figure 3 b).The interfacial convection is further inhibited by highly viscous dissipation. The transfer then becomes purely diffusive and of course slower. But if the molecules are not very surface active, they can easily desorb inducing a "contraction" of the interface and the latter step is not reached (figure 3c).
This is the case for a drop of water hanging in toluene containing acetone which "kicks" as long as the acetone is being transferred from toluene to water. But, when dodecyl ammonium chloride (4.10^{-3} mol.1^{-1}), a compound much more surface active than acetone and very similar to those which we used in our experiments is added, the drop does not start "kicking".
Indeed, we observed that no motions occurs if our system is made from pure nitrobenzene and an aqueous solution of $C_{16}Cl$. How can the presence of HPi a non surface active compound, bring about instability ? To answer this question requires more physicochemical information about the transfer of the solutes.

2.Physicochemical analysis

In order to obtain more information about the importance of the transfer according the nature of the ionic species, we have calculated the final equilibrium concentrations under typical initial conditions, computing them from electroneutrality and measured partition coefficients and dissociation equilibrium constant of HPi in nitrobenzene. The results are summarized and schematized in figure 4a. They show a full dissociation of HPi and

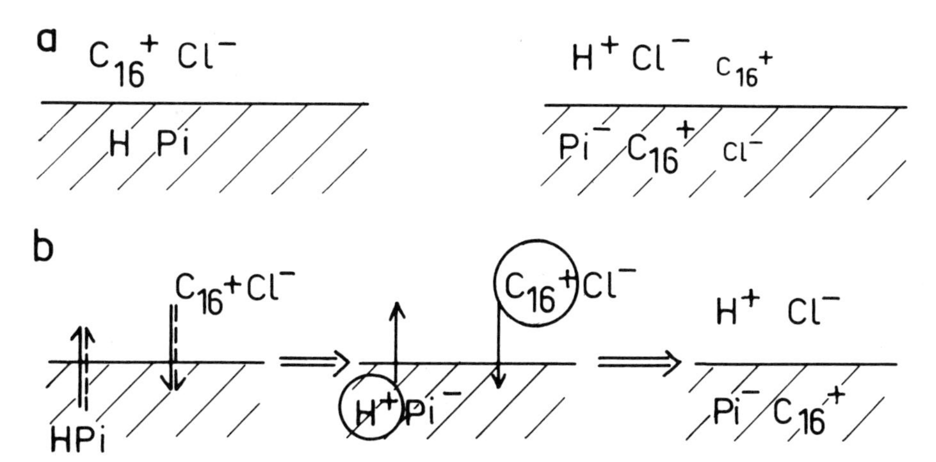

Figure 4. Transfer of HPi and $C_{16}Cl$ a) computed at the equilibrium b) assumed during the relaxation towards the equilibrium.

a nearly total exchange of H^+ with C_{16}^+ as a counter ion for Pi^-. This means that, although the driving force comes from the flux of $C_{16}Cl$ which passes from water towards nitrobenzene and the flux of HPi in the opposite direction, only C_{16}^+ passes through the interface because of the large difference in the partition coefficients. Indeed the counter ion of C_{16}^+ is Pi^- rather than Cl^- because the structure of the Pi^- ion is very similar to the nitrobenzene molecules. This is possible because at the same time and for similar reasons only H^+ ions pass through the interface while Pi^- ions remain in the organic phase. Thus a flux of C_{16}^+ can occur by means of two opposite fluxes (H^+ and C_{16}^+) (figure 4 b). A decrease of $C_{16}Cl$ and the formation of $C_{16}Pi$ result because of a counter ion exchange reaction

$$Pi_{nitro} + C_{16}Cl_{int} \rightarrow C_{16}Pi_{int} + Cl^-_{water}$$

Now $C_{16}Pi$ is also surface active. Therefore we have to consider its influence on the interfacial convection.
We have shown [19] that, actually, the amphiphilic C_{16}^+ ions only are surface active, but their interfacial density depends on the nature of the counter ion. Indeed, the structure of the Pi^- ion is very similar to the nitrobenzene molecules and the partition coefficient of $C_{16}Pi$ P_{water}^{nitro} is about 10^8 times larger than this of $C_{16}Cl$. Thus for the same water total concentration, there are two very different distributions of C_{16}^+ (figure 5) [20]. Now, the

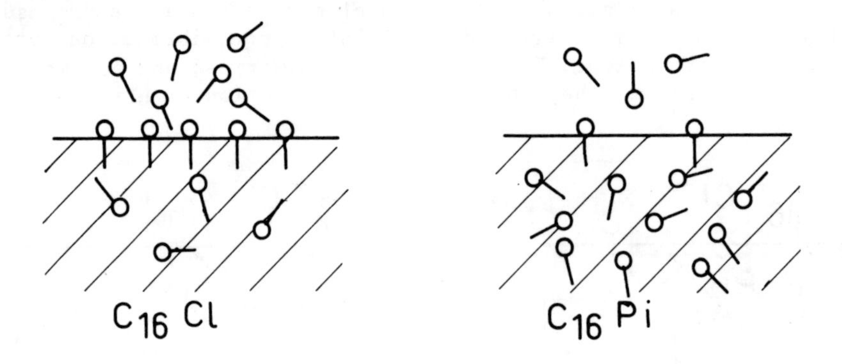

Figure 5. Distribtuion of $C_{16}Cl$ and $C_{16}Pi$ in the bulk and at the interface, at the equilibrium.

surface activity depends on the aqueous concentration of C_{16}^+ so that for a same initial concentration

$$[C_{16}Cl]_{int} \gg [C_{16}Pi]_{int}$$

and

$$\gamma\, C_{16}Cl \ll \gamma\, C_{16}Pi$$

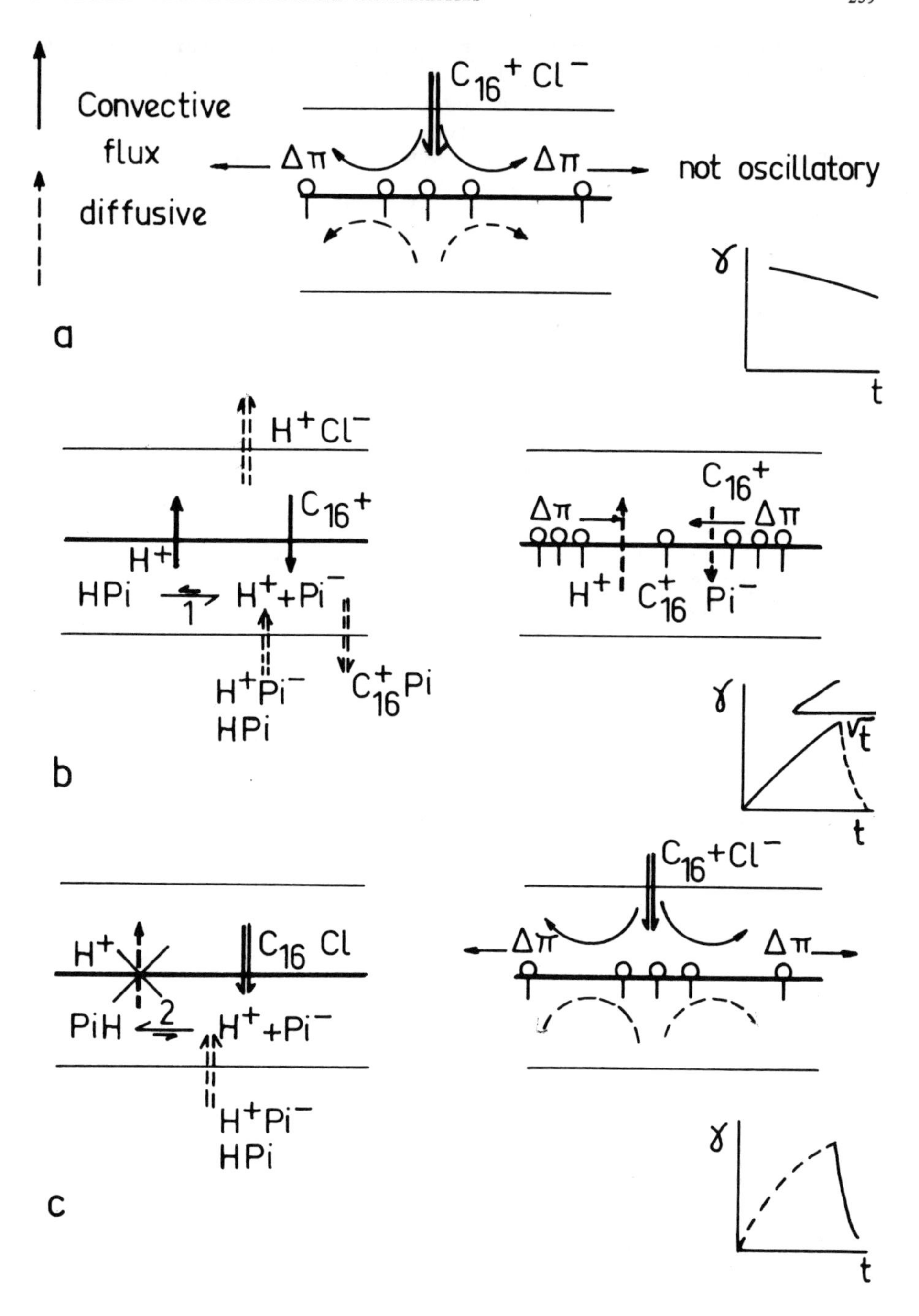

Figure 6. Coupling between hydrodynamical and physicochemical effects.

In conclusion if $C_{16}Cl$ is replaced by $C_{16}Pi$, the interfacial tension increases and C_{16}^+ ions desorb and leave the interface. At this step, we have to explain how can the concentration of $C_{16}Cl$ and $C_{16}Pi$ vary during the transfer. Several authors [21,22,16,23] studying similar interface have shown that the adsorption - desorption process is very fast as compared to diffusion convection. This implies that, at every time, the adsorbed monolayer is in equilibrium with the subjacent bulk layers and diffusion and convection are the limiting steps to the present transfer. Now we come to the coupling of these two effects.

3.Reaction mechanism

At the beginning, the driving force comes from the flux of $C_{16}Cl$ which passes from water towards nitrobenzene. Indeed the interfacial tension is lowered to $\gamma = 12.8$ mN/m which is near the value of the interfacial tension of such a system containing only $C_{16}Cl$ at the same initial concentration. This flux promotes at some point of the interface, eddies on either side of the interface as mentioned before while the interfacial tension decreases due to the adsorption of $C_{16}Cl$ (figure 6 a). In the second step the transfer is influenced by the counter ion exchange interfacial reaction. Thus, as explained before there are two convectives fluxes, a C_{16}^+ flux towards nitrobenzene and a H^+ flux towards water. Now, the nitrobenzene bulk sublayer depleted in H^+ by this convective flux is enriched only by a diffusive flux from the nitrobenzene bulk (figure 6b). Thus, the H^+ concentration in this sublayer is very low as compared to the concentration of the Pi^- ions for which no depletion occurs. Thus the displacement of the dissociation equilibrium of HPi occurs along the direction 1 as

$$HPi \underset{1}{\rightleftarrows} H^+ + Pi^-$$

according to the Le Chatelier law, which results in an accumulation of Pi^- ions in the sublayer. Therefore, as mentioned before, the interfacial tension should increase, proportionally to the time because the fluxes are convective (figure 6c). Then C_{16}^+ ions leave the interface, promoting a "clearing". Under these conditions $\Delta\pi$ will oppose the movements of the eddies while the fluxes become diffusive. The interfacial tension should increase at this moment, proportionally to the square root of time. We checked this point as shown in the figure 7 where the variations in γ during the ascending part of the oscillation are displayed as a function of time. It can be seen that the variation in γ is proportional to the time at the beginning of the oscillation, then to the square root of time, about fifteen seconds later. In the next step, the increase of the concentration of Pi^- ions relative to H^+ favors the displacement of the dissociation equilibrium according to the Le Chatelier law, along the direction 2 as

$$Pi^- + H^+ \overset{2}{\rightleftharpoons} HPi$$

It caused HPi to be recombined and H^+ consumed, so that H^+ is no longer available to pass through the interface thus preventing the interfacial exchange reaction. The flux of C_{16}^+ as $C_{16}Cl$ becomes predominant and the interfacial tension decreases while the interfacial convection occurs again as in the beginning . The development of eddies produces a hydrodynamical agitation which would restore the bulk concentration of all species on

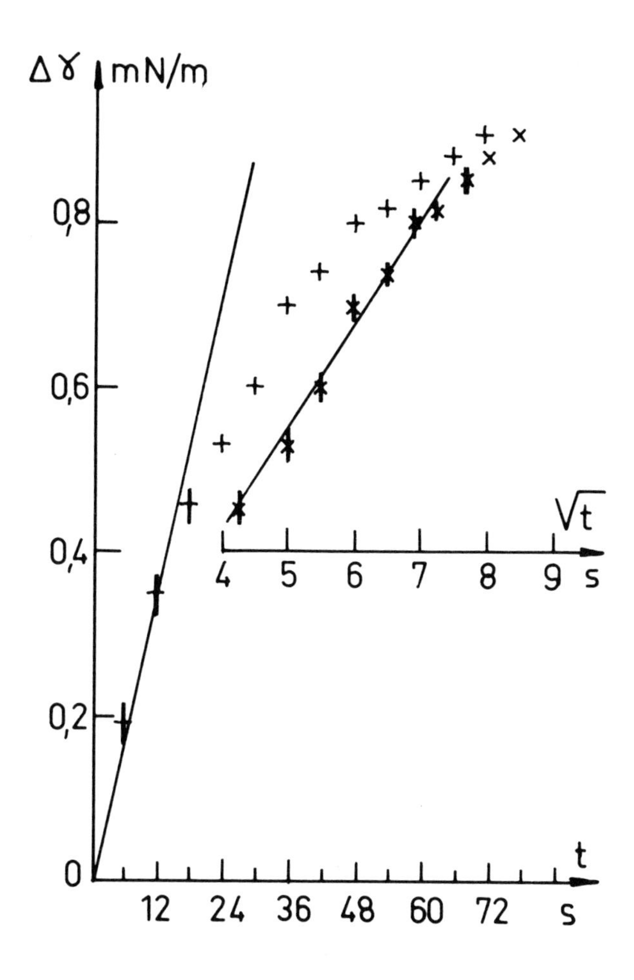

Figure 7. Variation of γ with the time and with the square root of time. Interface (nitroethane) HPi 1.25 10^{-3} mol.$^{-1}$ / (water) $C_{12}Br$ 5.10^{-3} mol.$^{-1}$.

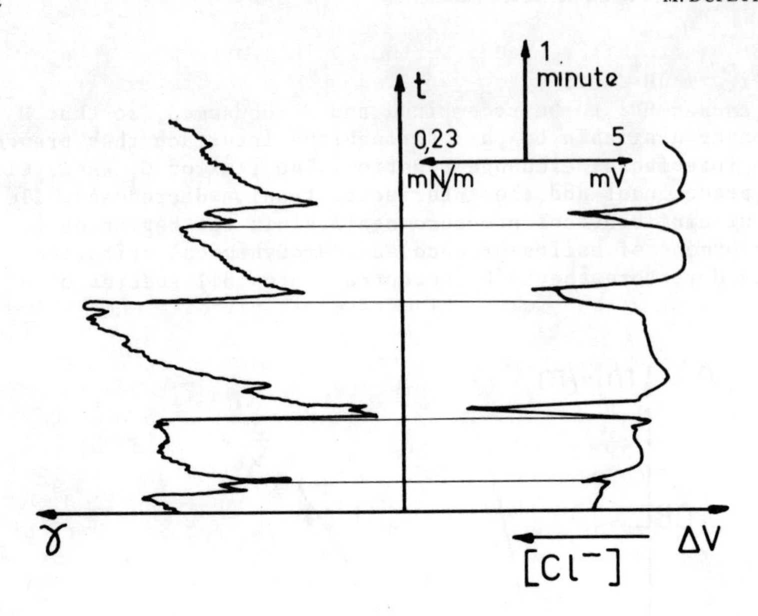

Figure 8. Variation of γ and ΔV with time

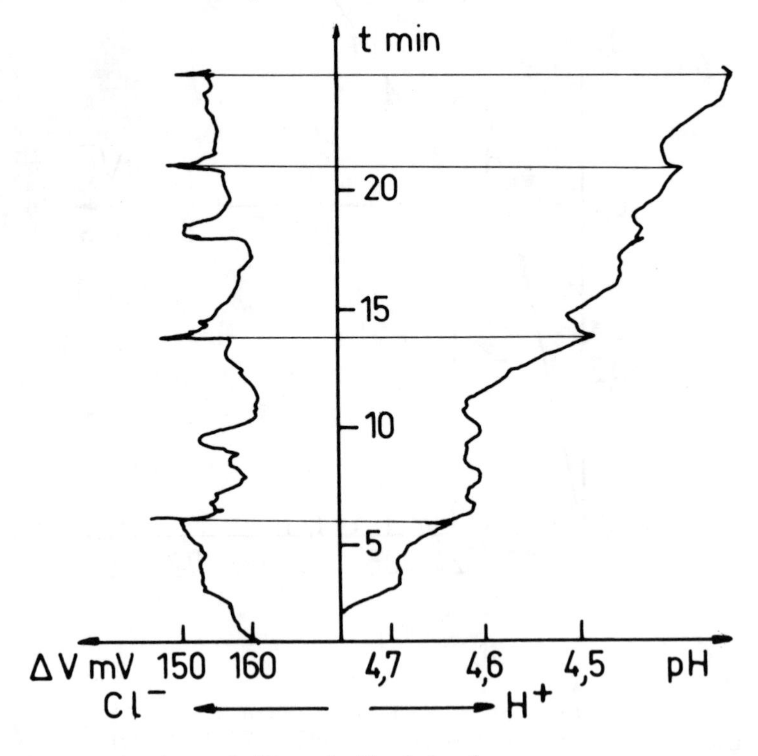

Figure 9. Variation of ΔV and pH with time

either side of the interface. So the cycle starts again.
There is some evidence which verified such a mechanism. If it is true :
1 - The ascending part of the oscillation is ruled by the transfer of H^+ through the interface which is controlled by the dissociation reaction of HPi along the direction 1, while during the descending part Pi^- ions are recombined along the direction 2. Therefore the possibility of such a dissociation reaction is fundamental. Now, we observed that if HPi is replaced by KPi, a compound very similar but fully dissociated in nitrobenzene, no motion occurs.
2 - During the descending part of the oscillation, an increase of the concentration of $C_{16}Cl$ is correlated with a decrease in the interfacial tension and in the H^+ concentration. We performed simultaneous measurements of Cl^- concentration - by means of a silver - silver chloride electrode - and of γ , on one hand, (figure 8) of pH on other hand (figure 9). It can be noticed that the expected correlations are verified - with some delay in the response of the silver chloride electrode relative to the variation in γ. This can be due to the bulkiness of the sensors which does not allow us to perform the measurement at exactly the same point on the interface.
3 - The descending part of the oscillation is initiated by a stop in the H^+ flux. If an injection of concentrated HCl is made with a micropipet very close to the interface, a decrease of both the period and the amplitude of the oscillation should be observed. In figure 10 it can be seen that, just after such an injection,

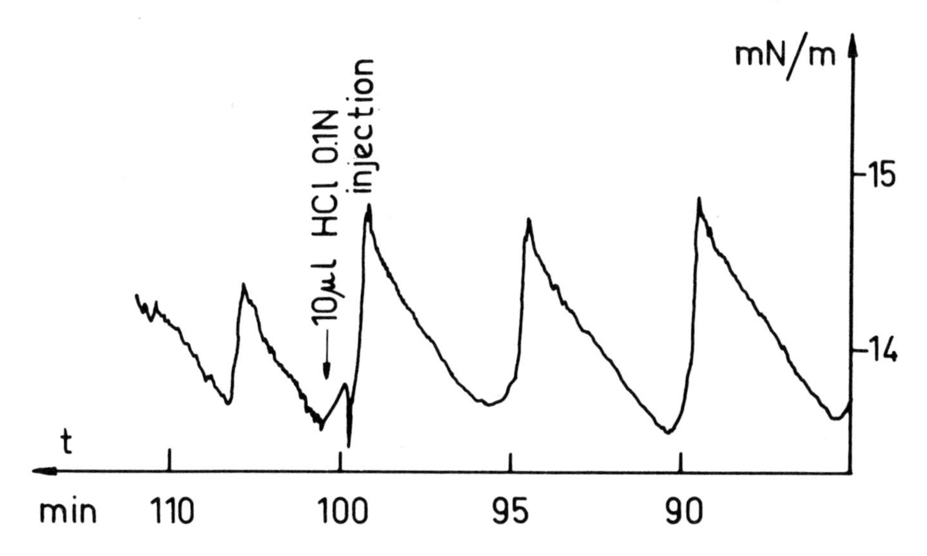

Figure 10. HCl injection effect in the aqueous phase. Interface HPi $3.5.10^{-4}$ $mol.l^{-1}$(nitrobenzene)/$C_{16}Cl$ 4.10^{-4} $mol.l^{-1}$(water)

both the period and the amplitude of the oscillation decrease.

CONCLUSION

This system is only one particular case of the coupling which can occur at a liquid - liquid interface. We can now try to draw some more general conclusions about such heterogeneous systems. If two bulk solution come in contact there are the possibility of transfer of material from one to the other and the existence of an interface which allows adsorption - desorption processes and eventually interfacial reactions. If the transferred compounds are surface - active, the transfer through the interface can be convective, if not, diffusion occurs.

This interfacial convection is possible during a notable time only if some process of desorption occurs. Now, the desorption can be spontaneous with a short chain surfactant, which is easily desorbable. In this case, the kinetics of transfer, adsorption and desorption have to be considered simultaneous in order to account for some instability, that is the same molecule will be adsorbed and desorbed. When the surfactant is not or only slightly desorbable - that is if the chains are long - the desorption cannot be spontaneous but it can happen that it is assisted by an interfacial reaction which transforms a very adsorbable species into an easily desorbable one as by our counter ion exchange reaction.

$$\underset{\text{nitro}}{Pi^-} + \underset{\text{adsorbable}}{\boxed{C_{16}\,Cl}} \longrightarrow \underset{\text{desorbable}}{\boxed{C_{16}\,Pi}} + \underset{\text{water}}{Cl^-}$$

In this case, it is not the same molecule which is adsorbed or desorbed. This is the precise point where chemistry is involved The kinetics of this reaction has to be considered, eventually taking into account some solvatation or reorientation processes during the transfer. Thus, the model becomes very complex due to the numerous species involved. It is for this reason that we did not succeed in solving the problem rigorously. Knowing the mechanism of the phenomenon, we are now able, to make the necessary approximations.

Finally, there is a possible coupling between a transfer and some interfacial reaction, permitting an instability to occur without any chemical reactions with complex kinetics as for chemical oscillations in homogeneous system. So it is possible that, contrary to these latter ones, of which only three really different examples are known, there are a lot of heterogeneous systems which can give interfacial convection leading to a more or less oscillating behaviour, going from liquid - liquid extraction to biochemical oscillations - or even transformation of chemical energy into mechanical one.

REFERENCES

1. Dupeyrat, M. and Michel, J., 1971, J. Exp. Suppl. 18, pp. 269
Dupeyrat, M. and Nakache, E., 1977, "Electrical Phenomena at the Biological Membrane Level", Roux,E. Ed., pp. 377-388, Elsevier Amsterdam
2. Sherwood, T.K. and Wei, J.C., 1957, Ind. Eng. Chem. 49, n°6, pp. 1030
3. Lewis, J.B., 1954, Chem. Eng. Sci. 3, pp. 248-260
4. Ward, A.F.H. and Brooks, L.H., 1952, Trans. Faraday Soc. 48, pp. 1124
5. Davies, J.T. and Wigill, J.B., 1960, Proc. Roy. Soc. A, pp. 255-277
6. Linde, H. and Kunkel, M., 1969, Wärme Stoffübertragung Bd 2, pp. 60
7. England, D.C. and Berg, J.C., 1971, AIChE J. 17, n°2, pp. 313
8. Levich, V.G. and Krylov, V.S., 1970, Annu. Rev. Fluid Mech. 1, pp. 293
9. Vidal, A. and Acrivos, A., 1968, I and EC Fund. 7, n°1, pp. 53
10. Palmer, H.J. and Bose, A., 1981, J. Colloid Interface Sci. 84, pp. 291
11. Pearson, J.R.A., 1958, J. Fluid Mech. 4, 489
12. Sternling C.V. and Scriven, L.E., 1959, AIChE J. 5, pp. 514
13. Ruckenstein, E. and Berbente, C., 1964, Chem. Eng. Sci. 19, pp. 329
14. Palmer, H.J. and Berg J.C., 1971, J. Fluid Mech. 47, n°4, pp. 779
15. Proctor Hall, T., 1983, Phil Mag. 36, pp. 385
16. Dupeyrat, M. and Nakache, E., 1980, J. Colloid Interface Sci. 73, pp. 332
17. Davies, T.V. and Haydon, D.A., 1958, Proc. Roy. Soc. A 243, pp. 492
18. Davies, T.V. and Rideal, E.K., 1963, "Interfacial Phenomena", pp. 322-360, Academic Press, New York, London
19. Nakache, E. and Dupeyrat, M., 1982, J. Chim. Phys. 79, pp.563
20. Nakache, E., Vignes-Adler,M. and Dupeyrat,M., J. Colloid Interface Sci., in press
21. Davies, J.T. and Rideal, E.K., 1963, "Interfacial Phenomena", pp. 180, Academic Press, New York, London
22. Ter Minassian-Saraga,L., 1955, J. Chim. Phys. 52, pp. 181
23. Van den Tempel,M. and Lucassen-Reynders, E.H., 1983, Adv. Coll. Int. Sci. 18, pp. 281

SHAPE INSTABILITIES OF MOVING REACTION INTERFACES[1]

J. Chadam[2]

IHES[3]
91440, Bures-sur-Yvette
France

A multi-dimensional generalization of the Stefan problem which includes surface tension is studied as a mathematical model for growth in a metastable medium. Planar solutions are shown to exist for all time only if the data is sufficiently small, otherwise the velocity of the front becomes infinite in finite time. Planar fronts which exist for all time are shown to be morphologically unstable without surface tension and to be stable with respect to perturbations of short wavelength when surface tension is included. Above a critical value of the surface tension, planar fronts are completely stable. The mathematical techniques used are a combination of soft analysis based on the maximum principle and functional analytic/integral equation type hard estimates.

1. INTRODUCTION

In recent years the shape instabilities which occur in solidification processes have attracted a great deal of attention (see, for example [1,2] and the references therein). Here we shall look at some of the mathematical questions which arise in the context of a simplified mathematical model. Specifically, if the interface separating the phases is nearly planar and given by $x = r(y,t)$, and the density of the melt is $u(x,y,t)$, then the governing equations are (see [3] for more details),

G. Nicolis and F. Baras (eds.), Chemical Instabilities, 247–262.

$$\frac{\partial u}{\partial t} = D\Delta u + v\frac{\partial u}{\partial x} \qquad x > r(y,t) \tag{1.1}$$

$$u = \gamma K \tag{1.2}$$

$$D\left(\frac{\partial u}{\partial x} - \frac{\partial u}{\partial y}\cdot\frac{\partial r}{\partial y}\right) + vu = k^{-1}\frac{\partial r}{\partial t} \tag{1.3}$$

$\Bigg\}\; x = r(y,t)$

$$u \to u_\infty \tag{1.4}$$

$$\frac{\partial u}{\partial x} \to 0 \tag{1.5}$$

$\Bigg\}$ as $x \to \infty$

$$u(x,y,0) = u_o(x,y) \quad \text{for} \quad x > r(y,0) = r_o(y) \tag{1.6}$$

where $D, v, \gamma, k, u_\infty$ are non-negative constants, and

$$K = -\frac{1}{2}\frac{\partial^2 r}{\partial y^2}\left(1 + \left(\frac{\partial r}{\partial y}\right)^2\right)^{-1/2} \tag{1.7}$$

is the mean curvature of the surface $x = r(y,t)$. These equations generalize the multi-dimensional version of the classical Stefan problem (which models the melting of ice) through the inclusion of flow in equations (1.1) and (1.3) and a surface tension effect in equation (1.2) making the equilibrium concentration on the interface shape dependent. The main difference, however, is that u_∞ is taken to be positive rather than negative as in the classical literature [4,5]. Thus in our problem the interface grows into a metastable phase (a typically unstable situation) as opposed to the classical problem in which the interface grows away from a stable phase (a typically stable situation).

The objective of this note is to make this last statement mathematically more precise and in the course of the analysis delineate and solve problems which do not appear in the classical treatment. In trying to examine the morphological stability of planar fronts one must first find a special planar solution and show it is marginally stable among all planar solutions before one can use it as the base of a perturbation expansion. In section 2 we examine the existence, uniqueness and asymptotic behavior of planar solutions to equations (1.1-6). The results are generally similar to the $v = 0$ case [6] but are more precise and more easily proved. If $ku_\infty > 1$, however, we show in section 3 that the velocity of the front blows up in finite time which is not the case for arbitrarily large but negative ku_∞ [6,7]. This gives a rather complete, fully non-linear treatment of the planar versions of (1.1-6). In section 4 we examine the stability of planar fronts with respect to transverse perturbations. If there is no surface tension (i.e. $\gamma = 0$) we find that all bumps are unstable while the inclusion of surface tension stabilizes the high frequency modes. In fact if γ is larger than some critical value (which depends on all the other parameters) then all modes are stabilized and otherwise a finite interval of small frequency remains unstable.

2. PLANAR PROBLEM - GLOBAL EXISTENCE AND ASYMPTOTICS.

The planar version of problem (1.1-6) is

$$\frac{\partial u}{\partial t} = D\Delta u + v\frac{\partial u}{\partial x} \qquad x > r(t) \tag{2.1}$$

$$\left.\begin{aligned} u &= 0 \qquad &(2.2)\\ D\frac{\partial u}{\partial x} &= k^{-1}\dot{r} \qquad &(2.3)\end{aligned}\right\} \quad x = r(t)$$

$$\left.\begin{aligned} u &\to u_\infty \qquad &(2.4)\\ \frac{\partial u}{\partial x} &\to 0 \qquad &(2.5)\end{aligned}\right\} \quad x \to \infty$$

$$u(x,0) = u_o(x) \quad \text{for} \quad x > r(0) \tag{2.6}$$

It is not difficult to see that this possesses a travelling front solution of the form

$$u_p(x,t) = u_\infty\left(1 - e^{-\left[\frac{(v+V)}{D}(x-Vt)\right]}\right) \tag{2.7}$$

$$r_p(t) = Vt \tag{2.8}$$

where the velocity of the front is given by

$$V = ku_\infty(1-ku_\infty)^{-1}v \tag{2.9}$$

In order to obtain the decay at infinity one must have $v+V = (1-ku_\infty)^{-1}v > 0$; i.e. $ku_\infty < 1$. It turns out that this condition for the special solution (2.7-9) is essential for all planar solutions in that we will show in section 3 that if it is not met then one can find initial data for which the velocity of the front will become unbounded in finite time. Here we shall demonstrate that if ku_∞ (and v/D) is sufficiently small (presumably $ku_\infty < 1$ is the best result) then solutions to problem (2.1-6) exist globally in time for arbitrary data $0 \leq u_o(x) \leq u_\infty$ and $r(0)$, and the position of the resulting front $r(t)$ remains in a bounded interval around $r_p(t)$ for all time.

<u>Definition 2.1</u>. The pair $u(x,t)$, $r(t)$ is a solution of problem (2.1-6) over the interval $0 < t < T$ if (i) $\partial^2u/\partial x^2$ and $\partial u/\partial t$ are continuous in $r(t) < x < \infty$, $0 < t < T$; (ii) u and $\partial u/\partial x$ are continuous in $r(t) \leq x < \infty$, $0 < t < T$; (iii) u is also continuous in $r(t) \leq x < \infty$, $0 \leq t < T$; (iv) $r(t)$ is continuously differentiable in $0 \leq t < T$ and (v) the equations (2.1-6) are satisfied with the limit in (2.4,5) being uniform on bounded t intervals.

Following Friedman [4, chap.8] one can turn the problem of seeking such solutions to the partial differential problem (2.1-6) into an equivalent existence problem for non-linear integral equations of Volterra type. Indeed one can write

$$\begin{aligned} u(x,t) = & \int_{r(o)}^{\infty} E(x-z,t)u_o(z)dz \\ & - \int_o^t E(x-r(s),t-s)\frac{\partial u}{\partial x}(r(s),s)ds \end{aligned} \tag{2.10}$$

where E is the fundamental solution of equation (2.1) given by

$$E(x-z,t-s) = \frac{1}{2\sqrt{\pi D}(t-s)^{1/2}} e^{-\frac{(x-z+v(t-s))^2}{4D(t-s)}} \tag{2.11}$$

Then, differentiating with respect to x, taking the limit as $x \to r(t)+0$, noting the jump discontinuity in the second integral, one obtains for $U(t) = \frac{\partial u}{\partial x}(r(t),t)$

$$U(t) = 2[\int_{r(o)}^{\infty} \frac{\partial E}{\partial x}(r(t)-z,t)u_o(z)dz - \int_o^t \frac{\partial E}{\partial x}(r(t)-r(s),t-s)U(s)ds] \tag{2.12}$$

$$= 2[\int_{r(o)}^{\infty} E(r(t)-z,t)u_o'(z)dz - \int_o^t \frac{\partial E}{\partial x}(r(t)-r(s),t-s)U(s)ds] \tag{2.13}$$

if u_o is differentiable and, as required by part (iii) of definition 2.1, $u_o(r(0),0) = 0$. Integrating equation (2.3) one obtains

$$r(t) = r(0)+kD \int_o^t U(s)ds \quad . \tag{2.14}$$

By a straightforward modification of the $v = 0$ proof [4,p.216] one obtains the basic result of this integral equation approach - that the differential equations (2.1-6) and the integral equations (2.13,14) are equivalent.

Theorem 2.1 [4,p.221] $u(x,t)$, $r(t)$ is a solution of problem (2.1-6) over the interval $0 < t < T$ if and only if $U(t)$ is a continuous solution of the integral equation (2.13) over $0 \leq t < T$, where $r(t)$ is given by equation (2.14).

The integral equation can be solved by a contraction mapping argument [4,p.222] in the space $C[0,T)$ with T sufficiently small. This local solution can be extended globally (i.e. $T \to \infty$) if $|U(t)|$ remains finite [4,p.223]. This is the only thing which requires checking in the present case.

Theorem 2.2. If $u_o \in C^1[r(0),\infty)$ with $0 \leq u_o(x) \leq u_\infty$, $u_o(r(0)) = 0$ and $u_\infty - u_o(x) \in L^1(r(0),\infty)$, then problem (2.1-6) has a unique global solution, provided that $k(u_\infty + \|u_o'\|_\infty)$ and v/D are small.

Proof. In the interval of existence $U(t)$ is bounded from below by the maximum principle. Specifically, $0 \leq u_o$ on $r(0) < x < \infty$ and $u = 0$ on $x = r(t)$, $0 \leq t < T$ imply that $u(x,t) \geq 0$ on $r(t) < x < \infty$, $0 < t < T$. Thus $U(t) = u_x(r(t),t) \geq 0$ for $0 < t < T$. The upper bound must be obtained from the integral equation

$$U(t) = \text{Initial term of (2.12 or 13)} \qquad (2.15)$$

$$-2\int_0^t - \frac{2(r(t)-r(s)+v(t-s))}{8\sqrt{\pi}D^{3/2}(t-s)^{3/2}} U(s) e^{-\frac{(r(t)-r(s)+v(t-s))^2}{4D(t-s)}} ds$$

Unfortunately, the last integral is positive so that it cannot be ignored as the in the classical, $v = 0$, melting proof [4]. Nor can the exponential term be ignored as in the $v = 0$, solidifying case [6, thm.2.3]. One must first estimate the quantity $r(t)-r(s)+v(t-s)$. Surprisingly, the result for $v \neq 0$ is easier to obtain and more precise than the best possible result available in the $v = 0$ melting case [5,p.184].

Proposition 2.3. With the hypotheses of theorem 2.2, if $s \leq t$ are in the interval of existence $[0,T)$ of problem (2.1-6) given by theorem 2.1, then

$$-A \leq r(t)-r(s)-ku_\infty v(1-ku_\infty)^{-1}(t-s) \leq A \qquad (2.16)$$

where A is a positive constant given by

$$0 \leq A \leq ku_\infty(1-ku_\infty)^{-1}\left[\int_{r(0)}^\infty \left(1-\frac{u_o(x)}{u_\infty}\right)dx + 2v^{-1}\right] \qquad (2.17)$$

Returning to the proof of theorem 2.2, and making very crude estimates to obtain the most direct, but certainly not the best possible argument, we have

$$U(t) \leq \text{const}(\|u_o\|_\infty + \|u_o'\|_\infty) \tag{2.18}$$
$$+ \frac{1}{2\sqrt{\pi}D^{3/2}} \int_o^t \frac{[kD\int_s^t U(w)dw + v(t-s)]}{(t-s)^{3/2}} U(s) e^{-\frac{(r(t)-r(s)+v(t-s))^2}{4D(t-s)}} ds$$

using both initial terms of (2.12) and (2.13) for the first term and (2.14) in the second. In order to retain the large $(t-s)$ behavior in the exponential we obtain, directly from (2.16),

$$[r(t)-r(s)+v(t-s)]^2 \geq \begin{cases} 0 & t-s \leq A/\gamma \\ [\gamma(t-s)-A]^2 & t-s > A/\gamma \end{cases} \tag{2.19}$$

where $\gamma = v(1-ku_\infty)^{-1}$; Thus if $W(t) = \sup_{0\leq s\leq t} |U(s)|$, one has

$$W(t) \leq c + \frac{kDW(t)^2 + vW(t)}{2\sqrt{\pi}D^{3/2}} [\int_{t-A/\gamma}^t (t-s)^{-1/2} ds +$$
$$+ \int_o^{t-A/\gamma} (t-s)^{-1/2} e^{-\frac{[\gamma(t-s)-A]^2}{4D(t-s)}} ds] . \tag{2.20}$$

where $c = \text{const}(\|u_o\|_\infty + \|u_o'\|_\infty)$. What is important is that the last two integrals are uniformly bounded, the first obviously by $2\sqrt{A/\gamma}$ and the second by $2\sqrt{\pi}D/\gamma$ after making the change of variables $w = (\gamma(t-s)-A)/[2\sqrt{D}(t-s)^{1/2}]$. Thus one obtains a quadratic inequality of the form $0 \leq c+bW+aW^2$ which, because of the continuity of W , guarantees that $W(t) \leq [-b-(b^2-4ac)^{1/2}]/2a$ if $b < 0$ and $b^2-4ac > 0$. The first follows from choosing v/D small and the second is obtained by taking $kc = k(\|u_o\|_\infty + \|u_o'\|_\infty)$ small.

All that remains then is to prove proposition 2.3. Consider the quantity $q(t) = \int_{r(t)}^\infty (u_\infty - u(x,t))dx$ which seems to be central in many of these calculations [6, thm.2.4] [7, thm.IV.A3]. Differentiating

$$\dot{q}(t) = (k^{-1}-u_\infty)\dot{r}(t)-vu_\infty \tag{2.21}$$

which upon integration gives

$$q(t)-q(s) = (k^{-1}-u_\infty)(r(t)-r(s))-vu_\infty(t-s) \tag{2.22}$$

Rearranging, one has

$$\begin{aligned} -k(1-ku_\infty)^{-1}q(s) &\leq r(t)-r(s)-ku_\infty v(1-ku_\infty)^{-1}(t-s) \\ &\leq k(1-ku_\infty)^{-1}q(t) \end{aligned} \tag{2.23}$$

since $q(t) \geq 0$ by the maximum principle. The result follows then by showing that $q(t)$ is uniformly bounded above.

To this end consider problem (2.1-6) in terms of $w = u_\infty - u$ and compare this with the solution to the problem

$$\frac{\partial z}{\partial \tau} = D\Delta z+v\frac{\partial z}{\partial x} \qquad x > r(t)\ ,\ 0 < \tau < t \tag{2.24}$$

$$z = u_\infty \qquad x = r(t)\ ,\ 0 < \tau < t \tag{2.25}$$

$$z(x,0) = w_o(x) = u_\infty-u_o(x)\ ,\ x > r(t) \tag{2.26}$$

Because $r(\tau) \leq r(t)$ for $\tau \leq t$, $w(x,\tau) \leq z(x,\tau)$ for $r(t) < x < \infty$, $0 < \tau \leq t$ by the maximum principle, so that

$$q(t) = \int_{r(t)}^{\infty} w(x,t)dx \leq \int_{r(t)}^{\infty} z(x,t)dx \tag{2.27}$$

and the last integral can be estimated because one has an explicit expression for $z(x,t)$. Indeed

$$\begin{aligned} z(x,t) &= \int_{r(t)}^{\infty} G(x-y,t)w_o(y)dy \\ &\quad + u_\infty\int_o^t \frac{\partial G}{\partial y}(x-r(t),t-s)ds \end{aligned} \tag{2.28}$$

where the Green function is given by

$$G(x-y,\tau-s) = E(x-y,\tau-s)[1-e^{-\frac{(x-r(t))(y-r(t))}{D(\tau-s)}}] \tag{2.29}$$

and E is defined in equation (2.11). Integrating expression (2.28) from $r(t)$ to ∞, the first term can be dominated by $\int_{r(t)}^{\infty} w_o(y)\,dy \leq \int_{r(0)}^{\infty}(u_\infty - u_o(y))\,dy = q(0)$ because $\int_{-\infty}^{\infty} G(x-y,t)\,dy = 1$. For the second term

$$
\begin{aligned}
\frac{\partial G}{\partial y}(x-r(t),t-s) &= E(x-r(t),t-s)\left[\frac{x-r(t)}{D(t-s)}\right] \\
&= -2\,\frac{\partial E}{\partial x}(x-r(t),t-s) - \frac{v}{D}\,E(x-r(t),t-s) \qquad (2.30)\\
&\leq -2\,\frac{\partial E}{\partial x}(x-r(t),t-s)
\end{aligned}
$$

Thus the second term can be estimated by

$$
\begin{aligned}
& u_\infty \int_{r(t)}^{\infty}\left[\int_o^t -2\,\frac{\partial E}{\partial x}(x-r(t),t-s)\,ds\right]dx \\
&= \frac{u_\infty}{\sqrt{\pi}} \int_o^t \frac{e^{-\frac{v^2(t-s)}{4D}}}{\sqrt{D}(t-s)^{1/2}}\,ds \\
&= \frac{u_\infty}{\sqrt{\pi}}\,\frac{4}{v}\int_o^{\frac{v\sqrt{t}}{2\sqrt{D}}} e^{-a^2}\,da \leq 2\,\frac{u_\infty}{v} \qquad (2.31)
\end{aligned}
$$

which concludes the proof.

The above is intended to show that the study of the multi-dimensional problem (1.1-6) _with_ curvature terms cannot be easy in view of the difficulties encountered in proving the existence of global solutions for the planar version (2.1-6) in which the curvature term _disappears_. One might ask whether these difficulties arise only because of the method used. While it is true that more refined estimates (resulting in a more complicated proof) would prove existence for a wider class of data, the next section shows that one cannot obtain global existence for all data thus making a concrete distinct between melting problems [7,thm. IV.A3] and the solidification problem treated here.

3. FINITE-TIME BLOW-UP.

As already suggested by the special solution (2.7-9) we conjecture that $ku_\infty < 1$ suffices for global existence. Here we show that if this condition does not obtain however, then one can have the velocity of the planar front become infinite in finite time.

Theorem 3.1. Suppose $u_o \in C^3(r(0),\infty)$, $u - u_o \in L^1(r(0),\infty)$ $0 \leq u_o \leq u_\infty$, $u_o(r(0)) = 0$, $u_o'' \leq 0$ and $u_o''(r(0)) = -[kDu_o'(r(0))+v]\, u_o'(r(0))$. If $ku_\infty > 1$, then there exists a $0 < T < \infty$ such that

$$\lim_{t\to T-0} U(t) = +\infty .$$

Proof. We shall combine equation (2.21) with a lower bound for $\dot{r}(t)$ to obtain a differential inequality for $q(t)$. To this end recall that $0 \leq u(x,t) \leq u_\infty$ in $r(t) < x < \infty$ and $u_x(r(t),t) = (kD)^{-1}\dot{r}(t) \geq 0$ by the maximum principle for t in the interval of existence (to be so understood throughout the proof). Similarly $u_{xx}(x,t) \leq 0$ for $r(t) < x < \infty$ because u_{xx} satisfies equation (2.1), $u_{xx}(0,x) \leq 0$ and, by differentiating (2.2), one finds that $u_{xx}(r(t),t) = -(kD)^{-1}\dot{r}(t)[\dot{r}(t)+v] \leq 0$. Thus $q(t)$ can be bounded from below by the area of the triangle in figure 1. to obtain

$$q(t) \geq \frac{kDu_\infty^2}{2\dot{r}(t)} \tag{3.1}$$

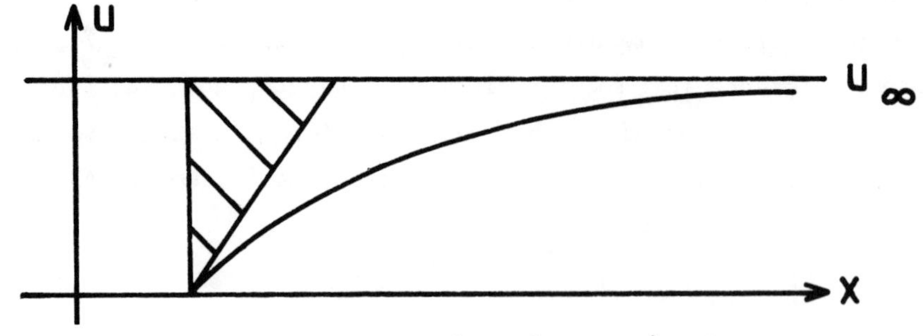

Figure 1. Graph of u as a function of x at time t.

Combining (2.21) and (3.1) one obtains

$$\frac{d}{dt}(q(t)^2) = 2q\dot{q} \leq Du_\infty^2(1-ku_\infty)-vu_\infty q$$
$$\leq Du_\infty^2(1-ku_\infty) \tag{3.2}$$

using $1-ku_\infty < 0$ and $q(t) \geq 0$. Integrating, one obtains

$$q^2(t) \leq q^2(0)+Du_\infty^2(1-ku_\infty)t \tag{3.3}$$

since $1-ku_\infty < 0$, this is clearly impossible after

$$T = q^2(0)[Du_\infty^2(ku_\infty-1)]^{-1} \tag{3.4}$$

Before this critical time one has from (3.1) and (3.3)

$$kDU(t) = \dot{r}(t) > \frac{kDu_\infty^2}{2}(q^2(0)-Du_\infty^2(ku_\infty-1)t)^{-1/2} \quad . \tag{3.5}$$

4. SHAPE INSTABILITIES OF PLANAR FRONTS.

In section 2 we showed that for small data global solutions to the planar version of equations (1.1-6) (i.e. equations (2.1-6)) exist and that the special solution (2.7-9) with the same asymptotic value u is marginally stable in this class (i.e. take $s = 0$ in inequality (2.16)). Thus planar solutions of problem (1.1-6) will lose stability to a more complicated developing shape only at those modes which are shown to be unstable by the linearized analysis about the special solution (2.7-9). We shall show that if $\gamma = 0$, every shape perturbation is unstable; i.e. diffusion causes all of the amplitudes to grow. The addition of a surface tension effect (i.e. $\gamma \neq 0$) causes the high frequency perturbations to be stabilized leaving only an interval $(0,m_c)$ unstable where the critical frequency depends on all the parameters. As γ increases this unstable interval shrinks until for all γ larger than some critical value, all modes are stable.

We begin by scaling equations (1.1-6) not only to reduce the algebra but also to point out explicitly that there are only two parameters in the problem. If $x' = (v+V)D^{-1}(x-Vt)$, $y' = (v+V)D^{-1}y$, $t' = (v+V)^2D^{-1}t$, $u' = u/u_\infty$ and $r' = (v+V)D^{-1}(r-Vt)$, where V is defined in equation (2.9), then equations (1.1-6) become (dropping the primes)

$$\frac{\partial u}{\partial t} = \Delta u + \frac{\partial u}{\partial x} \qquad x > r(y,t) \tag{3.6}$$

$$\left.\begin{aligned} u &= \frac{-g}{2h(1-h)}\frac{\partial^2 r}{\partial y^2}\left(1+\left(\frac{\partial r}{\partial y}\right)^2\right)^{-3/2} && (3.7)\\ \frac{\partial u}{\partial x} - \frac{\partial u}{\partial y}\frac{\partial r}{\partial y} + (1-h)u &= h^{-1}\frac{\partial r}{\partial t} + 1 && (3.8)\end{aligned}\right\} \quad x = r(y,t)$$

$$\left.\begin{aligned} u &\to 1 && (3.9)\\ \frac{\partial u}{\partial x} &\to 0 && (3.10)\end{aligned}\right\} \quad x \to \infty$$

$$u(x,y,0) = u_o(x,y) \quad \text{for} \quad x > r(y,0) = r_o(y) \tag{3.11}$$

where $h = ku_\infty$ and $g = \gamma kUD^{-1}$. In these variables the special solution (2.7-9) is

$$u_p(x,t) = 1-e^{-x} \tag{3.12}$$

$$r_p(t) = 0 \tag{3.13}$$

Writing the solution of problem (3.6-11) as

$$u_\varepsilon(x,y,t) = u_p(x) + \varepsilon u(x,y,t) + O(\varepsilon^2) \tag{3.14}$$

$$r_\varepsilon(y,t) = 0 + \varepsilon r(y,t) + O(\varepsilon^2) \tag{3.15}$$

the linearized equations in u,r are

$$\frac{\partial u}{\partial t} = \Delta u + \frac{\partial u}{\partial x} \qquad x > 0 \tag{3.16}$$

$$\left.\begin{aligned} u &= -r - \frac{g}{2h(1-h)} \frac{\partial^2 r}{\partial y^2} && (3.17)\\ \frac{\partial u}{\partial x} &= r + \frac{g}{2h} \frac{\partial^2 r}{\partial y^2} + h^{-1} \frac{\partial r}{\partial t} && (3.18) \end{aligned}\right\} \quad x = 0$$

$$u, \frac{\partial u}{\partial x} \to 0 \qquad x \to \infty \tag{3.19}$$

$$u(x,y,0) = u_o(x,y) \qquad x > 0 \tag{3.20}$$

Because equations (3.16-20) are linear there is no loss in generality in considering solutions of the form $u(x,y,t) = U_m(x,t)\cos my$ and $r(y,t) = r_m(t)\cos my$. Dropping subscripts one obtains

$$\frac{\partial U}{\partial t} = \frac{\partial^2 U}{\partial x^2} - m^2 U + \frac{\partial U}{\partial x} \qquad x > 0 \tag{3.21}$$

$$\left.\begin{aligned} U &= \left(-1 + \frac{gm^2}{2h(1-h)}\right) r && (3.22)\\ \frac{\partial U}{\partial x} &= \left(1 - \frac{gm^2}{2h}\right) r + h^{-1}\dot{r} && (3.23) \end{aligned}\right\} \quad x = 0$$

$$U, \frac{\partial U}{\partial x} \to 0 \,, \qquad x \to \infty \tag{3.24}$$

$$U(x,0) = U_o(x) \qquad x > 0 \tag{3.25}$$

Theorem 4.1. For arbitrary $U_{mo} \in C^1 \cap L^1(0,\infty)$ and $r_m(0)$, equations (3.21-25) have a unique global solution with U_m analytic and r_m continuously differentiable. Moreover the amplitude (more precisely, the real part of $r_m(t)$ behaves asymptotically according to the following list.

i) $m = 0$: r_m is uniformly bounded

ii) $m \neq 0$, a) $g = 0$: r_m is exponentially increasing

b) $g \neq 0$, $0 < g < 2(1-h)$: there exists a critical frequency $m_c (< [2h(1-h)g^{-1}]^{1/2})$ such that in $0 < |m| < |m_c|$, r_m is exponentially increasing, r_{m_c} is uniformly bounded, and in $m_c < |m|$, r_m is exponetially decreasing.

c) $g \neq 0$, $2(1-h) < g$: r_m is exponentially decreasing.

<u>Proof</u>. The existence can be established by applying Volterra integral equation techniques to the integral version of problem (3.21-25) or by Laplace transform methods [6,thm.IV.B1]. Here we shall concentrate only on the asymptotic behavior which can be observed directly from the dispersion relations obtained in the second method. One can derive them formally but rapidly by postulating exponential solutions of the form $U_m(x,t) = u_m(x)e^{st}$ and $r_m(t) = r_m(0)e^{st}$. Substituting these into (3.21-24) one obtains

$$(s+m^2+1/4)^{1/2} = [s + h/2 - \frac{g(1-2h)m^2}{4(1-h)}][h - \frac{gm^2}{2(1-h)}]^{-1} \quad (3.26)$$

When $m = 0$, $s = 0$ is the only root and it is a simple zero except at the limiting case $h = 1$ which is excluded. This establishes part i). When $g = 0$ we obtain

$$(s+m^2+1/4)^{1/2} = \frac{s}{h} + \frac{1}{2} \quad (3.27)$$

which can be seen graphically to have a positive root for all $|m| \neq 0$ because $h > 0$. When $g \neq 0$ and $|m| \neq 0$ one proceeds by examining for which parameter values the s-intercept on the right is above that on the left. This obtains for $g(1-h)^{-1} > 2$ when $m^2 < 2h(1-h)g^{-1}$ and for $g(1-h)^{-1} < 2$ when $m_c^2 < m^2 < 2h(1-h)g^{-1}$, the root of

$$(m^2+1/4)^{1/2} = [\frac{h}{2} - \frac{g(1-2h)m^2}{4(1-h)}][h - \frac{gm^2}{2(1-h)}] \quad (3.28)$$

(which exists when $g(1-h)^{-1} < 2$) being the definition of the critical frequency m_c . Parts b) and c) follow then by listing all the cases obtained by intersecting this classification with that determined by whether the slope of the line on the right side of (3.26) is positive or negative (i.e. $2h(1-h)g^{-1} > m^2$ or $< m^2$ respectively). One must check explicitly however that when the line has positive slope and intercept above the parabola, then the real parts of all roots are negative.

A full non-linear analysis of any problem of this type has not as yet been carried out. Indeed the weakly non-linear analysis for this specific problem promises to be interesting not only because many modes can be unstable simultaneously but also because they might interact non-trivially with the marginally stable planar development. This last phenomenon does not arise in the classical Wollkind and Segal analysis [8] because coupling problem (2.1-6) to a thermal field effectively translates the interval of unstable frequencies to the right so that it is disjoint from $m = 0$ which becomes a stable mode.

1. Research supported in part by a grant from the NSF-USA and the DOE-USA.

2. The author would like to thank the director, Professor N. Kuiper, and the members of the IHES for their kind hospotality while this work was done.

3. Permanent address : Mathematics Department, Indiana University, Bloomington, Indiana 47405, USA and Geo-Chem Research Assocs., 700 N. Walnut St., Bloomington, Indiana 47401, USA.

REFERENCES

1. Langer, J.S., 1980, Rev. of Mod. Phys. 52, pp. 1
2. Ockendon, J., 1980, Linear and non-linear stability of a class of moving boundary problems, Proc. Seminar on Free Boundary Problems, Magnese, E. ed., Pavia 1979, Inst. Naz. di Alta Mat., Rome
3. Chadam, J. and Ortoleva, P., 1983, The stability effect of surface tension on the development of a free boundary in a planar Cauchy-Stefan problem, accepted for publication in JIMA
4. Friedman, A., 1964, "Partial Differential Equations of Parabolic Type", Prentice Hall, Edgewood Cliffs, New Jersey
5. Rubinstein, L.I., 1971, The Stefan Problem 27, Translations of Mathematical Monographs, AMS, Providence Rhode Island
6. Baillon, J.B., Bertsch, M., Chadam, J., Ortoleva, P., Peletier, L.A., 1983, Existence, uniqueness and asymptotics for planar, supersaturated solidifying Stefan-Cauchy problems, to appear Proc. of Brézis-Lions Seminar, Collège de France

7. Auchmuty, G., Chadam, J., Merino, E., Ortoleva, P. and Ripley, E., 1983, Geo-Chem. Res. Assoc. Report, June
8. Wolkind, D.J. and Segel, L.A., 1970, Phil. Trans. Roy. Soc. 268, 351, London

MORPHOLOGICAL STABILITY DURING UNIDIRECTIONAL SOLIDIFICATION : INFLUENCE OF MELT RHÉOLOGY AND MARANGONI CONVECTION

B. Billia, A. Sanfeld[1], A. Steinchen[1] and L. Capella

Laboratoire de Physique Cristalline (E.R.A. n° 545) - Faculté de St-Jérôme - 13397 Marseille Cedex 13, France

During unidirectional growth of a binary alloy from its melt dissipative structures are formed. A viscoelastic melt and Marangoni convection at the solid-liquid interface are studied in order to predict some new possibilities of instability.

1. INTRODUCTION

Since the Constitutional Supercooling Criterion [1] and the analysis of Mullins and Sekerka [2] the study of the morphological stability of a planar solidification front during unidirectional growth of a binary alloy has been extensively developed [3]. Most of the authors consider that the melt is motionless so that heat and matter are transported by diffusion only. Convection in the melt can yet be actually important [4,5]. The influence of natural convection has recently been taken into account for a viscous liquid phase [6,7].

We analyse natural convection for some viscoelastic rheological behaviors for the melt and then Marangoni convection. A prospective investigation is achieved so as to bring out new effects on the limit of morphological stability of a plane solid-melt interface. The analysis is made for a locally non-convective reference state.

2. CONSTRAINED UNIDIRECTIONAL GROWTH OF BINARY ALLOYS

2.1. Bridgman growth

An isotropic solid alloy is formed from its melt at a rate V

G. Nicolis and F. Baras (eds.), Chemical Instabilities, 263–270.

(see Figure 1). Liquid (ℓ) and solid (s) are infinite along the growth direction z. A heating device settles the solidification velocity and the thermal gradients in liquid and solid, G_ℓ and G_s. For the reference stationary state the solid-melt interface moves at constant velocity V_o, the solute mass density in solid $\rho^s_{2,0}$ is equal to that in the bulk liquid ρ^∞_2. Upon solidification the solid acts as a sink or a source of solute depending on the partition coefficient k (ratio of the solute weight fractions in solid and liquid, at the planar front), what results in a jump on the solute mass density profile (see Figure 1.b).

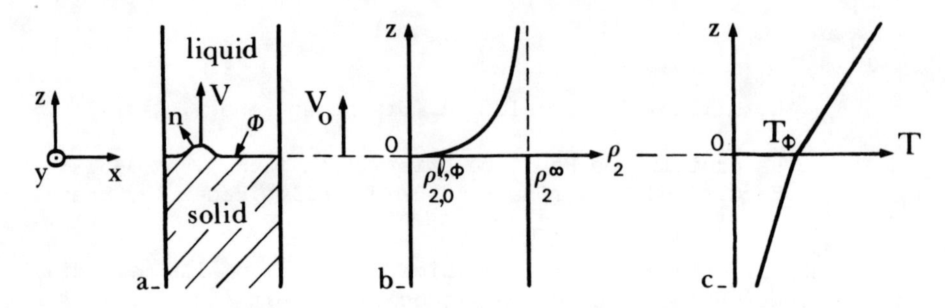

Figure 1. Unidirectional solidification. a : Planar front with a shape fluctuation - b : Solute mass density profile (k > 1) - c : Temperature profile.

Three parameters control an experiment : V_o, G_ℓ and ρ^∞_2. Two of them being held constant, e.g. G_ℓ and ρ^∞_2, a bifurcation is observed for a critical velocity V_c. Below V_c the front is planar, beyond V_c cellular or dendritic dissipative structures appear [8].

2.2. Basic equations for the bulks

Neglecting solute diffusion in solid and thermodiffusion in liquid, the following processes are taken into account in a frame attached to the reference planar front :

<u>Heat diffusion in solid</u>.

$$\partial T_s/\partial t = D^s_{th} \nabla^2 T_s + V_o \partial T_s/\partial z \qquad (1)$$

where T_s and D^s_{th} are the temperature and thermal diffusivity in solid.

Close to the melting point a solid is not totally rigid so that the possibility for some convective motion has strictly to be considered. Nevertheless we assume that these flows are slow enough for the related terms to be neglected.

Heat transport in liquid.

$$\partial T_\ell/\partial t + (\vec{v}_\ell+\vec{V}_o).\vec{\nabla}T_\ell = D^\ell_{th}\ \nabla^2 T_\ell + V_o\ \partial T_\ell/\partial z \quad (2)$$

where T_ℓ, D^ℓ_{th} and $\vec{v}_\ell$ are the temperature, thermal diffusivity and fluid velocity in liquid.

Solute transport in liquid.

$$\partial \rho^\ell_2/\partial t + (\vec{v}_\ell+\vec{V}_o).\vec{\nabla}\rho^\ell_2 = D\nabla^2\rho^\ell_2 + V_o\partial\rho^\ell_2/\partial z \quad (3)$$

where ρ^ℓ_2 and D are the solute mass density and diffusion coefficient in the melt.

Mass transport in liquid. For an incompressible melt the equation of continuity reduces to :

$$\vec{\nabla} \cdot \vec{u}_\ell = 0 \quad (4)$$

where $\vec{u}_\ell = \vec{v}_\ell + \vec{V}_o$ is the fluid velocity in the laboratory frame.

The motion of a volume element is described by the momentum balance equation which reads in the Boussinesq approximation :

$$\rho^{\ell,\phi}_o\ \partial\vec{u}_\ell/\partial t + \rho^{\ell,\phi}_o\ \vec{v}_\ell.\vec{\nabla}\vec{u}_\ell = -\vec{\nabla}\ p^\ell + \mathrm{Div}\ \bar{\bar{\pi}}_\ell + \rho^\ell\vec{g} \quad (5)$$

where p^ℓ is the hydrostatic pressure, $\bar{\bar{\pi}}_\ell$ the deviatoric part of the stress tensor and $\vec{g}$ the gravitational field. The liquid mass density ρ^ℓ is related to its value at the planar front $\rho^{\ell,\phi}_o$ by a linearized equation of state :

$$\rho^\ell = \rho^{\ell,\phi}_o \left[1 - \alpha_T\ (T_\ell - T_{\phi,0}) - \alpha_\rho(\rho^\ell_2 - \rho^{\ell,\phi}_{2,0}) \right] \quad (6)$$

with α_T and α_ρ the thermal and solutal coefficients of expansion of the fluid.

The expression of $\bar{\bar{\pi}}_\ell$ depends on the rheological behavior of the melt. For the different viscoelastic behaviors we consider it reads :

$$\text{Newton :}\ \bar{\bar{\pi}}_\ell = \eta_\ell\dot{\bar{\bar{\varepsilon}}}_\ell \quad (7.a)$$

$$\text{Maxwell:}\ \bar{\bar{\pi}}_\ell - \tau_\ell\dot{\bar{\bar{\pi}}}_\ell = \eta_\ell\dot{\bar{\bar{\varepsilon}}}_\ell \quad (7.b)$$

$$\text{Kelving-Voigt :}\ \bar{\bar{\pi}}_\ell = E_\ell\ \bar{\bar{\varepsilon}}_\ell + \eta_\ell\ \dot{\bar{\bar{\varepsilon}}}_\ell \quad (7.c)$$

where $\bar{\bar{\varepsilon}}_\ell$ is the deformation tensor, η_ℓ the viscosity coefficient, E_ℓ the elastic modulus and τ_ℓ the relaxation time. The dot means the time derivative. For small deformations :

$$\dot{\varepsilon}_{\ell}^{ij} = (\partial u_i/\partial j + \partial u_j/\partial i)/2 \qquad (8)$$

where i,j are the coordinates x,y,z.

2.3. Boundary conditions

The temperature, solute and velocity fields in the bulks must satisfy the following boundary conditions at the solid-liquid interface :

Surface mass balance.

$$(1 - \rho_o^{s,\phi}/\rho_o^{\ell,\phi})\vec{V}.\vec{n} = \vec{u}_\ell|_\phi.\vec{n} \qquad (9)$$

where the surface mass density is considered as a constant.

Surface heat balance.

$$(K_S \vec{\nabla} T_S - K_\ell \vec{\nabla} T_\ell)_\phi.\vec{n} = L\vec{V}.\vec{n} \qquad (10)$$

where K_ℓ and K_S are the thermal conductivities in liquid and solid. L is the latent heat of fusion per unit volume.

Surface solute balance. The variation of the solute mass density in the interphase must be taken into account for it may be important for dilute alloys [9].

$$\partial\Gamma_2/\partial t = D\vec{\nabla}\rho_2^\ell|_\phi.\vec{n} + \rho_o^{s,\phi}(\rho_2^{\ell,\phi}/\rho_o^{\ell,\phi} - \rho_2^{s,\phi}/\rho_o^{s,\phi})\ \vec{V}.\vec{n} \qquad (11)$$

Normal surface momentum balance (Laplace condition). There is no transport of the matter in the surface in the direction normal to the front for a continuous renewal of atoms takes place at the solid-liquid interface. We thus obtain for the Laplace condition :

$$(p^\ell - p^S)_\phi = \sigma\kappa + \pi_\ell^{zz} \qquad (12)$$

where p^S is the pressure in solid, σ the solid-liquid interface tension and κ the interfacial curvature. π_ℓ^{ij} are the components of tensor $\bar{\bar{\pi}}_\ell$.

Tangential surface momentum balance. Despite the possibility of convective motions in the interphase we neglect them and get :

$$\partial\sigma/\partial i + \pi_\ell^{iz} = 0 \qquad (i = x,y) \qquad (13)$$

These two relations form the Marangoni condition. They have to be completed by an equation of state for the surface tension :

$$\sigma = \sigma_o + \beta_T(T_\phi - T_{\phi,o}) + \beta_\Gamma(\Gamma_2 - \Gamma_{2,o}) \qquad (14)$$

where the coefficients β_T and β_Γ are constants.

Growth kinetics. Around the reference state the kinetic law can be written [10] :

$$(\vec{V}-\vec{V}_o).\vec{n} = \nu \left[T_{Eq} - T_\phi - (T_{Eq} - T_\phi)_o \right] \qquad (15)$$

where the differential kinetic coefficient ν is a positive quantity. We have for the equilibrium temperature :

$$T_{Eq} = T_M + m\,\rho_2^{\ell,\phi} + T_M(p^\ell - p^S)_\phi/L \qquad (16)$$

with T_M the melting temperature of pure solvent and m the liquidus slope.

Moreover we assume that temperature is continuous at the front and the partition coefficient k constant.

3. INFLUENCE OF MELT RHEOLOGY AND SOLID SHRINKAGE

A linear stability analysis is performed in terms of normal modes. For illustrative purpose and so as to be able to proceed analytically we here restrict ourselves to growth under zero gravity and assume that the principle of the exchange of stabilities holds. The viscoelastic effects appear in the momentum balance equation (Eq. 5) and in the Laplace condition (Eq. 12). The latter contribution has been previously neglected [6,7].

3.1. Newton or Maxwell behavior

For wavenumbers ω for which the thermal steady state approximation is valid [11], we obtain for the non-oscillatory marginal state the Mullins-Sekerka condition with additional terms related to the mass density difference between liquid and solid :

$$(K_\ell G_\ell + K_S G_S)/(K_\ell + K_S) + \omega^2 T_M \sigma'/L - mG_c(p_2 - \rho_o^{s,\phi} V_o/\rho_o^{\ell,\phi} D)/$$

$$\left(p_2 - (1-k)\rho_o^{s,\phi} V_o/\rho_o^{\ell,\phi} D\right) - \omega m G_c(\rho_o^{s,\phi} - \rho_o^{\ell,\phi})\left[(p_2-\omega)(p_1+\omega) + (p_2 - p_1\right.$$

$$\left. -\omega - \rho_o^{s,\phi} V_o/\rho_o^{\ell,\phi} D)\rho_o^{s,\phi} V_o/\rho_o^{\ell,\phi} D\right] / \left[\rho_o^{s,\phi}(p_2 - (1-k)\rho_o^{s,\phi} V_o/\rho_o^{\ell,\phi} D)\right.$$

$$\left.(p_1^2 - \omega^2 + p_1 \rho_o^{s,\phi} V_o/\rho_o^{\ell,\phi} D)\right] = 0 \qquad (17)$$

where :

$$p_1 = \rho_o^{s,\phi} V_o / 2\rho_o^{\ell,\phi} \eta_a^{\ell} + \left[(\rho_o^{s,\phi} V_o / 2\rho_o^{\ell,\phi} \eta_a^{\ell})^2 + \omega^2 \right]^{1/2} \qquad (18.a)$$

$$p_2 = \rho_o^{s,\phi} V_o / 2\rho_o^{\ell,\phi} D + \left[(\rho_o^{s,\phi} V_o / 2\rho_o^{\ell,\phi} D)^2 + \omega^2 \right]^{1/2} \qquad (18.b)$$

with η_a^{ℓ} the kinematic viscosity coefficient. G_c is the solute mass density gradient in the melt at the planar front and σ' a generalized surface tension :

$$\sigma' = \sigma - 2\eta_a^{\ell} (\rho_o^{s,\phi} - \rho_o^{\ell,\phi}) V_o \qquad (19)$$

For most alloys there is a shrinkage upon solidification so that $\rho_o^{s,\phi} > \rho_o^{\ell,\phi}$. In this case the last term in Eq. 19 is destabilizing so that the generalized capillary effect could be destabilizing whereas the pure one is always stabilizing. This generalized effect would then alter the region of absolute stability predicted by the Mullins-Sekerka theory [2]. This cannot happen for ordinary growth conditions but might be possible for high growth velocities like those which are encountered during rapid solidification [12] or for alloys for which σ approaches zero, perhaps under the action of a surfactant. After straightforward transformations one can show that the last contribution in Eq. 17 is stabilizing for $\rho_o^{s,\phi} > \rho_o^{\ell,\phi}$, and usual growth conditions ($\omega \gg V_o/D$).

3.2. Kelvin-Voigt behavior

In this case the condition for the non-oscillatory marginal state is simply $\rho_o^{s,\phi} = \rho_o^{\ell,\phi}$. For such an alloy the planar front is always stable ($\rho_o^{s,\phi} > \rho_o^{\ell,\phi}$) or unstable ($\rho_o^{s,\phi} < \rho_o^{\ell,\phi}$) what has never been observed. Alternately it might be possible either that the marginal state is oscillatory or that there exists no melt with a Kelvin-Voigt rheology close to the melting point, i.e. with a pronounced elastic behavior.

4. INFLUENCE OF MARANGONI CONVECTION

Under zero-gravity conditions it is necessary to take into account the time variation of Γ_2 in Eq. 11 so as to have enough coupling between Marangoni convection and unidirectional solidification. We have thus to seek for an oscillatory marginal state.

For systems with equal densities $\rho_o^{s,\phi}$ and $\rho_o^{\ell,\phi}$ we derive the dispersion relation for two limiting cases depending on the time frequency α_i for the evolution of the normal mode ω.

<u>Large wavenumbers</u> : $\alpha_i / D\omega^2 \ll 1$. We obtain for the real part

of the marginality condition :

$$\left[\bar{\omega}-(1-k)V_o/D\right]\left[(K_\ell G_\ell+K_S G_S)/(K_\ell+K_S) + \omega^2 T_M\sigma/L - mG_c(\bar{\omega}-V_o/D) / \left(\bar{\omega}-(1-k)V_o/D\right)\right] - \alpha_i^2 L\left[-m\beta_T/\beta_\Gamma + \left(1+(K_\ell+K_S)\omega/\nu L\right) /2\left((V_o/2D)^2+\omega^2\right)^{1/2}\right]/\omega D(K_\ell+K_S) = 0 \quad (20.a)$$

and for the imaginary part :

$$\left[-m\beta_T/\beta_\Gamma+1/2\left((V_o/2D)^2+\omega^2\right)^{1/2}\right](K_\ell G_\ell+K_S G_S)/(K_\ell+K_S) + mG_c\left[1-\left(1+(2\omega D/V_o)^2\right)^{-1/2}\right]D/V_o + D\left(\bar{\omega}-(1-k)V_o/D\right)\left[L/\omega (K_\ell+K_S)+1/\nu\right] + \omega^2 T_M\sigma/2L\left[(V_o/2D)^2+\omega^2\right]^{1/2} = 0 \quad (20.b)$$

where $\bar{\omega}$ equals p_2 (Eq. 18.b) with $\rho_o^{s,\phi} = \rho_o^{\ell,\phi}$

For rapid growth kinetics ($\nu \to +\infty$) Eqs. 20 cannot be together satisfied out of the domain of morphological instabilities given by the Mullins-Sekerka criterion. Therefore Marangoni convection does not affect the limit of morphological stability of the planar front. For very sluggish growth kinetics ($\nu \to 0$) Marangoni convection should appear and influence the morphology of the solid-melt interface.

Small wavenumbers : $\alpha_i/D\omega^2 \gg 1$. A preliminary study suggests that Marangoni convection might be important during growth of binary alloys when the coefficient β_Γ tends to zero. This could be realized when σ is independent of Γ_2 or when the surface tension of solvent and solute are nearly the same.

Acknowledgements: The authors gratefully acknowledge financial assistance from the NATO (Grant n° 220.81).

1 - Permanent address : Service de Chimie-Physique II, Université Libre de Bruxelles, Bruxelles, Belgium.

4. REFERENCES

1. Tiller, W.A., Jackson, K.A., Rutter, J.W. and Chalmers, B. 1953, Acta Met. 1, pp. 428-437
2. Mullins, W.W. and Sekerka, R.F. 1964, J. Appl. Phys. 35, pp. 444-451
3. Delves, R.T., in *Crystal Growth*, Pamplin, B.R. Ed. 1974 (Pergamon Press, Oxford) pp. 40-103
4. Pimputkar, S.M. and Ostrach, S. 1981, J. Crystal Growth 55, pp. 614-646

5. Hurle, D.T.J. 1981, Adv. Colloid Interface Sci. 15, pp. 101-130
6. Coriell, S.R., Cordes, M.R., Boettinger, W.J. and Sekerka, R.F. 1980, J. Crystal Growth 49, pp. 13-28
7. Hurle, D.T.J., Jakeman, E. and Wheeler, A.A. 1982, J. Crystal Growth 58, pp. 163-179
8. Billia, B. 1982, Dissertation, Marseille
9. Favier, J.J. 1982, J. Electrochem. Soc. 129, pp. 2355-2360
10. Coriell, S.R. and Sekerka, R.F. 1980, in *Rapid Solidification Processing Principles and Technologies II*, Mehrabian, R., Kear, B.H. and Cohen, M. Eds (Claitor, Baton Rouge) pp. 35-49
11. Sekerka, R.F. 1967, in *Crystal Growth*, Steffen Peiser, H. Ed (Pergamon Press, Oxford) pp. 691-702
12. Rimini, E. 1984, This book, pp. 367-385.

MODE SELECTION ON INTERFACES

H. Müller-Krumbhaar

Institut für Festkörperforschung,
Kernforschungsanlage Jülich,
D-5170 Jülich, Fed. Rep. Germany

Driven nonlinear systems often tend to develop spatially periodic patterns. The underlying mathematical models usually permit a continuous set of linearly stable solutions. As a possible mechanism of selecting a specific pattern the principle of marginal stability is presented, being applicable to situations, where a propagating front leaves a periodic structure behind. We restrict our discussion to patterns on interfaces which are more easily accessible than three-dimensional structures, for example in hydrodynamic flow. As a concrete system a recently analyzed model for dendritic solidification is discussed.

G. Nicolis and F. Baras (eds.), Chemical Instabilities, 271–285.

1. Introduction

A typical driven system producing complicated spatial patterns is the growth of a crystal into a supercooled liquid [1]. The interface free energy tending to minimize the total area of the interface between solid and liquid tries to keep the interface flat. Small protuberances of the interface towards the supercooled liquid on the other hand have an advantage over the flat part in dissipating latent heat of freezing into the supercooled liquid, thus tending to advance faster than the rest of the interface. This effect acts destabilizing. At short wavelengths $\lambda < \lambda_s$ the stabilizing effect due to surface tension dominates, at longer wavelengths $\lambda > \lambda_s$ the destabilizing effect due to the driving diffusion field dominates. Hence the flat interface is unstable against perturbations with wavelengths $\lambda > \lambda_s$, with maximum instability at about $\lambda \sim \sqrt{3}\,\lambda_s$.

This consideration [1], however, does not tell us, which pattern finally will be formed. As an illustration we show in fig. 1) model calculations of the formation of complicated interface patterns, starting from an initially (t=0) flat interface with two protuberances (further expalnations in section II).

Experimentally this situation is not easily realized because of the necessary control of external boundary conditions. A technically simpler experiment is concerned with the three-dimensional growth of dendritic crystals. There, a spherical container filled with a transparent liquid is supercooled and seeded at the center by a small crystal through a capillary tube [2]. The crystal grows in dendritic form under dissipation of latent heat of freezing into the supercooled liquid (fig. 2, from ref. [2]).

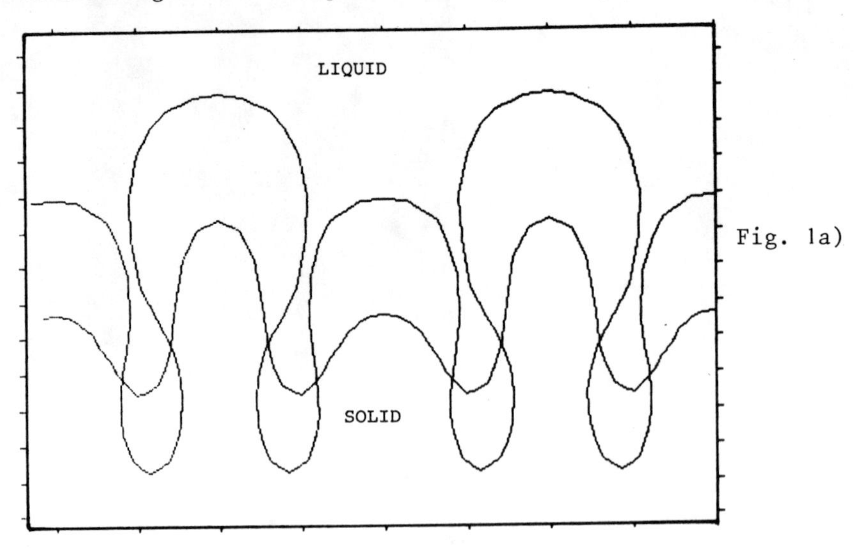

Fig. 1a)

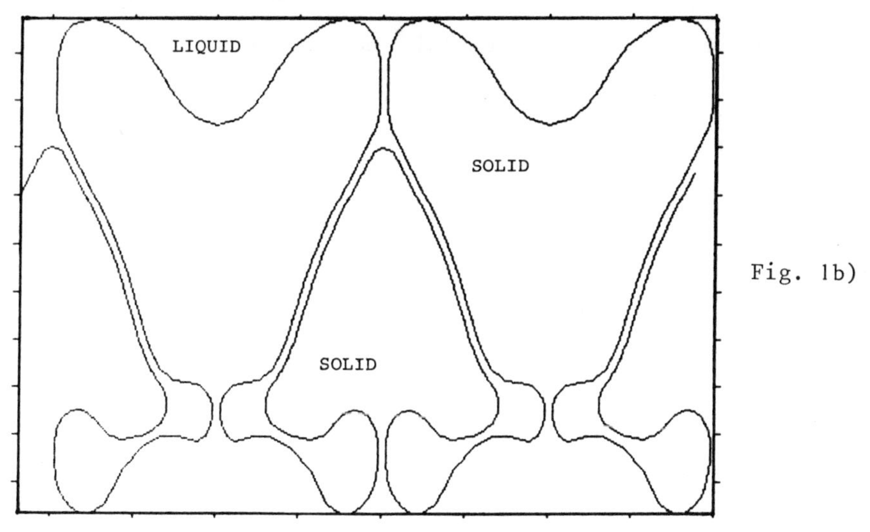

Fig. 1b)

Model calculations as in fig. 1 give similar results, fig. 3. A seed with initially three symmetric protuberances evolves into a feathery structure. Note that, while in the experiment there are small crystalline anisotropies stabilizing the growth directions, the model calculations are fully isotropic (apart from the initial perturbation). Small differences in the three dendritic main-branches are due to amplification of numerical discretization and round-off errors, caused by the inherent physical instability (further remarks in section II).

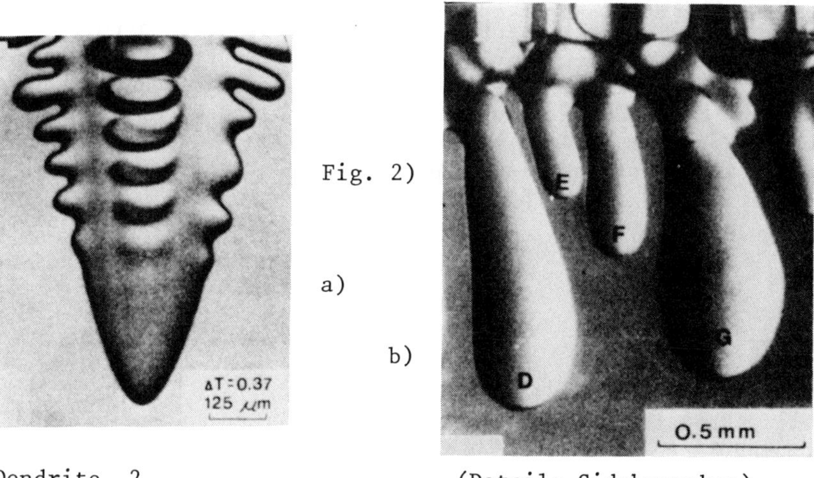

Fig. 2)

a)

b)

Dendrite 2 (Detail: Sidebranches)

Because of limited space the theory here can only be sketched. The propagation of the interface is driven by a difference $\Delta\mu$ in chemical potentials between solid and liquid across the interface. A local equation of motion [3] was given as

$$v \equiv \frac{\partial \xi}{\partial t} = -\kappa \frac{\delta G}{\delta \xi} \qquad (1)$$

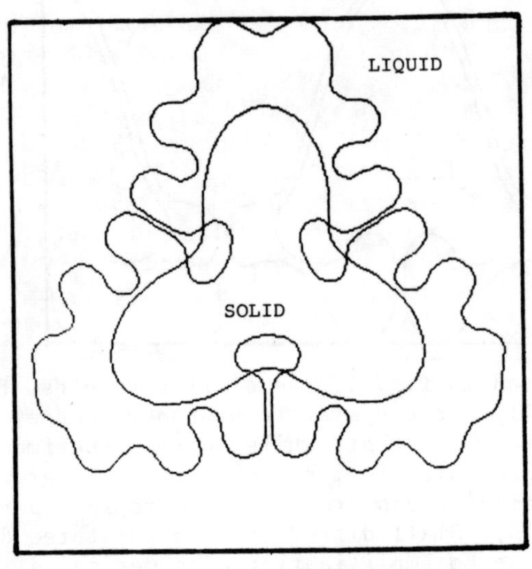

FIG. 3)

where $\delta\xi$ is the (local) normal displacement of the interface, δG is the associated change in Gibbs free energy, κ is a (anisotropic) mobility coefficient and v is the resulting normal growth rate. The derivative $\delta G/\delta\xi$ in (1) gives for the isotropic case: $(\Delta\mu-\gamma/R)$. Here, γ is the interface free energy and R is the local radius of (mean curvature). Eq. (1) now constitutes only part of the problem. The driving force $\Delta\mu$ is not directly accessible by experiment, but is related to the deviation of the interface temperature T_s from the equilibrium temperature T_o at the melting point: $\Delta\mu \sim T_o-T_s$. Eq. (1) thus represents only a (moving) boundary condition to the diffusion equation

$$\frac{\partial T}{\partial t} = D \nabla^2 T \qquad (2)$$

for heat in the fully three dimensional system. In principle the problem may be reduced to an equation for the interface alone by using Greens function techniques [4]. The resulting equation, however, is then non-local in space and time (long-range, resp. memory effects). Even though memory-effects can often be neglected, the long range character of the diffusion field cannot easily be ignored. On the other hand they sometimes can be integrated out for highly symmetric structures like a growing sphere, where the quasi-stationary approximation (instantaneous adjustment of the diffusion field to the interface structure) gives

$$v \sim R^{-1} \ (1 - R_c/R) \qquad (3)$$

for the normal growth rate, with $R_c = \gamma/\Delta\mu$ being the radius of the critical nucleus, the prefactor R^{-1} being proportional to the gradient of the diffusion field at the interface. R is the radius of the sphere.
Despite these difficulties we may still be able to work with approximate local equations, as long as the long range effects provide only quantitative corrections to the systems behavior and do not change the analytical structure qualitatively. For our concrete model of dendritic growth [6] we have argued along these lines, to obtain a local nonlinear partial differential equation. In the next section the crude model will be introduced and its limitations will be discussed. In the last section the resulting principle of marginal stability will be presented and relations to other ideas for pattern selection and generalizations will be outlined.

2. Dendritic Crystallization

The overall shape of a dendritic crystal branch is approximately parabolic (fig. 2) and thus can essentially be parametrized by the radius R of curvature at the tip. For the overall growth in axial direction we may approximately ignore the shaft thus obtaining eq. (3) as a crude approximation for the dependence of the axial growth rate upon the size of the primary dendrite. In the following we will allow for local deviations $R(x)$ $0<x<\infty$ of the surface from the average (parabola-like) amplitude $R_{average} = r_o$ = constant, writing $R(x)=r_o+\tilde{r}_1(x)$. Note that the spatial coordinate x denotes the distance from the tip down the shaft in the coordinate system moving at axial velocity v with the dendrite. A detailed linear stability analysis of the full diffusion problem [6] suggests an equation of the structure

$$\frac{\partial R(x,t)}{\partial t} = - v \frac{\partial R}{\partial x} - R \cdot \frac{\partial^2 R}{\partial x^2} - \frac{\partial^4 R}{\partial x^4} \qquad (4)$$

for the initial time-development of a spatial perturbation on the paraboloidal dendrite surface R=const. The first term on the right hand side is just the transformation from the laboratory system of reference into the moving coordinate system. The second term is the diffusional destabilizing term. The instability is more pronounced the broader the unperturbed dendrite. The third term represents stabilization by surface tension. In this equation the simplification of neglecting the quantities of long ranged terms present in the full analysis has already been incorporated [6]. The reason for this approximation is, that at least as long as the ratio $p = r_o/\ell$ ($\ell=2D/v$ being the diffusion length) is non-zero, the Greens-function has an exponential cutoff and consequently the true linear spectrum [6] is well approximated by eq. (3) [7]. The boundary condition at x=0 was assumed to keep all odd derivatives $\partial^n R/\partial x^n = 0$, for symmetry reasons. Eq. (3) still has a defect, insofar, as due to the only quadratic nonlinearity it is asymptotically ($R(x)\to\infty$) not stabilized. In reality, however, the local radius of curvature cannot become smaller than the critical radius $R_c = \varepsilon = \text{const.}$ (3). We therefore insert a term $(1-R_c\cdot K)$ into eq. (3), where K is the local curvature, approximated by $K \approx -\dfrac{\partial^2 R}{\partial x^2}$:

$$\frac{\partial R}{\partial t} = -v\frac{\partial R}{\partial x} - R\frac{\partial^2 R}{\partial x^2}\left(1+\varepsilon\frac{\partial^2 R}{\partial x^2}\right) - \frac{\partial^4 R}{\partial x^4} \qquad (5)$$

This is our crude nonlinear model for dendritic growth [7]. A slight generalization would be

$$v_\perp = -v\frac{\partial R}{\partial x} + R\cdot K(1-\varepsilon K) + \frac{\partial^2 K}{\partial x^2} \qquad (5a)$$

which appears more natural as it describes the displacement of the surface in locally normal direction rather than normal to the original paraboloid, in accordance with ref. [3]. At this stage, however it is not useful, since eq. (5a) does not prevent self-intersection of the surface in contrast to eq. (5). The problem of neglecting the long-ranged terms, as is outlined later in this section is also present here.

Eq. (5) was integrated on the computer by discretizing the x-coordinate. A small perturbation $r_1(x)$ was inserted onto a constant value of $R = r_o + r_1(x)$, some distance away from the tip. In a short time it produces sidebranches of asymptotically constant amplitude propagating down the dendrite shaft in the moving coordinate system. This corresponds to stationary sidebranches in the laboratory system of reference (fig. 4a).

The tip region on the left is still unperturbed. If the original parabolic dendrite $R=r_o$ is too narrow, $r_o<r^*$, the front of the sidebranching pattern cannot keep up with the tip-speed and an unperturbed dendrite would remain for time $t\to\infty$. If on the other hand $r_o>r^*$, the front propagates up to the tip, fig. 4b), and the tip settles in a complicated limit-cycle behavior, producing side-branches at constant frequency (fig. 4c). Note that the curvature at the tip $R(0)^{-1}$ settles around some unique value, independent of the details of the initial conditions.

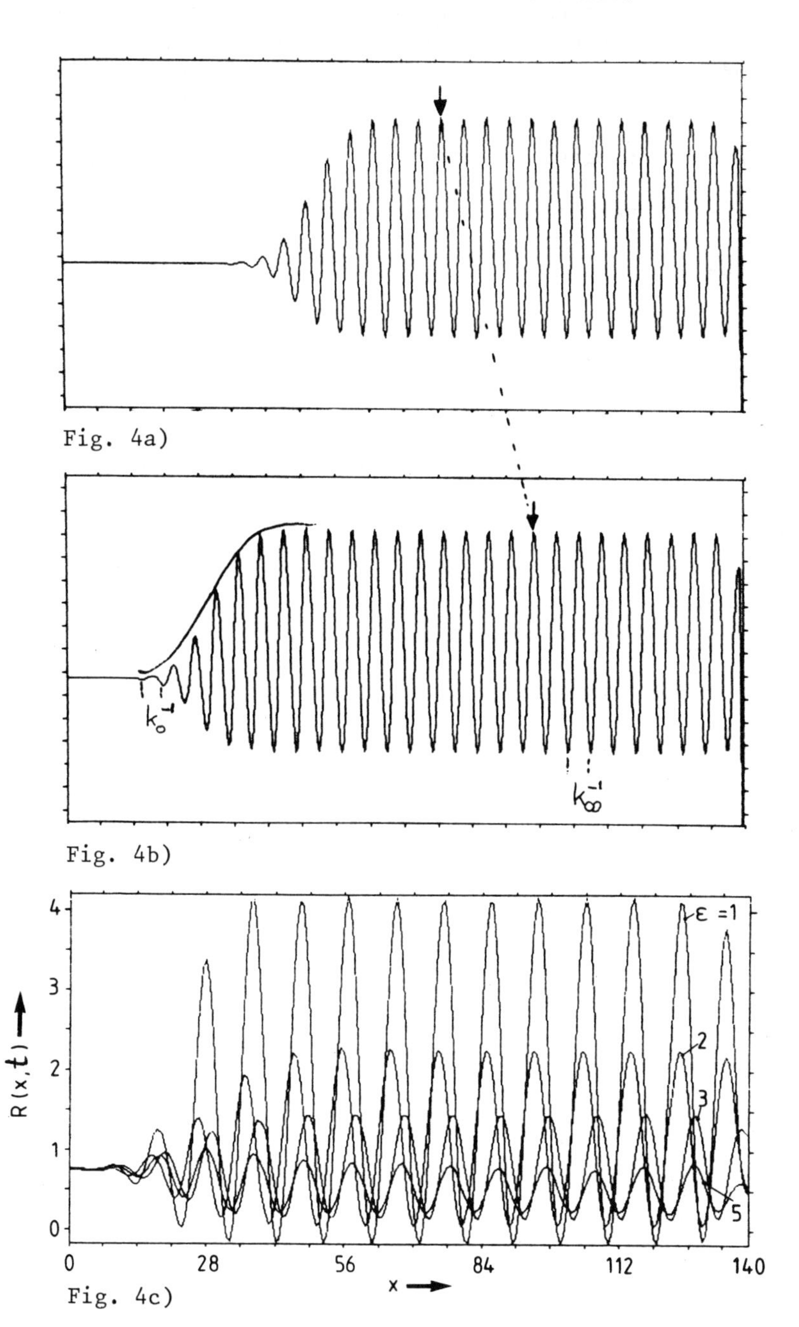

Fig. 4a)

Fig. 4b)

Fig. 4c)

From this numerical analysis we find, that within numerical precision ($\approx 1\%$) the tip radius $R(0,t)$ performs very small oscillations near the value $R(0,t) \approx r^*$, i.e. just at the width of the dendrite, where the speed of the propagating sidebranch-front equals the axial velocity of the tip. Moreover, the result is practically independent of the parameter $\varepsilon(\approx 1..2)$ which governs the sidebranch amplitude. This behavior is also found in experiments [2]. Our explanation for this is summarized as the marginal stability principle, to be discussed in the next section.

Comparing our results [5,6] with experiments, we find excellent quantitative agreement over five orders of magnitude in growth rate. The basic success for this agreement seems to be the applicability of linear analysis (and, consequently, marginal stability; section 3). But this concerns only the basic selection of the average size and the resulting growth rate. Experiments show [2], that the sidebranches do not simply grow out maintaining the originally generated periodic spatial pattern. In fact they compete with each other, the fastest ones developing secondary sidebranches like independent primary dendrites. We therefore would like to write a local equation for the interface-displacement not referring to a specific coordinate system as still done in eq. (5), where all quantities refer to parabolic coordinates.

Quite obviously, all we formally have to do is giving an expansion of eq. (1) in local quantities, e.g. curvature and local direction (to incorporate anisotropy effects). Neglecting anisogropy we may write: $v = \sum_{n=o}^{\infty} a_n K^n$ with arbitrary expansion coefficients a_n. This is just a formal extension of eq. (5). But we also need the third term in eq. (5) in order to stabilize the system against short wavelength fluctuations. This leads us to an equation of the form

$$v = q_o \{1 + K - K^3 + \nabla^2 K\} \qquad (6)$$

where constants mostly have been dropped, the second derivative $\nabla^2 K$ being taken along the surface; q_o will be discussed below. The first term provides a constant drift, the second term destabilizes a flat interface, the third term gives a limitation to the curvature K and the last term corresponds to the third term in eq. (5). Sofar, this local equation still has a defect, since it ignores the competition between protrusions of the surface a distance apart.

This is exactly the point, where nonlocal effects enter. A first crude estimate for these effects is obtained as follows. Take a corrugated, on the average planar, surface. Between two distant points, let the straight linear distance be denoted by Δ, the distance along the corrugated surface by S. Define an asymptotic ratio by $q_\infty = \lim_{\Delta \to \infty} (\Delta/S)$. Now, if we neglect local effects, we note, that the amount of latent heat generated for a constant normal displacement of the interface is proportional to q_∞^{-1}, compared with a non-corrugated interface. The average growth rate thus is reduced by q_∞. A simple local form q_o for this factor may be written as

$$q_o^{-1}(S_o) = \int dS \frac{|S-S_o|}{\Delta(S-S_o)} / \int dS, \qquad (7)$$

where S and S_o denote points along the interface (for simplicity we consider only a one-dimensional interface in a two-dimensional system). Obviously, eq. (7) has the correct limiting properties for $\Delta/S = q_\infty = \text{const}$. Moreover, this specific form ensures, that $q_o \to 0$ if two positions of the surface are tending to touch.

Eq. (6) with definition (7) was inplemented on the computer by discretizing S with automatically adjusted density of points along the surface, using the program generated in [3]. The calculation of fig. 1, 3 took about 8 minutes each. The precision was tested on circular shapes to be within 0.1 percent. The breaking of the three-fold symmetry in fig. 3 is due to an amplification of discretization and roundoff-errors by the second term in eq (6), since no measures where taken to maintain the symmetry in the automatic adjustment of grid-point densities. In fig. 1 periodic boundary conditions were used at the open ends, the initial condition being two protrusions of equal shape.

Note that the generated surface-profile is not selfintersecting due to (7). Specifically in fig. 1 one clearly sees the effect of competition between different side branches. Coarsening of the side-branches is not observed, because of the crude approximations in (6). We leave this complicated subject to further analysis and return to simpler equations of type (5) in the following section.

3. The Principle of Marginal Stability

A) Propagating Pattern Selection

The marginal stability criterion presented below applies to situations, where a periodic pattern is produced by propagation of a wave front, fig. 4a,b). We will look first at cases where spatial boundaries are sufficiently far away to be negligible. The left part in fig. 4a is still flat, while on the right of the front a periodic pattern is formed.

It is suggestive to assume, that the front propagation plays an essential role in this situation. Sufficiently far to the left the amplitudes of the spatial oscillations are very small. Therefore we may assume, that in this range we may linearize eq. (5) around the constant $R = r_o$ amplitude. We have performed a linear stability analysis (4) looking for eigenfunctions propagating to the left. The structure of the eigenfunctions about the constant component r_o is [6]

$$y_k(x,t) = \exp\{q_k[x+(v_k-v)t]\}\cos\{k[x+(u_k-v)t]\} \qquad (8)$$

in the coordinate system, moving at velocity v to the left. This characterizes a periodic pattern with wavevector k moving to the left at velocity u_k. It is multiplied by an envelope, decreasing exponentially to the left and moving at velocity v_k to the left. q_k is the slope of the envelope. The dependence of $q_k \cdot v_k$ and u_k upon k is defined by the spectrum of the linearized equation (4):

$$y_k(x,t) = \exp(\Omega_k t)\exp(ikx) \qquad (9)$$

with $\mathrm{Re}[\Omega_k] = q_k \cdot (v_k - v)$ and $\mathrm{Im}[\Omega_k] = k(u_k - v)$.

The important point now is, that $\mathrm{Re}[\Omega_n]$ has a quadratic maximum for some value k_o. By adjusting v (=choosing the speed of the moving frame of reference properly, we may set this maximum to zero, such that $\mathrm{Re}[\Omega_n] \sim -(k-k_o)^2$. Quite obviously this value of $v=v_o$ corresponds to the speed of the fastest propagating mode. This is now what we conjecture to be the operating point of "marginal stability" [6], where the spectrum has its extreme value at $\mathrm{Re}\,\Omega_k = 0$.

Let us assume for the moment that the conjecture be correct. With another mild assumption we then can predict the spacing of the periodic structure formed behind the front (fig. 4a,b). Far to the right, the pattern cannot propagate effectively, but only change its wavevector via phase-diffusion. Thus sufficiently far to the right of the front the pattern has come to rest. Its wave-vector k_∞ can be deduced directly from the wavevector k_o at the far left ahead of the front (corresponding to marginal linear stability) under the assumption, that the number of nodes of the pattern (inflection points of the curve) are conserved. This is a very intuitive assumption, since a change in the number of nodes represents a large local perturbation, which is not observed during the evolution of the pattern in fig. 4a,b). The same observation was recently made with other nonlinear equations of similar type [9]. Comparing the number of nodes flowing towards the moving front at a rate $(v_{k_o} - u_{k_o})$ with the number flowing (asymptotically) away from it at a rate v_{k_o}, we obtain the asymptotic wavenumber k_∞ as

$$k_\infty = k_o\,(1 - U_{k_o}/V_{k_o}) \qquad (11)$$

directly from the linear analysis of eq. (4). Thus, if the marginal stability principle is correct, one obtains the periodicity of the pattern by simple linear analysis plus conservation of nodes.

Sofar we could not give a rigorous proof for this principle of marginal stability. It is, however, also correct for a system with a simple wavefront with no periodic pattern selected [10]. There it correctly predicts the asymptotic speed of the wavefront selected.

To understand the origin of this mechanism we will shortly discuss an approximate solution [7] to the nonlinear system (5). At the far right (fig. 4) the pattern looks sinusoidal. We substitute as a two-mode approximation

$$R(x,t) = r_o(t) + \tilde{r}_1(x,t), \qquad \tilde{r}_1 = r_1(t)\cos(k[x-vt]) \tag{12}$$

In the full nonlinear system (4) with the above-mentioned boundary conditions at x=0 the first term produces a drift of the pattern towards the right. Furthermore, as mentioned, this term stabilizes the system against sidebranch-fluctuations for $r_o<r^*$. We model this by rewriting (5) as

$$\frac{\partial R}{\partial t} = -\left(\frac{1}{2}r^*\right)^2 \tilde{r}_1(x,t) - R\frac{\partial^2 R}{\partial x^2}\left(1+\varepsilon\frac{\partial^2 R}{\partial x^2}\right) - \frac{\partial^4 R}{\partial x^4} \tag{13}$$

with R defined by (12). For $r_o < r^*$ the amplitude $r_1(t)$ decays with time.

This allows us, to rewrite (13) in two-mode approximation as two coupled nonlinear differential equations for $r_o(t)$ of the structure:

$$\frac{dr_o}{dt} \sim r_1^2\,(1-\varepsilon\, r_o) \tag{14}$$

$$\frac{dr_1}{dt} \sim r_1\,(r_o-r^*) - \varepsilon\, r_1^3$$

Irrelevant constants have been dropped. The flow diagram is sketched in fig. 5).

One sees that for $r_o<r^*$ small amplitudes r_1 tend to decay, $r_1 \to 0$. But if due to external fluctuations, e.g. by thermal noise, sidebranches r_1 are permanently excited, a drift in r_o towards r^* will be observed. This is due to the nonzero component of the

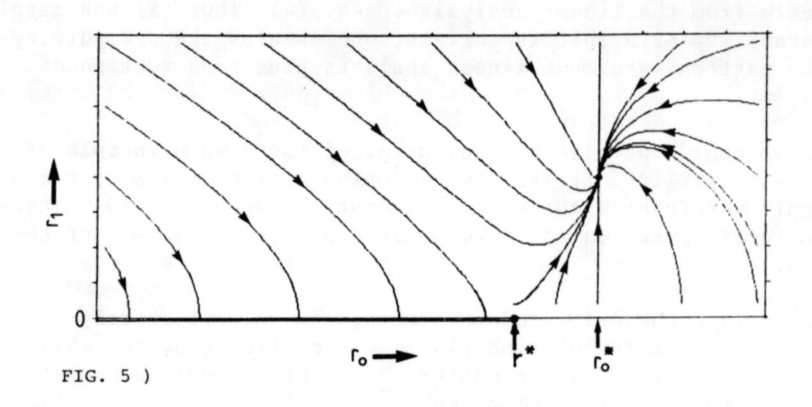

FIG. 5)

flow lines in $+r_o$-direction. Thus, narrow dendrites $r_o<r^*$ are slowly broadened. Ultimately, the trajectories end up at r_o^*, at a slightly larger value than the point of marginal stability $r_o = r^*$, $r_1=0$. The point r_o^*, depending on ε, was numerically indiscriminable from r^*. So, for our system (5) with mentioned boundary conditions, the point of marginal stability is not exactly the operating point, but it plays an important role in the selection mechanism.

On the other hand, for the simpler system with a propagating front without periodic pattern [10] it turns out, that the fixed point at r_o^* coincides with the end-point r^* of the line of fixed point $r_o < r^*$, $r_1=0$. In this case the marginal stability criterion gives exactly the operating point of the system.

An alternative way of understanding the marginal stability mechanism is the philosophy, that the fastest mode of the set (8) dominates the systems behavior. This seems to be true in cases with periodic patterns [7,9]. In the simpler system [10], however, the mode marginally stable against perturbations (of finite support) is the slowest one. And again, the full nonlinear analysis there shows, that it is ultimately being selected [10].

We conclude, therefore, that a satisfactory criterion for mode selection requires a nonlinear analysis as sketched in the flow diagram fig. 5). For cases analyzed sofar, however, the marginal stability principle appears to play a dominant role.

B) Generalization and Alternatives

The selection mechanism described in subsection A) relies on the following fact. Assume a continuous set of stable solutions to be given, characterized by the parameter r_o. If in an (N+1)-mode approximation the flow lines starting from $r_o(t=0)<r^*$ have a component $dr_o(t)/dt>0$ for small but arbitrary $r_n(t=0)\neq 0$ $(n=1,2,..,N)$ and if for initial conditions $r_o(t=0)>r^*$ they have a component $dr_o(t)/dt<0$, then the mode characterized by $r_o=r^*$ is the ultimately selected mode. Constructing arbitrary differential equations for $r_n(t)$ with such a distinguished value of $r_o=r^*$ we note that the point $r_o=r^*$, $r_n=0(n=1,...N)$ is "generically" not marginally stable, if it is not the end point of a line of fixed points, Insofar it is surprising, that a recent analysis of cellular interfaces seems to recover a marginal point [11], while from arguments above it is "generic", that this is the point being selected.

An alternative way of looking at mode selection is, to relax the condition of mode-conservation by appropriately chosen boundary conditions [12]. Pure stability arguments, also including the possible readjustment of wave numbers, are then to be used as selection criteria.

For reaction-diffusion-type equations of the form (i, j=1, 2...,M):

$$\frac{dY_i(x,t)}{dt} = D_{ij}\, Y''_j + N(\{Y_j\}), \tag{15}$$

N being a nonlinear combination of Y_j's, one obtains the criterium [12]

$$\int dx\, \tilde{Y}'_i\, D_{ij}\, Y'_j = 0, \tag{16}$$

where the prime denotes the spatial derivative, and $\tilde{Y}'_i$ is the left eigenfunction (Re $\Omega=0$) of the operator corresponding to a linearization around stationary stable solutions of (15). The philosophy here is, that condition (16) ensures vanishing of the right-hand-side of (15) even under slow spatial variation of the wave number. To conclude these considerations it appears, as if there is no unique selection criterium applicable to arbitrary cases. We feel, however, that a classification of both experimental situations and corresponding selection might be possible in the future. In contrast to equilibrium thermodynamics we therefore cannot simply talk about "states", but also have to worry about "pathways".

As a last point we shortly give a few remarks about "collective behaviour" of (linearized) modes near marginal stability. For the left of the propagating front (fig. 4a,b) we have argued, that in our case the fastest mode dominates the systems behavior. But the nonlinear terms in eq. (5) will also tend to couple other modes with the asymptotically dominating one. A simple estimate of what could happen is obtained in the following way.

Assume that at t=0 the structure of the profile ahead of the interface can be described by the superposition

$$y(x,t) = \int dk\ \rho(k)\ \ y_k(x,t) \tag{17}$$

with $y_k(x,t)$ from eq. (8) and $\rho(k)$ being an arbitrary weight function. Assume further, that the most important effect of the nonlinearities in eq. (4) is, to provide a nonvanishing, time-independent weight $\rho(k)$ also at later times $t>0$. (In reality, phase shifts will enter the problem). The most important part of the spectrum is near $k \approx k_o$, where the system is least stable. As an explicit choice for $\rho(k)$ near k_o we may assume either $\rho(k) \approx \rho_o$ = constant or $\rho(k) \sim |k-k_o|^{-\lambda}$, $0<\lambda<1$. The latter choice is suggested by the increasing susceptibility of the system to perturbations for modes with $\Omega_k \lesssim 0$ as $k \to k_o$. Simple saddle point-integration recovers the structure of the dominant mode (8) with $k=k_o$, but with a correction to the velocity

$$V_{eff} = V_{k_o} - O(\frac{1}{t}) \tag{18}$$

This crude analysis [8], based on the asymptotic degeneracy of the continuous spectrum as $k \to k_o$, indicates the existence of long-time tails in the mode-selection problem, with corresponding long-time tails in the resulting pattern (11). Thus even for long duration of experimental measurements the predicted pattern (11) eventually may not be reached. This behavior resembles features in dynamic critical phenomena and in the microscopic dynamics of simple fluids.

In conclusion we think, that despite the difficulties of these questions a number of ideas outlined in this lecture are promising enough to expect visible progress in the not too distant future.

References

1. W.W. Mullins, R.F. Sekerka, J. Appl. Phys. 35, 444 (1964)
2. S.C. Huang, M.E. Glicksman, Acta Met. 29, 701 717 (1981)
3. H. Müller-Krumbhaar, T. Burkhardt, D. Kroll, J. Crystal Growth 38, 13 (1977)
4. J.P. v.d. Eerden, in "Modern Theory of Crastal Growth", A.A. Chernov, H. Müller-Krumbhaar eds. (Springer Verlag, Heidelberg, 1983)
5. J.S. Langer, Rev. Mod. Phys. 52, 1 (1980)
6. J.S. Langer, H. Müller-Krumbhaar, Acta Met. 26, 1681, 1689, 1697 (1978); 29, 145 (1981)
7. J.S. Langer, H. Müller-Krumbhaar, Phys. Rev. A27, 499 (1983)
8. H. Müller-Krumbhaar (unpublished)
9. G. Dee, J.S. Langer, Phys. Rev. Letters 50, 383 (1983)
10. D. Aronson, H. Weinberger; Adv. Math. 30, 33 (1978)
11. M. Kerszberg, Phys. Rev. B27, 3909 (1983)
12. L. Kramer, E. Ben-Jacob, H. Brand, M.C. Cross, Phys. Rev. Lett. 49, 1891 (1982)

References

1. [illegible], R.A. [illegible], J. Appl. Phys. 47, [illegible]
2. [illegible] Krumhansl, [illegible] 29, 701-715 (1981)
3. [illegible] Krumhansl, [illegible] Burkhardt [illegible] J. Crystal [illegible] 14 (1977)
4. [illegible], in "[illegible] Theory of Crystal [illegible]" [illegible] Krumhansl [illegible] (Springer Verlag, [illegible] 1983)
5. [illegible] Mod. [illegible] 1 (1980)
6. [illegible] Krumhansl, Acta [illegible] (1981)
7. [illegible] Phys. Rev. [illegible] (1983)
8. [illegible]
9. [illegible] Phys. Rev. Letters 50, [illegible] (1983)
10. [illegible] Metall. [illegible] (1978)
11. [illegible] Rev. [illegible] 800 (1977)
12. [illegible] Phys. [illegible] (1982)

PART IV

GEOLOGY

THE SELF ORGANIZATION OF LIESEGANG BANDS AND OTHER PRECIPITATE PATTERNS

P. Ortoleva

Departments of Chemistry and Geology
Indiana University, Bloomington, Indiana 47405 and
Geo-Chem research Assoc.,
700 North Walnut St., Bloomington, Indiana 47401

Precipitation/dissolution kinetics coupled to transport can lead to a number of interesting pattern forming processes, the most well studied being the Liesegang bands. Here we review these phenomena and a number of models that have been set forth to explain them. Geological applications of the theory will be discussed.

I. BRIEF SURVEY OF PRECIPITATE PATTERNING PHENOMENA

In the century of research on Liesegang banding and other precipitate patterning phenomena, many effect have been noted [1-5]. A partial list is as follows.

(i) Liesegang bands occur when coprecipitates are interdiffused. Alternating bands of more and less (or no) precipitate may result.

(ii) Regular band induction in originally uniform sols occurs when they are subjected to an overall composition or temperature gradient or subject to a chemically active site or boundary condition involving the solutes into which the precipitate particles dissolve.

(iii) Secondary banding occurs when one of the bands as in (i) or (ii) breaks up into bands on a finer scale.

(iv) Band spacing may increase or decrease along the cross-diffusion gradient or under imposed electric fields may be constant. Profiles and regularity of bands may vary along the gradient also.

(v) Patterns in an experiment an in (i) may form an a tube wall in an interesting corksrew geometry.

G. Nicolis and F. Baras (eds.), Chemical Instabilities, 289–297.

(vi) Mosaic, concentric ring, spiral and banded patterns may occur spontaneously as a uniform sol ages.

These phenomena have been observed in laboratory experiments and, in some cases, in geochemical systems [1,2,6,7] (also Boudreau, these proceedings).

II. BANDING IN A GEOCHEMICAL REDOX FRONT PROBLEM

Consider the effect of imposing oxygenated waters on a sandstone aquifer containing pyrite [7,8]. A very schematic representation of this process is

$$X + P \rightleftarrows F + T \tag{1}$$

$$X + F \rightleftarrows G. \tag{2}$$

Here P and G are the minerals pyrite and goethite while X, F and T represent the mobile oxygen, Fe^{2+} and thiosulfate species. Initially in the aquifer G = 0 and upon mobilization of F, a supersaturation with respect to goethite may occur so that G is nucleated. The interesting aspect of this situation in the present context is that the G content behind the advancing pyrite redox front can be oscillatory.

This system has been simulated as follows. Consider a system in the spatial (r) interval $0 < r < \infty$. A flow through the porous medium of velocity v is imposed. Taking rates W_1 and W_2 for the P and G kinetics respectively we assume the following reaction transport model :

$$\frac{\partial X}{\partial t} = D_X \frac{\partial^2 X}{\partial r^2} - W_1 - W_2 \tag{3}$$

$$\frac{\partial P}{\partial t} = - W_1 \tag{4}$$

and similarly for F, T and G. The initial data is $P(r,0) = P_0$, $G(r,0) = 0$. Rates W_1 and W_2 were modeled to account for equilibrium, nucleation of G and the dependence of the rates on particle size via

$$W_1 = k\, P^{2/3}\, [KX - FT] \tag{5}$$

$$W_2 = q\, (G + g)^{2/3}\, [FX - Q] \quad . \tag{6}$$

The quantity g allows for nucleation of G. If $F \cdot X > Q_n$, Q_n being the critical value of the goethite equilibrium constant Q for nucleation, then g is a small number g_0 whereas g vanishes otherwise. The 2/3 power law reflects the expected proportionality of the dissolution rate on grain surface area.

The driving force for the overall process P G comes from the boundary conditions at the "inlet" r = 0. We take

$X(0,t) = X_M$, $F(0,t) = F_M$ and $T(0,t) = 0$.

A series of numerical, finite difference simulations were carried out to investigate the behavior of the system as a function of X_M, F_M and P_0 for given values of the other parameters. An interesting case in the present context is shown in Fig. 1.

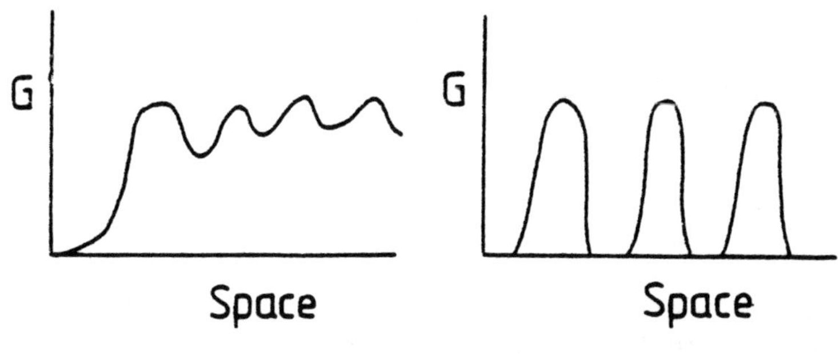

Figure 1. Figure 2.

Note that in this figure the space between the G concentrations are not void of precipitate whereas changing parameters (such as increasing q) beyond critical values can open up a gap between bands where G vanishes over an interval as in Fig. 2.

The model is essentially that of Ostwald which emphasized a cycle of supersaturation, nucleation and depletion of co-precipitates to explain Liesegang bands [9]. However this theory, especially as formulated mathematically by Prager [10] emphasizes the presence of a gap between bands. This occurs in the model presented here as a special limit of parameter values but is not required. The model presented here as well as teh Ostwald-Prager theory are designed with phenomenon (i) in mind. They clearly cannot explain the other phenomena, however.

III. COMPETITIVE PARTICLE GROWTH THEORY

At the suggestion of the author, Flicker and Ross [11] allowed a uniform PbI_2 sol to age to see if a symmetry breaking instability would occur - see also [12,13]. The system did indeed pattern, displaying phenomenon (vi). To explain this phenomenon a new mechanism of precipitate pattern formation, the "competitive particle growth" (CPG) model was introduced [4,5,12, 14] and found strong experimental support [4,5,12-15] .

CPG theory is based on the observation that because of surface tension, larger particles have a lower free energy (and hence are less soluble) than smaller ones. Thus a large particle will grow at the expense of its smaller neighbors by lowering the local concentrations to its "solubility" and below that of its smaller neighbors. This effect is clearly an accelerating one. CPG theory then makes the assertion that because of the competitive effect, particles in one region can grow at the expense of particles in neighboring regions driven by their difference in local average particle size.

The simplest formulation of the CPG model is to assume that the particle size distribution is sufficiently narrow that it is acceptable to speak of the local particle radius $R(\vec{r},t)$ for all particles in a small volume element about point $\vec{r}$. R is taken to satisfy

$$\frac{\partial R}{\partial t} = k\,[\,c - c^{eq}(R)\,] \tag{7}$$

where k is a rate coefficient (that may depend on R) and we have assumed growth occurs from a monomer of concentration c which attains an equilibrium value $c^{eq}(R)$ for a radius R particle. In cases as in the previous section where growth is from coprecipitates, as in Reaction (2), the law can take a form like $k\,[\,XY - K^{eq}(R)\,]$ for the growth process $X + Y \rightarrow$ particles.

The monomer concentration c is taken to satisfy a simple reaction-transport equation :

$$\frac{\partial c}{\partial t} = D\,\nabla^2 c - 4\pi n p R^2\,\frac{\partial R}{\partial t} + W \tag{8}$$

where n is the particle number density, p is the molar density and W is the rate of monomer production (if any) from cross diffusion reactions; W may be calculated explicitly for a case like $X + Y \rightarrow$ monomer when $W = \varepsilon^{-1} XY$ and $\varepsilon \rightarrow 0$. In that case W is a δ-function source with a weight that can be calculated from matched asymptotic methods [14]. For cases when the mechanism of particle growth is by direct addition of X and Y (say), then similar equations, without the source term, can be written down for X and Y.

Instability of the uniform sol is predicted by a linearized analysis of (7,8) about the uniform state $R = \bar{R}$ (arbitrary), $c = c^{eq}(\bar{R})$. Instability is of the svegliabile type and is discussed in the other article by the author in this volume. Numerical simulations show that these equations have much structure as can be seen by the sample run shown in Fig. 3.

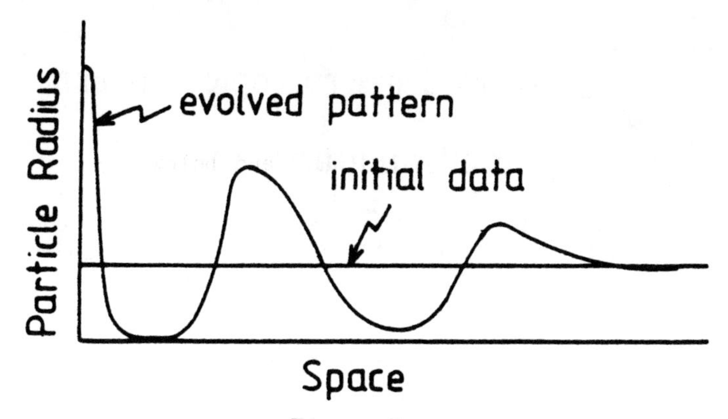

Figure 3.

Here we see, schematically, results of a numerical simulation of the CPG model equations (7,8). An imperceptible initial bump in the R-profile at left is amplified into a large scale pattern. Note the nascent first satellite that never quite makes it, being eliminated by the strong central maximum. It is found that the spacing between the bands is essentially constant, in the dimensionless variables used in [15]. In the latter study a case is made for the existence of the invariant first satellite induction length. In other cases, quite extremely different from that of Fig. 3, the invariant first satellite induction length also is observed although the first satellite band "comes of age" (grows above the initial average value of R). Simulations with a monomer source to account for the cross diffusion of coprecipitates were undertaken : one example is seen in Fig 4. Many examples with a range of pattern types (depending on parameter choices) are given in [14] for both the cross-diffusion and electrically driven cross flux cases. Simple CPG theory does not include nucleation; it can only explain phenomena (ii),(iii), (v) and (vi). According to CPG, secondary banding is the result of a slight gradient in R across a given band while the spontaneous self organization of the uniform sol results from CPG amplification of omnipresent macroscopic variations from uniformity. The dynamics of this growth of macroscopic length and amplitude scale patterning from the initially unpatterned, uniform sol has been studied by two-dimensional light absorption techniques and anlyzed in terms of the kinetics of the correlations recently [13]. In Fig. 4 we see a correlation function for a precipitate pattern of PbI_2 that formed spontaneously in a petri dish from a uniform sol (for which the correlation function c vanishes) - see Ref. [13] for details.

IV CPG THEORY AND PARTICLE SIZE DISTRIBUTION

What is needed to study all aspects of precipitation pat-

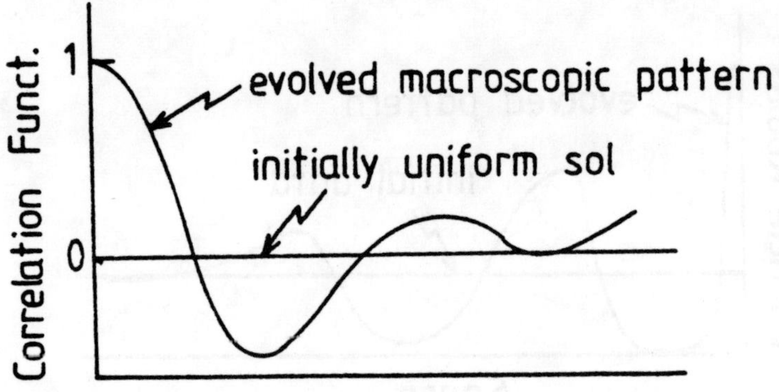

Figure 4.

terning in a realistic way is a theory that incorporates the following processes.

(i) The nucleation of particles
(ii) Local Ostwald ripening whereby the particle size distribution is always evolving to a state of fewer but on-the-average larger particles.

A first attempt at this was accomplished in [16] where the Lifshitz-Slyozov theory was adopted to allow for spatial variations including the concept of the local particle size distribution. In that work, however, undue emphasis was placed on the diffusion of the particles and the concept of forming macroscopic length scale patterns via the svegliabile mechanism was not realized (see a discussion in the other paper of the author in these proceedings).

The statistical aspects of nucleation and ripening require that precipitate patterning phenomena be described by a particle size distribution formalism. Recently we have carried out a simulation of the spatio-temporal evolution of the size distribution function as follows - see Ref. [17] for details. A semimicroscopic description was adopted whereby space is divided up into boxes containing varying numbers of particles of a range of sizes. In the spirit of ensemble theory, the numbers and sizes in each box for the given member system of the ensemble was chosen by picking particle positions and sizes at random , weighted with a given size distribution $f(\vec{r},R,t=0)$ ($f\, d^3r\, dR$ is the number of particles in d^3r about point $\vec{r}$ with radius R in the interval $(R,R+dR)$). The N particles so placed with initial radii R_i^o were grouped according to spatial cells in which they resided and into size classes c = 1,2,...M. So as to minimize the calculations, all particles in a given class i were taken to have the same size $R_i(\vec{r}_\alpha)$ and labeled with a spatial cell index α in which they resided. Letting $n_i(\vec{r}_\alpha)$ be the number of particles of size class i in the cell about point α, we simulated the evolution of the system with the model equations

$$\frac{\partial R_i}{\partial t} = k\,[\,c - c^{eq}(R_i)\,] \tag{9}$$

$$\frac{\partial c}{\partial t} = D\,\nabla^2 c - 4\pi p \sum_{i=1}^{M} \frac{\partial R_i^3}{\partial t} + W, \tag{10}$$

the obvious generalization of (7,8) to the system of many particles sizes. Simulations were carried out using numerical integration techniques.

Construction of the particle size distribution is straightforward via the simple algorythm

$$f(R,\vec{r},t) = \frac{1}{\Delta R_i} \sum_i n_i(\vec{r}) \quad ; \quad R < R_i < R + \Delta R \tag{11}$$

(although certain statistical and numerical techniques are actually used to greatly improve on this "straightforward" approach in [17]). The results of these simulatons are most interesting as follows.

As the width of the initial particle distribution increases many of the features of the patterns predicted by the monodisperse CPG theory - namely band intensity or sharpness - are lost. Also a "bump" in the local average size only can induce a given number of satellite bands before homogeneous ripening shuts off the CPG feedback (since the particles encoutered far from the initial bump by te advancing banding zone increases with distance from that initial bump). Thus if homogeneous ripening is very fast compared to the satellite band induction time, banding is reduced to a few satellites or even totally eliminated.

Another important consequence of monodispersal is the existence of greedy giants [12]. These giants are rare particles from the tail of the initial size distribution that can have significant macroscopic consequences - an interesting case where fluctuations may dominate over the mean field (monodisperse) theory of the simple CPG theory of the previous section. A case in point is the initial bump on the average particle size induction of satellite band induction run of Fig. 5. In that run we see that there are some large particles in the region between the central initial bump and the first satellite. Apparently these greedy giants have taken advantage of the elevated monomer concentration induces by the dissolution of the more average sized particles resulting from the satellite induction process induced by the central initial bump. The possibility of this phenomenon depends sensitively on the tail of the initial size distribution of the homogeneous sol.

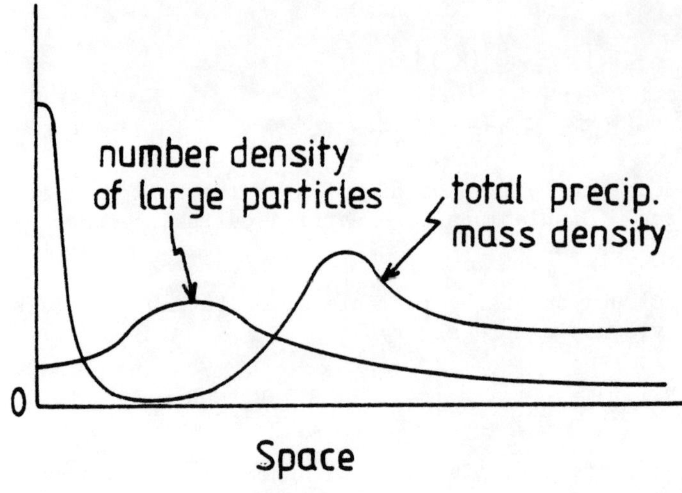

Figure 5.

The tail of the distribution can have another profound effect on system evaluation - tail flutter pattern induction. In any volume element there are increasingly fewer particles the further the size is in the initial size distribution's tail by definition of the latter. Thus spatial fluctuation in the tail of the distribution - called tail flutter here - is expected to be very large on a scale less than or on the order of the typical distance between these particles. Since the uniform sol is unstable to pattern formation of the svegliabile type, these large scale tail flutter perturbations can set off the pattern growth and strongly effect the length scale. One expects from these arguments that in the uniform PbI_2 sol mosaic pattern process, there should be greedy giants observed in the otherwise precipitate free space between the concentrate precipitate areas. This is the other manifestations of nonmonodispersal are all observed in precipitate patterning experiments. Thus the evidence is strong that the CPG model can explain a host of precipitate patterning effects in non-Ostwald-Prager systems.

REFERENCES

1. Hedges, E. and Myers, J.E., 1926, "The problem of Physico-Chemical Periodicity", Longmans-Green, New York
2. Liesegang, R.E., 1913, "Geologische Diffusionen", Steinkopf, Dresden and Leipzig
3. Stern, K.H., 1967, Nat. Bureau of Standards (U.S.) Spec. Pub. n°292
4. Ortoleva, P., 1978, in "Theoretical Chemistry" IV, Eyring, H. ed., Academic Press, New York

5. Kai, S., Mitler, S.C. and Ross,J., Measurements of Temporal and Spatial Sequences od Events in Periodic Precipitation Processes (preprint)
6. Merino, E. and Ortoleva, P., in "The Self Organization of Geo-Chemical Periodicity" (book manuscript in preparation)
7. Ortoleva, P., Auchmuty, G., Chadam, J., Hettmer, J., Merino, E., Moore, C., Ripley, E., Redox Front Propagation and Banding Modalities, Physica D (submitted for publication)
8. Merino, E., Moore, C., Ripley, E., Auchmuty, G., Chadam, J., Hettmer, J. and Ortoleva, P., Kinetic Modeling of Redox Roll Fronts and Their Instabilities (submitted for publication)
9. Ostwald, W., 1925, Kolloid-Z 36, pp. 380
10. Prager, S., 1956, J. Chem. Phys. 25, pp. 279
11. Flicker, M.R. and Ross, J., 1974, J. Chem. Phys. 60, pp. 3458
12. Feinn, D., Ortoleva, P., Scalf, W., Schmidt, S. and Wolff, M. 1978, J. Chem. Phys. 67, pp. 27
13. Lecuna, J., Johnson, J., Ortoleva, P. and Sporleder, R., Development of Macroscopic Correlation in Precipitating Systems, J. Chem. Phys. (submitted for publication)
14. Feeney, R. and Ortoleva, P., Experimental Studies of Precipitating Banding (in preparation)
15. Feeney, R., Strickholm, P. and Ortoleva, P., Quantitative Comparison of Precipitation Banding Experiments and Competitive Particle Growth Theory (in preparation)
16. Lovett, R., Ross, J. and Ortoleva, P., 1978, J. Chem. Phys. 69, pp. 947
17. Strickholm, P. and Ortoleva, P., Particle Size Distribution Effects in the CPG Model of Liesegang Banding, J. Chem. Phys. (submitted for publication) ;
Ortoleva, P., 1982, Zeit. Für Fisick B 49, pp. 149

EXAMPLES OF PATTERNS IN IGNEOUS ROCKS

Alan E. Boudreau

Department of Geology, University of Oregon, Eugene, OR 97403 (Current: Department of Geological Sciences, AJ-20, University of Washington, Seattle, WA 98195)

The Stillwater Complex is a layered igneous intrusion situated in south-central Montana. Locally, the sequence of layered rocks show many features analogous to Liesegang phenomenon. Similarities with pattern formation in crystallizing salt solutions suggests that some types of igneous layering may be the result of an orderly aging (or Ostwald ripening) of the crystal assemblage.

1.1 Introduction

Layered igneous intrusions are a class of igneous rocks which formed by the slow cooling of a large body of magma at some depth within the crust of the earth. The resulting rock is not homogeneous in texture or composition, but instead shows a layered segregation of mineral phases on scales ranging from millimeters to kilometers. Fine-scale layering (or inch-scale layering) is characterized by a layered segregation of mineral phases on a millimeter to centimeter scale. Fine-scale layering may exhibit a geometric regularity which is strikingly similar to banding produced in Liesegang experiments. An example of such layering is shown in figure 1.

The similarity of fine-scale layering with Liesegang phenomenon has led to the suggestion that the igneous layering is the result of a periodic nucleation mechanism. In one such model, two different rate laws, one for mass diffusion and one for heat transport, govern the evolution of gradients at the crystallization front and lead to supersaturation and nucleation which is periodic in space and time [1]. This paper will present some evidence that fine-scale layering may instead be the result of pattern

G. Nicolis and F. Baras (eds.), Chemical Instabilities, 299–304.

Figure 1. A broken outcrop of fine-scale layering in a norite of the Banded zone of the Stillwater Complex. Hammer at lower right provides scale.

development during the orderly aging (or Ostwald ripening) of the crystal assemblage, a mechanism which has increasing theoretical and experimental support [2,3,4].

2.1 Fine-scale layering in the Stillwater complex

The Stillwater Complex is a large layered intrusion situated in south-central Montana. Its radiometric age has been measured at 2.7×10^9 yrs [5]. The currently exposed portion of the complex, roughly 7 x 50 km, is only a small part of the intrusion's original dimensions.

Within the Stillwater Complex, fine-scale layering is only locally well developed. It is found associated with chromite-bearing rocks and, in the most extensive developments, with layered norites associated with anorthosites in the middle portions of the complex. Where best developed, as in figure 1, the layering may be present over a stratigraphic distance of several tens of meters.

The most notable feature of the layered norites (aside from the layering) is the coarse grain texture of the rock. It is found to be the case that where layering is present as distinct layers which are continuous over the length of an outcrop, and when the individual layers are more nearly monomineralic, the rock is considerably coarser grained than the unlayered, medium grained rocks which are typical of most of the complex. This textural difference is shown in figure 2.

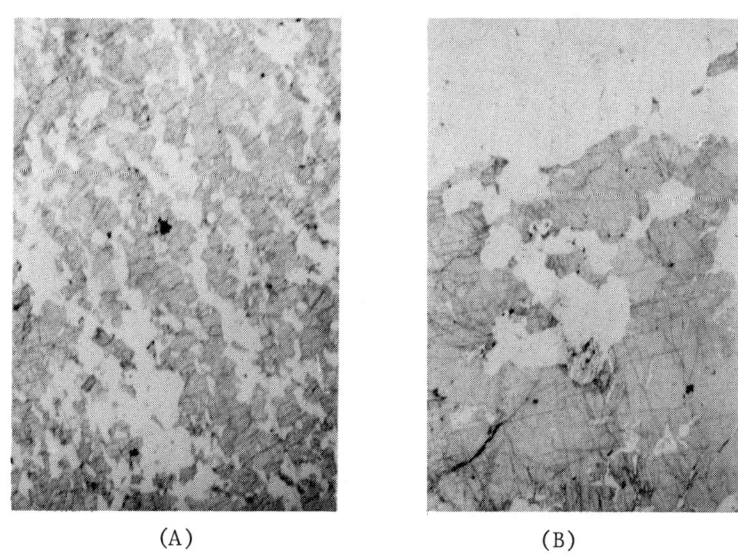

(A) (B)

Figure 2. Photomicrograph of an unlayered gabbronorite (A), which shows a moderate mineral lamination, and a coarse grain layered norite (B). Both sections are approximately 3 cm long.

At best, the unlayered rocks show a moderate lamination caused by a parallel alignment of mineral grains. In contrast, figure 2 (B) shows the coarse texture found in norites in which fine-scale layering is well developed. In the layered rocks, the crystals which comprise the individual layers form a near-continuous network of interlocking grains. Although individual layers appear ragged in thin section, the layering is uniform on the scale of an outcrop.

Another common feature of fine-scale layering is the common occurrence of two different length scales in the spacing of the individual layers. One such example is again shown in figure 1. In this rock, the dark pyroxene bands occur as regularly-spaced paired layers. It is also found that some individual layers in the banded chromite-bearing rocks may consist of two or more finer layers (figure 3). These features are similar to the secondary banding reported in Liesegang experiments, where it has been shown that individual bands can actually be composed of a series of finer bands [6].

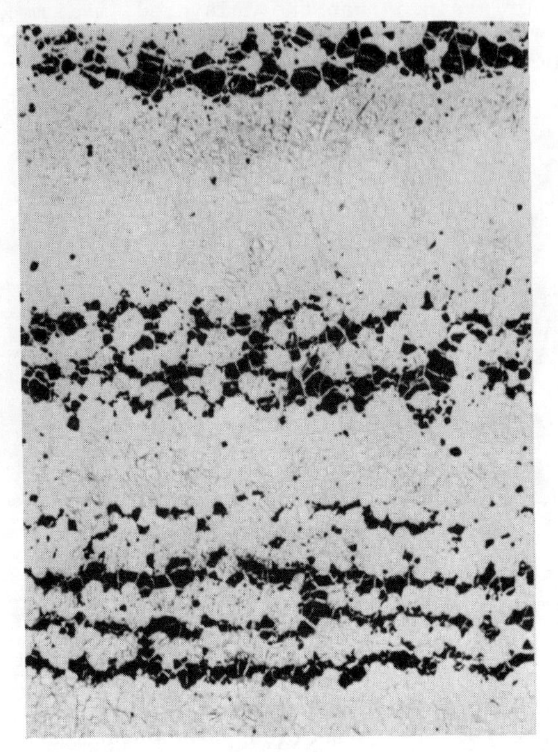

Figure 3. Chromite layers (black) developed within an olivine cumulate host (now partially serpentinized). Photomicrograph is 3.2 cm long.

Yet one other secondary feature found in fine-scale layering is the presence of a mosaic or polygonal pattern in the crystal arrangement when individual layers are viewed along the bedding plane. Shown in figure 4 is one such pattern of olivine crystals seen on a bedding surface where parting in the rock has fortuitously run parallel to layering . The pattern has been enhanced by low temperature hydrothermal alteration of the original igneous minerals. Patterns similar to these have been produced in crystallizing aqueous salt systems. In these experiments, a mosaic pattern can be formed by allowing a thin layer of crystallites in a gel solution to simply age [2].

Figure 4. Mosaic pattern on bedding surface of a banded troctolite from the Stillwater Complex.

3.1 Layer formation during aging.

Recent experimental and theoretical work have demonstrated that Liesegang-type pattern behavior need not be the result of a nucleation mechanism [2,3,4,7]. Instead, an orderly pattern can evolve in systems which begin as an unpatterned assemblage of crystallites. The most readily visualized explanation is that pattern behavior results from a feedback mechanism during aging of the crystals. Because larger crystals require lower solution concentrations than do small crystals (which arises from the surface free energy contribution to the overall free energy of a crystal), regions of larger crystals will grow at the expense of nearby, smaller crystals. Dissolution of unfavored grains becomes even more rapid as the size disparity increases. Thus, any homogeneous assemblage of crystals is inherently unstable to minor deviations in crystal grain size. In an advancing front of crystallization, one would expect a gradient in crystal size whereby crystals would be progressively larger away from the nucleation front. This suppiles the asymmetry such that competitive crystal growth can result in a layered segregation of mineral phases. This process would also enhance any banding present because of non-uniform nucleation. Furthermore, such a process can be envisioned to work on successively smaller scales and cause the segregation of primary layers into finer, secondary bands.

The coarse rock textures seen in the layered norites and the presence of analogous secondary structures suggests that the mechanism as developed for pattern growth in aqueous salt systems was operative during the formation of layered igneous rocks. Layering produced by the destruction of mineral grains will result in fewer and fewer grains as the pattern evolves, and hence the coarse rock texture is an expected consequence of layer growth. Such a view is radical in that igneous textures which are generally considered to have been present when the rock first crystallized actually may have evolved only slowly over time. In a work to be presented elsewhere, calculations will be presented which imply that unusual crystallization conditions must have existed during the formation of fine-scale layering, which is supported by the relative rarity of this type of layering.

4.1 References

1. McBirney, A. R., and Noyes, R. M., 1979, J. Petrol. 20, pp. 487-554.
2. Feinn, D., Ortoleva, P., Scalf, W., and Wolff, M., 1978, J. Chem. Phys. 69, pp. 27-39
3. Lovett, R., Ortoleva, P., and Ross, J., 1978, J. Chem. Phys. 69, pp.947-955.
4. Kai, S., Muller, S. C., and Ross, J., 1982, J. Chem. Phys. 76, pp. 1392-1406.
5. DePaolo, D. J., and Wasserburg, G. J., 1979, Geochim. Cosmochim. Acta 43, pp. 999-1008.
6. Ramaiah, K.S., 1939, Proc. Indian Acad. Sci. Sect. A 9, pp. 467-478.
7. Flicker, M. and Ross, J., 1974, J. Chem. Phys. 60, pp. 3458-3465.

SURVEY OF GEOCHEMICAL SELF-PATTERNING PHENOMENA

Enrique Merino

Department of Geology, Indiana University, Bloomington, Indiana 47405, U.S.A.

Self-patterning phenomena involving chemical reactions take place in all rock types--sedimentary, igneous, and metamorphic. Because apparent examples of geochemical self-organization cannot as a rule be directly proven by experiment, it is often difficult even to establish that they are true cases of self-patterning, let alone identify the mechanisms involved. Examples (with variable certainty) are: (1) Iron oxide can form concentric bands in chert nodules in limestones, and regularly spaced, cigar-shaped, lumps in shales. (2) Agates (aggregates of microcrystalline quartz fibers) may display all of the following: (a) banding looking remarkably like Liesegang banding; (b) adjacent bands that consist alternatingly of non-twisted and twisted quartz fibers; and (c) a distinctive chevron interference pattern, visible observing the agate between crossed polarizers. (3) Precambrian iron formations display alternating bands on one or more size scales, often with great lateral extent. (4) Mississippi-Valley-type ores often display striking alternating bands ($\sim$ 1 cm) of any one or two of the minerals fluorite, dolomite, galena, sphalerite, and others; grain-size banding is also found. (5) Sets of evenly-spaced seams and stylolites, with spacings of up to tens of centimeters, are common in limestones and many other sedimentary rocks. (6) The conspicuous, roughly regular, light and dark mineral banding (schistosity, cleavage, or foliation) in most regionally metamorphosed rocks. (7) The regular banding exhibited by some contact metamorphic rocks (skarns). (8) Inch-scale, planar bands with considerable lateral extent are found in basic complexes such as the Stillwater in Montana and the Skaergaard in Greenland. (9) Many granitic rocks contain "orbicules" (roughly spherical, crystalline aggregates up to several tens of centimeters across) consisting of concen-

G. Nicolis and F. Baras (eds.), Chemical Instabilities, 305–328.

tric, alternating bands (∿ 0.1-1 cm) of two minerals. (10) Crystals of plagioclase feldspar with oscillatory compositional zoning can occur in intermediate volcanic rocks.

INTRODUCTION

Self-organization in geological systems has been recognized sporadically over the last several decades. In 1922 for example Hedges and Myers [1] mentioned natural occurrences of gold/quartz interbandings and attempted to account for them as cases of self-patterning; in 1918 Davis [2], to explain the formation of the bedded cherts of the Franciscan Formation of California [3], tried to show experimentally that silica-gel/clay systems, initially uniformly mixed, can resolve themselves spontaneously into alternating layers of silica gel (which, through dehydration and ripening, would later become microcrystalline quartz or chert) and clay; and in 1932 Hartman and Dickey [4] suggested, based on their experiments, that Liesegang-type interdiffusion could cause the characteristic banding of Precambrian iron formations. Earlier, Liesegang himself [5,6] had speculated that agate banding could form by diffusion of metals through lumps of silica gel in rhyolitic and andesitic volcanic rocks, and Knopf [7] had suggested "phenomena similar to those operative in the formation of Liesegang's rings" to account for alternatingly banded spherical structures in metamorphic rocks.

Perhaps because theoretical physico-chemical explanations of Liesegang periodic precipitation did not improve much from the early work [1,8] to that of Prager [9], and also because geological thinking has been rightly permeated since the 17th century by the stratigraphic principle of superposition--which states that any layer in a sequence is older than the one on top of it--and because of this principle's unspoken implication that each layer basically "knows" or "cares" nothing, from a genetic standpoint, about the layers that underlie and overlie it, it has been difficult to imagine that layers can exist that <u>are</u> linked genetically to each other and that, therefore, fall outside of the field of applicability of the principle of superposition. Layers generated by self-organization are of this kind. The recent theoretical progress made by physical chemists in explaining the dynamics of self-organization--a progress we do not intend to review here (but see, for example, [10-12] and the proceedings of this and previous workshops on non-equilibrium phenomena)--along with several very recent articles in the geological literature [13-21] point to the widespread occurrence of what could be called geochemical self-organization.

The purpose of this contribution is to describe several likely or certain cases of geochemical self-patterning and to suggest directions for future work. No attempt has been made at compiling a complete bibliography; the references given are preferably recent ones. As shown by the cases discussed below, self-patterning involving chemical reactions takes place in all rock types--sedimentary, metamorphic, and igneous. The cases presented below cover a wide spectrum of size scales, and all appear to involve first-order phase transitions. Some of the cases have been only geological curiosities until now, though to the extent that they are true cases of self-patterning, they illustrate the complexity and variety of the mechanisms operating in rocks and thus should gain importance for geologists.

Because apparent examples of geochemical self-organization cannot as a rule be repeated experimentally, the first difficult task is to establish that they are true cases of self-patterning, that is, cases where the rock or system was indeed unpatterned to begin with. Establishing that a given textural pattern is a case of true self-organization should be done in general by a combination of careful field, petrographic, and chemical observations.

The second task is to choose the mechanisms involved in generating a given case of self-organization so that they are consistent with field evidence and with the overall processes known independently to have operated in the genesis of the encasing rocks.

IRON OXIDE CONCENTRATIONS IN SEDIMENTARY ROCKS

Concentrations of ferric oxide (hematite, goethite) can be found in a variety of sedimentary rocks forming spatial patterns --concentric bands, planar bands, or columns. Hematite concentric bands are often present in chert concretions and nodules in limestones (Fig. 1a), and must then be a case of true self-organization (and one that took place during diagenesis), because the very concretions are themselves of diagenetic origin. Concentric or planar bands of hematite can also occur in clastic rocks such as the Baraboo Quartzite of Wisconsin and the zebra rocks from East Kimberley, Western Australia (Fig. 1b, see details in the caption; see ref. [21a]). In both of these cases the bands cut across stratification and must be true self-patterning, probably caused by a combination of water flow and reaction. Another clear case of (diagenetic) self-organization is the cigar-shaped columns of hematite in the Chino Valley Dolomite of Arizona (Fig. 2), which are normal to bedding and are arranged in a loose hexagonal pattern.

These hematitic patterns are minor features with little geological interest, but their genesis, once well understood, would illustrate the complexity of the kinetic interactions among reaction and transport in rocks, and would lend them considerable geochemical significance. In quantitative, kinetic modelling now in progress on the propagation of redox fronts through a rock we [48a] predict the formation of ferric hydroxide (goethite) bands through the instabilities arising in the interaction of flow of oxygenated water and dissolution/precipitation reactions in sandstones. Mechanisms of this kind may well be responsible for the hematitic patterns described above.

AGATE

Agates are rounded, banded aggregates (∿ 10 cm across) of microcrystalline quartz fibers typically found in the vesicles of rhyolitic or andesitic lavas. Agates have had little geological interest other than mineralogical, but their genesis poses fascinating problems.

Many agates simultaneously display the following associated features: (1) banding with regularly increasing spacing that looks much like Liesegang banding--see Fig. 3; the banding consists of thin brownish ribbons leaving between them colorless bands; both ribbons and bands are made up of quartz fibers oriented at right angles to the bands. All the fibers are elongated across the *c* axis of quartz [22,24]. (2) Systematically, the quartz fibers that make up the colorless bands have grown *twisted*, whereas those making up the brown ribbons are *un*twisted --see Figs. 4,5 and refs. [22,23]; and (3) typically agates have a distinctive chevron interference pattern easily visible under crossed polars. The reason for this is that in any given band of twisted fibers, each one is bodily rotated (around its length) by a constant angle with respect to the fiber immediately to its left. Every ten or twenty fibers the direction of this angle changes. The result is that the locus of points at which the *c*-axis is normal to a thin section (and at which, therefore, the quartz is at extinction viewed between crossed polars) is a narrow band that zig-zags with respect to the bands of fibers (Fig. 5).

Even leaving aside the problem of how and when silica, presumably uniform in concentration in the original lava flow, is "focused" at some points that end up becoming agates, there are basically two incompatible views on the origin of the banding: it results either from successive precipitation (out of groundwater) of silica bands in an empty space, or from Liesegang-type diffusion within an initially uniform, silica-rich lump, gelatinous or not, already accumulated in that space. The first alternative is very unlikely, because it would amount to the great coincidence that the successive layers had *ab-initio* the three textural regularities described above (regularly increasing spacing, bands that consist alternatingly of non-twisted and twisted quartz fibers, and chevron interference patterns). The second alternative, which was already proposed by Liesegang himself [5,6] and, probably independently, by Jones [25], has fascinating, but unexplored, implications on crystal growth, and on control of crystal structure and crystal habit by trace elements.

PRECAMBRIAN BANDED IRON FORMATIONS

The Precambrian banded iron formations are huge, flat accumulations, (∿ 500 km across and ∿ 500 m thick) mainly of silica and iron minerals arranged in bands, present in the Precambrian cores of all continents. They are the world's main ore of iron. The consensus is that they formed by precipitation out of shallow water bodies at earth's surface temperatures, and that somehow their origin appears to be tied to the composition and evolution of the earth's early atmosphere. Beyond this, everything else is controversial: How did such extreme chemical winnowing take place over whole basins up to hundreds of kilometers across? Why did these rocks not form again since the Precambrian? How did they become banded?

No attempt is made here to summarize the enormous literature on Precambrian iron formations. A recent comprehensive statement on their occurrence and origin is for example that of Mel'nik [26]. The one property of iron formations I am discussing here is their banding, which often occurs on two or even three size scales: microbands of submillimeter thickness, mesobands about 1 cm thick (these are the ones after which these formations are called "banded"), and macrobands meters thick. Good illustrations are given in refs. [27,28]. Several ideas exist to account for the banding: (1) Trendall [28] has proposed that there are cyclic periods of microbands, mesobands and macrobands, and that they coincide respectively with yearly, 25-year, and 600-to-1750-year cycles, though no one has explained which mechanisms connected those sun cycles to iron-formation genesis, and no one has clarified why the two longer cycles do not appear to have left their imprint in any other Precambrian or younger rocks. See also [58]. (2) Mel'nik [26] points out the possibility that the banding arose through the mineral-solubility changes associated with the Ostwald ripening, or aging, of mineral grains. (3) A widespread notion is that the banding simply reflects periodic, basin-wide changes in the water chemistry--a view that does actually little more than transfer the banding problem to that of explaining why such periodic changes in water chemistry should have occurred in the first place. (4) A fourth approach to account for the origin of some of the banding (R. M. Garrels, pers. comm., 1983) is that it is caused by the evaporation of one batch of aqueous solution at a time (which left an ABCDE mineral sequence), followed by the coming into the basin of a second batch of solution and by its evaporation (leaving another ABCDE sequence), and so on. The banding produced should then be of the type ABCDEABCDEABCDE..., but only exceptionally of the type ABABAB.... This hypothesis could be supported by stable-isotopic data and by field and petrographic data on banded iron formations that have undergone little metamorphism. (5) A fifth hypothesis is that the banding is not primary at all (that is, it did not arise at the time of deposition of the chemical sediments), but that it came about during diagenesis of chemical sediments that were initially essentially uniform in mineral composition and texture. The banding would have resulted from Liesegang-type diffusion (a possibility supported by experiments [4,29]) or from self-organization driven by Ostwald ripening (a possibility supported by theoretical and experimental work [30,49]) or from a combination of the two.

Leaving aside the huge-scale self-organization implicit in winnowing out just silica and iron from the six or eight most abundant components of the crust, there is therefore the possibility that the banding of iron formations (or at least one of the two or three banding scales often present) constitutes a case of self-patterning. It is a subject of great geological and geochemical importance. Much more research on it (petrographic, chemical, isotopic, field, experimental, and theoretical) would be welcome.

BANDED MISSISSIPPI-VALLEY TYPE ORES

Mississippi-Valley-type ores are economically important concentrations mainly of sphalerite, galena, and fluorite that form in many sedimentary basins during diagenesis at temperatures $< \sim 150^{o}C$, and often in association with dolomitization of the encasing limestone rock. Their genesis poses intriguing problems [e.g., 51,58] regarding metal sources, metal transport, factors controlling the mechanisms and location of precipitation, and the spatial association of the ores with dolomitized limestones.

In some districts [e.g., 52,53] the ores are strikingly banded on a scale of $\sim$ 1 cm (Fig. 6). The bands are sometimes asymmetric, or "polar", and they may consist of different minerals or of different crystal sizes of the same mineral.

There is little question that these alternating bands are a case of self-patterning, because both the identity of the minerals in question and the fact that their crystals are perfectly interlocking indicate that they precipitated _in situ_, and that the bands are not inherited from a preëxisting structure. Based on microscopic evidence, the precipitation of these crystals is thought to have happened some times in voids and other times replacing previously existing crystals of other minerals. The genesis of these bandings is really not understood [51-54].

STYLOLITES

Many sedimentary rocks display stylolites, which "...are partings between blocks of rock which exhibit complex mutual column and socket interdigitation" [31]. They are roughly planar, extend laterally for up to tens of meters, may form any angle to the bedding, and are widespread especially in monomineralic rocks such as limestones, cherts, and quartzites [32]. The interdigitations may be from a few micrometers to several tens of centimeters long, often have striations on their sides, and may or may not be perpendicular to the main plane of the stylolite. It has been clear for a long time [33] that each stylolite is the seam left by pressure dissolution (or dissolution caused by stress) of a region of an already-lithified sediment. Also, because fossil shells can sometimes be seen half dissolved by a stylolite but are otherwise undeformed (Figs. 28-30 in [33]), there is little doubt that the pressure dissolution took place in the <u>elastic</u> range of deformation.

The genesis of stylolites presents at least four intriguing problems: (1) How does a block of monomineralic rock manage to respond to an overall applied stress by generating within itself regions of <u>high</u> solubility (which become stylolites) and regions of <u>low</u> solubility (which become well cemented)? (2) Stylolites often occur at <u>regular</u> distances in a thin section, hand specimen, or outcrop (refs. [34-37] and many more). Another example is shown in Fig. 7. How are <u>regularly-spaced</u> stylolites generated? (3) In all probability, stylolites develop, under the action of stress and dissolution, from point-like textural heterogeneities. How can such punctual regions extend themselves laterally and become planar, even at an angle to the bedding [38]? (4) How are the column-and-socket interdigitations generated?

Stylolitized rocks, therefore, consist of planar, alternating regions of dissolution and precipitation. Since this reorganization took place after lithification and cannot have been inheritied from initial features of the sediment (or else the stylolites would never form at an angle to the bedding), stylolitized rocks constitute a certain case of self-organization. Elsewhere in this volume Ortoleva summarizes a quantitative kinetic theory [19] that can account for this spatial self-organization and that is consistent with isotopic [39] and petrographic [40] observations. This theory is based on a stress-driven feedback between porosity and pore-fluid composition whereby grains in a small region of initially higher-than-average porosity bear a higher-than-average local stress; this increases the local average solubility, which in turn causes the local average porosity to increase even more. (The terms "local" and "small region" refer here to volumes containing, say, 50 to 150 grains. As it stands, therefore, the theory cannot predict the pressure-solution/precipitation behavior of each grain in a rock, but only that of "small regions" several grain-lengths across.) The theory shows also that each local porosity maximum not only self amplifies but also induces a regularly spaced sequence of satellite maxima. This feedback between texture and pore-fluid composition is probably also the one involved in driving metamorphic differentiation and in particular differentiated layering (see below and refs. [17,20]). Predictions (based on the kinetic theory and on transport, thermodynamic and textural data) of the time of formation and spacing of stylolites are respectively 6000 years to 1.7 million years and 0.03 to 3 cm for quartzites and cherts, and 1 day-10^9 years and 0.001-50 cm for limestones.

The order-of-magnitude agreement between the predicted spacings and observed ones is good, for both limestones and quartzites. The order of magnitude of the time of formation of stylolites in quartzites appears reasonable. The range of predicted stylolite formation times for limestones is extremely wide because both the solubility and the dissolution rate constant for calcite may vary over many orders of magnitude depending on pH, P_{CO_2}, ionic strength, and inhibitors; at its lower end, the predicted formation time, even if it is too low by two or three orders of magnitude, points to the possibility of experimental production of stylolites in reasonable times.

DIFFERENTIATED LAYERING IN METAMORPHIC ROCKS

Metamorphic rocks form by heating (without melting) and deformation of preëxisting sedimentary or igneous rocks. Most metamorphic rocks the world over are immediately recognizable in the field by a conspicuous, roughly regular, light and dark mineral banding known, depending on the scale, as cleavage, schistosity, or foliation (see for example p. 229 in ref. [41]). This banding can often be shown, by one or another of the criteria listed below, not to have been inherited from the preëxisting rock, but to have developed by chemical differentiation. It therefore constitutes an excellent case of self-patterning. Criteria to recognize differentiated layering [17] are: (1) the banding cuts across bedding of the original sedimentary rock [42,21] and Fig. 8; (2) the banding cuts across mineral grains that are themselves metamorphic [43]; (3) the banding has a scale that is independent of the scale of bedding of the parent rock, which may be inferred from its chemical composition [44]; (4) individual bands that, on close examination, turn out to be very flat, discontinuous lenses [44]; (5) individual bands that do not correspond chemically to the composition of any ordinary sedimentary rock [45]; and (6) "microscopic examination may prove that a layered structure closely resembling bedding in the field crosses an earlier _s_-surface itself characterized by layering on a fine scale" [44].

The bands are usually alternatingly light (quartz and feldspars) and dark (layer silicates, amphiboles, pyroxenes, garnet, etc.), and often display preferred orientation of mineral grains. Because metamorphic rocks display abundant field and petrographic evidence of deformation, metamorphic banding has always been attributed to stress or to the strain produced by it.

We have proposed recently a quantitative kinetic theory able to account for the generation of evenly-spaced differentiated layering through the combined action of dissolution, diffusion of solute species along grain boundaries, and reprecipitation; the dissolution can be driven directly by applied stress [17] or by varying concentrations of dislocations, themselves caused by the applied stress (ref. [20] and article by Ortoleva in this volume). The stress both drives the changes in local texture and is modified by them. Several kinds of feedback between the rock's local texture (taken to be a collection of variables including the size, shape, orientation, number density, and dislocation density of all minerals) and the intergranular-fluid composition are described in [20].

Much work is needed to elucidate the effects on metamorphic differentiation of rock plastic flow and fluid infiltration (see section on skarns below). Reaction rates and diffusion coefficients are largely unknown for geological systems under metamorphic conditions (see Brady [46]). Experimental generation of mineral bands out of initially uniform aggregates of two or more kinds of crystals (even if these are not rock-forming minerals but substances with faster dissolution/precipitation kinetics) should be attempted.

SKARNS

Skarns are rocks produced by the chemical interaction (at hundreds of degrees C and 1 or 2 kb) of limestones or marbles with aqueous solutions that are thought to flow through them, introduce in them new components (mainly Si, Al, Fe and Mg), and often produce in them striking sets of alternating mineral bands. This is in contrast with metamorphic differentiation, where (see above) the banding is thought to come about by diffusional redistribution of the initial components, without flow and without introduction of other components.

The mineral bands often found in skarns (see for example [16,47,48] and Fig. 9) are alternatingly light (plagioclase, garnet, fluorite, etc.) and dark (hornblendes, pyroxenes, magnetite, vesuvianite, quartz, etc.); have spacings and thicknesses on the order of 0.1 to 1 cm; and can be planar or spherical. Banded skarns are another sure case of geochemical self-organization.

In quantitative kinetic modelling work in progress on flow/reaction situations we [48a] predict the formation of mineral bands, a result that máy be significant in regard to the genesis of skarns. Needed are more theoretical studies of coupled reaction and flow of (sub or supercritical) aqueous fluids through porous media, experimental measurement of relevant physical and chemical parameters under skarn-forming conditions, and additional detailed mineralogic studies of skarn occurrences.

LAYERED IGNEOUS INTRUSIONS

The layering of large igneous intrusions such as the Skaergaard Complex of Greenland, the Bushveld Complex in South Africa, and the Stillwater Complex of Montana is in some places perfectly repetitive and displays strikingly constant spacing (∿ 1 cm) over large horizontal extents as well as a mosaic pattern within the plane of individual layers; see, for example [13,18] and refs. therein and also the article by Boudreau in this volume. The common explanation for this layering is based on settling of crystals through the melt combined with intermittent gravity currents over the floor of the magma chamber. Many observations, however, contradict these ideas [13] and, furthermore, there is theoretical [49,30,18,13] and experimental [50] work suggesting that the bands were produced by double-diffusion in a compositionally and thermally inhomogeneous magma chamber and/or by competitive particle-growth mechanisms active during Ostwald ripening of crystalline aggregates. It seems clear that these rocks constitute outstanding examples of self-patterning involving first-order phase transitions, and that the convergence of theoretical, experimental, mineralogical and field work is still in the future.

ORBICULAR ROCKS

Igneous rocks such as granites and diorites (and even more basic plutonic rocks) sometimes contain ovoid bodies called orbicules (see Fig. 10 and plate 10B in [13]), which are typically several centimeters to as much as one meter in diameter and consist of concentric, alternating, light-and-dark mineral shells. In some cases at least, such bodies seem to have formed by rhythmic selective crystallization of minerals around a nucleus of foreign rock [47, p. 163]. Light colored shells consist predominantly of crystals of plagioclase along with a few crystals of a dark mineral like hornblende or biotite; dark-colored shells consist predominantly of biotite or hornblende with minor plagioclase. The distribution of any given mineral is not necessarily symmetric about the center of each shell: often, in fact, the volume fraction profile for the dark mineral along a radius is often sharp on the outside but tails off on the inside of the shell. It has been proposed [13] that orbicules form through the counter-diffusion of heat and silicate components under conditions of rapid growth. See also [7,55,56]. Orbicules are certain cases of self-organization and deserve more theoretical, petrographic, chemical, and field work.

OSCILLATORY BANDING IN PLAGIOCLASE CRYSTALS

Plagioclase feldspar is one of the most abundant minerals in the earth's crust, and can form under a broad range of conditions in sedimentary, metamorphic, and igneous rocks and from aqueous solutions or melts. It constitutes a complete solid-solution between the end members albite ($NaAlSi_3O_8$) and anorthite ($CaAl_2Si_2O_8$). Plagioclase displays interesting properties in its crystalline structure, twinning, ion ordering, absorption, and optical behavior [59,60]. Also, some igneous plagioclase crystals are often zoned in composition. This zoning can be normal (= monotonic variation from anorthitic cores to albitic rims), reverse (albitic cores to anorthitic rims), and, of direct interest here, oscillatory (see Fig. 11).

In oscillatory zoning, which appears to be limited to intermediate igneous rocks, the mole fraction of the anorthite end-member goes repeatedly up and down from core to rim. The amplitude of these spatial compositional oscillations is generally between 5 and 15 mole percent, and the wavelength is between 10 and 100 μm [14,59]. Recently, a quantitative kinetic theory has been proposed [14; see also Ortoleva's article in this volume] and enlarged upon [15,57] to explain the genesis of oscillatory zoning through the interaction of crystal growth with diffusion of ions from the bulk melt to the region of melt from which the crystal directly grows. In the theory [14], the rate of attachment of "molecules" of each end-member to the surface of the growing plagioclase crystal is taken to depend on the composition of the adjacent melt.

SUMMARY

I have briefly described above ten good examples, some certain and some probable, of geochemical self-organization in sedimentary, metamorphic, and igneous rocks. The order of magnitude of their length scale is most commonly in the range 10^{-2} to 10^{2} cm. They all appear to involve first-order phase transitions. Much more work is needed on all the cases mentioned here, especially combining geochemical, mineralogical and field evidence with dynamical modeling of the mechanisms deemed to be appropriate in each case. Quantitative dynamical theories of this kind are available only for a few of these instances of self-organization.

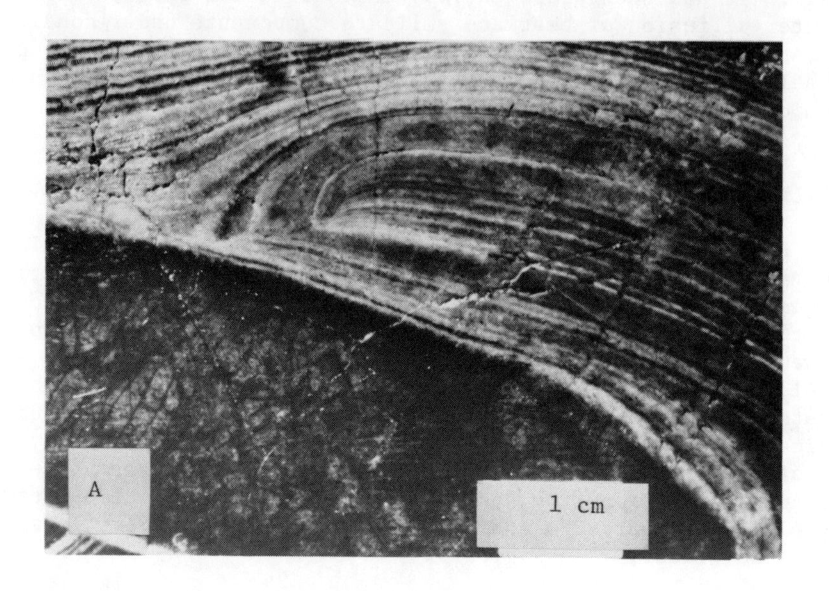

Various spatial patterns of ferric oxide.

Figure 1.A Concentric bands of hematite in a chert nodule in the Madison Group Limestone of Montana. The bands are ~ 1 mm thick. The nodule (~ 40 cm across and ovoid in shape) consists mostly of chert, which is an aggregate of microcrystalline quartz chemically precipitated in situ from aqueous solutions, at temperature $< \sim 50°C$. The chert nodule photographed here precipitated not in a void, but occupying the space formerly occupied by limestone. The hematite-bearing layers (dark) contain actually only a few percent of scattered hematite in the form of crystals $\ll 1\ \mu m$ in size.

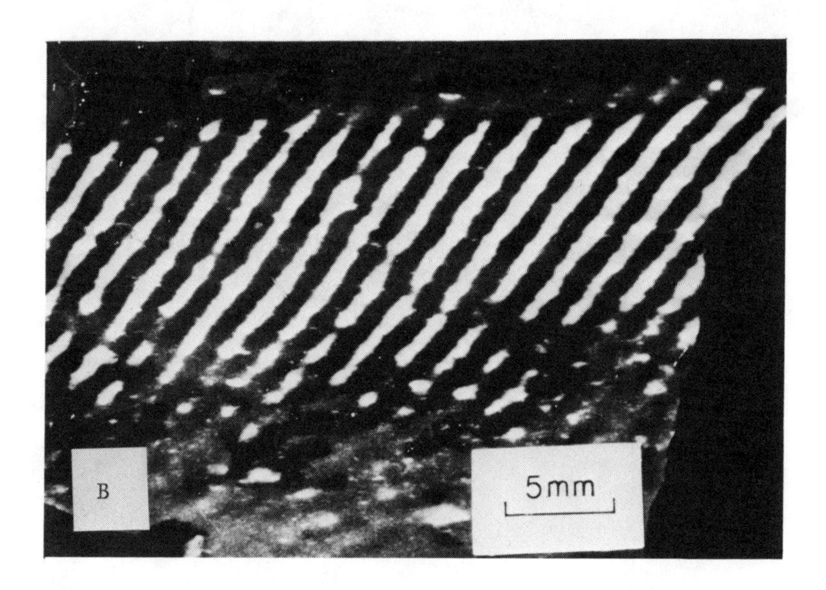

Figure 1.B Zebra mudstone from Argyle Downs Station, East Kimberley, Western Australia. See [21a]. The dark bands are portions of the mudstone containing 5-6% chemically-precipitated hematite. The light bands (∿5 mm thick) contain only ∿1% hematite. In other places in the area the rock is not banded but spotted : "As one moves farther south-west these spots become more numerous and coalesce. Most stages can be found between the spotted rock and the well known striped red and white rock." [21a, p. 57]. Sample provided by A.F. Trendall, Department of Mines, Perth, Western Australia.

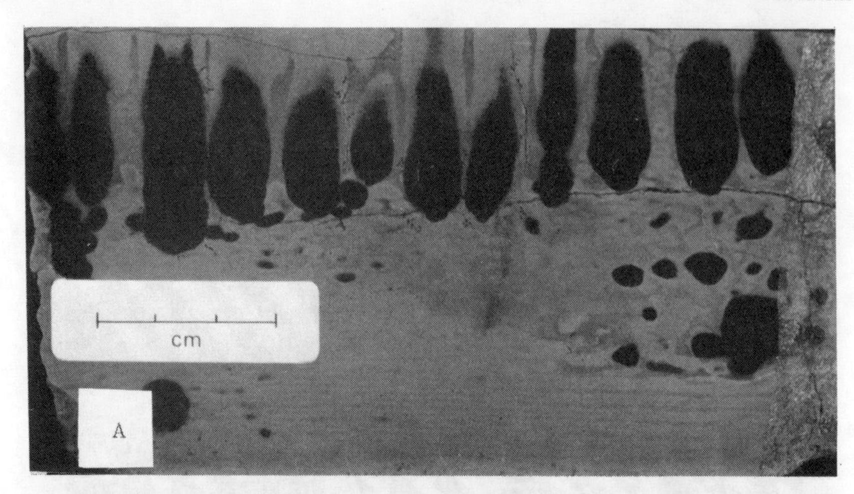

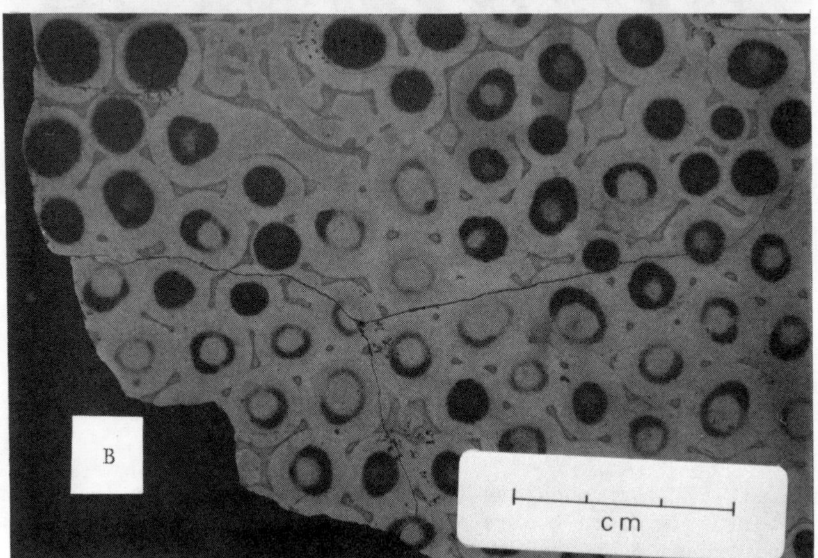

Figure 2. Hematite columns in the Chino-Valley Dolomite of central Arizona. The columns are cigar shaped, ∿ 0.5-1 cm thick, 1-1.5 cm apart, and perpendicular to the stratification, and they occur near the top of the formation, just under an unconformity, and a few centimeters above a region of the formation containing dozens of evenly spaced bands parallel to the stratification. Ferric oxide concretions are also present. The columns contain ∿20% hematite. The columns, bands, and ferric oxide concretions are all chemically precipitated in situ from aqueous solutions at temperatures < 50°C. Samples and description provided by Richard Hereford, U.S. Geological Survey, Flagstaff, Arizona.

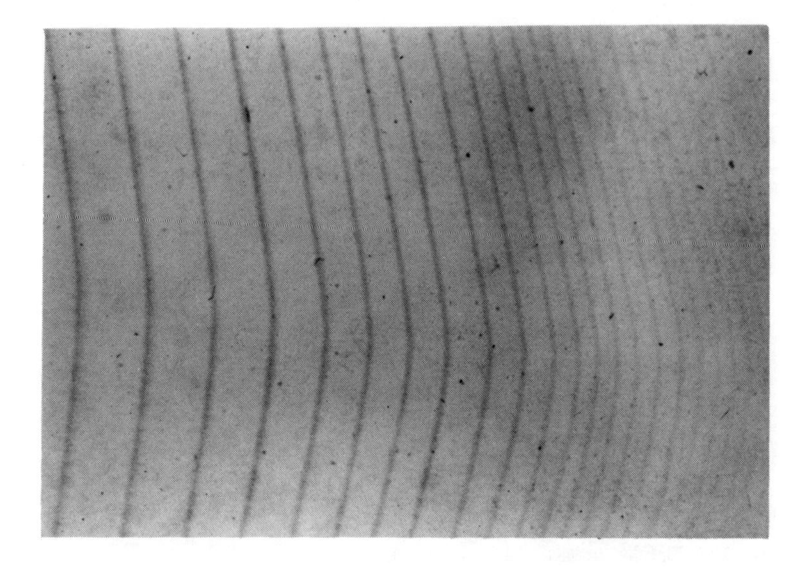

Figure 3. Banding in an agate from Chihuahua, Mexico. Both the dark, thin bands and the light spaces between them consist of small quartz fibers oriented at right angles to the bands, and with their c-axes across their lengths. The dark bands are here 0.03 to 0.07 mm wide. The quartz fibers are also about 0.03 to 0.07 mm long. The spacing increases evenly from ∿0.1 mm to 1 mm or more, and, aside from looking like Liesegang banding, coexists with the two regularities illustrated in Figs. 4 and 5. View under the plane-polarized light. Width of photograph ∿2 mm.

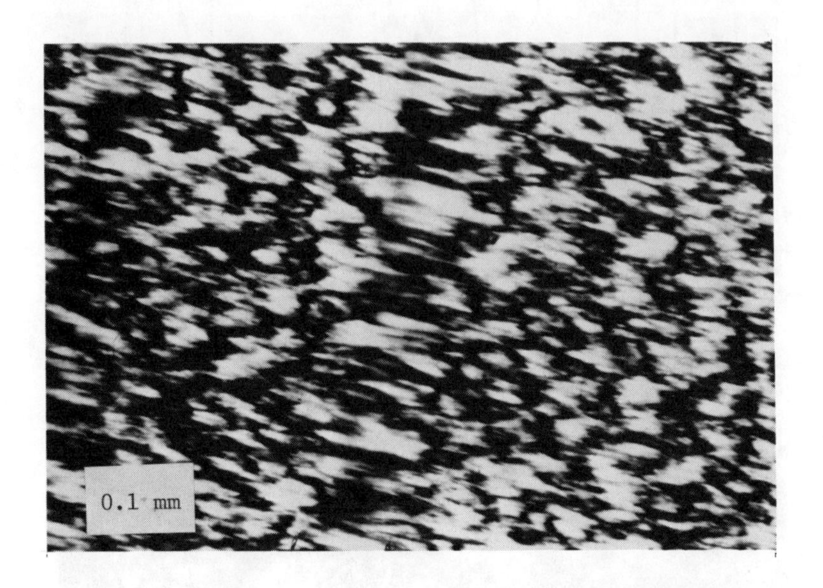

Figure 4. Same agate of Fig. 3, viewed under higher power and between crossed pollars. Running up and down through the center of the picture is one of the thin bands that appear dark in Fig. 3. The thin band is seen here to consist of untwisted quartz fibers -- untwisted because the birefringence is ∿ constant along the length of each fiber. The two regions to the right and left (which correspond to the light bands on Fig. 3) consist of fibers that have grown twisted, as shown by the continuous change in birefringence along any one quartz fiber. The chevron interference pattern visible especially in the lower right portion of the picture (see also Fig. 1 in ref. [22]) comes about because adjacent twisted fibers are oriented forming a given angle (around their axis) with respect to each other -- see text and Fig. 5.

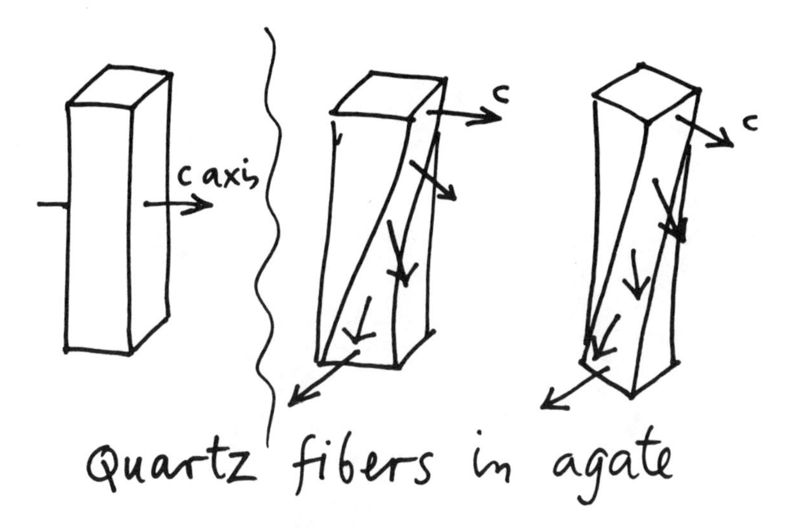

Figure 5. Sketch of some crystallographic characteristics of quartz fibers in the agate of Fig. 3 and 4.

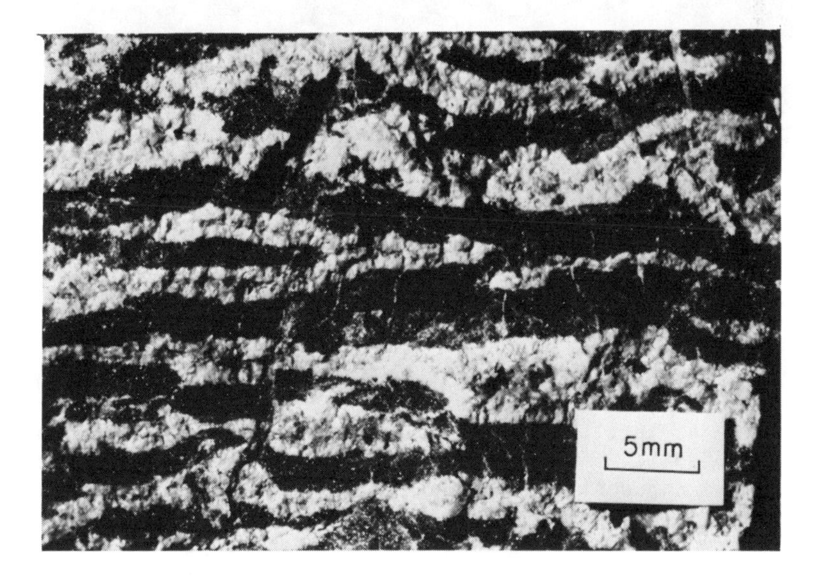

Figure 6. Centimeter-scale banding in a Mississippi-Valley type or from Orgiva, southern Spain. Both the dark and light bands shown here consist of dolomite (the dark bands contain also a small amount of scattered iron-oxide particles). The main difference between the bands is the dolomite crystal size -- large ($\sim$1 mm) in the light bands, small ($\sim$0.1 mm) in the dark ones. Examples of interbandings of two or more minerals are given in refs. [52,53], among others.

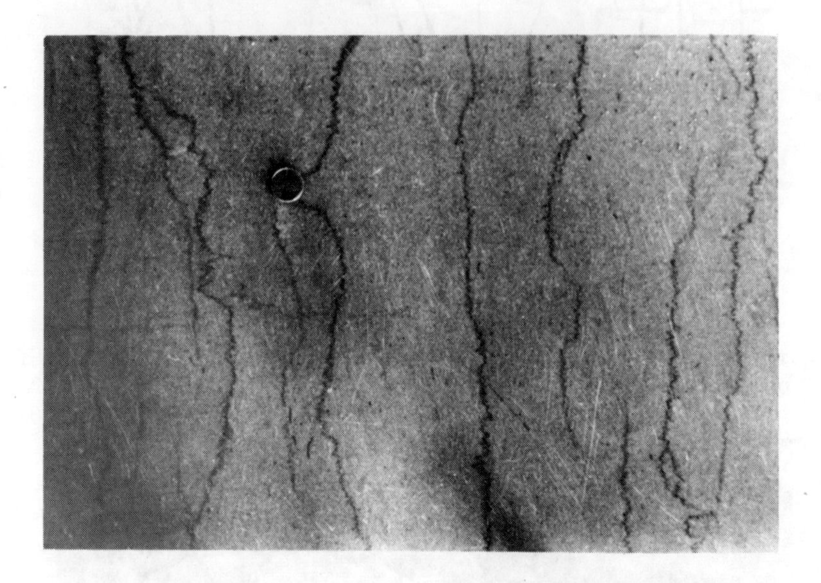

Figure 7. Evenly-spaced stylolites in the Salem Limestone of southern Indiana. Each stylolite is the seam of insoluble particles left by the stress-dissolution of a certain thickness of calcium carbonate. In contrast, in the region between any two adjacent stylolites, precipitation of calcite in pores has taken place rather than dissolution. A texture/pore-fluid chemistry feedback model has been proposed [19] to account for the generation of sets of stylolites in sedimentary rocks. Coin is 2 cm across.

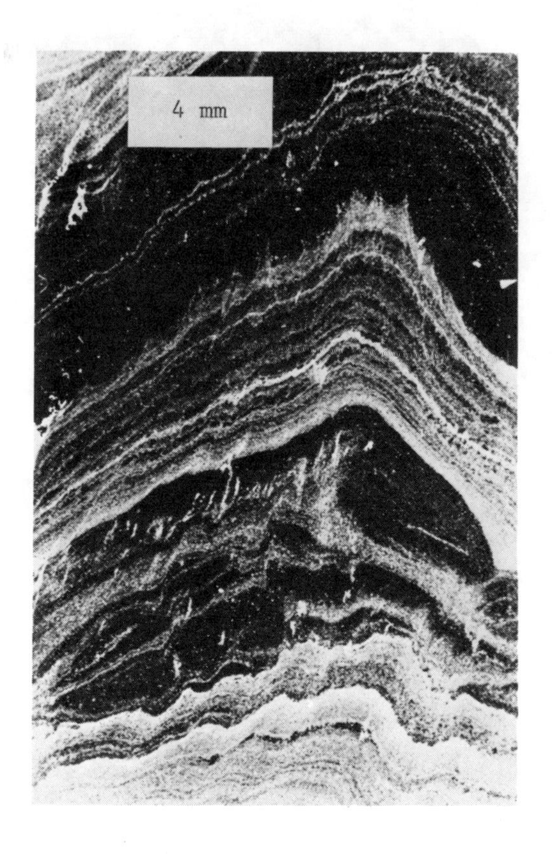

Figure 8. Incipient metamorphic banding (here approximatively vertical) forming a large angle to the sedimentary stratification (here running sideways) in the Sandsuck Formation, Ocoee Gorge, Ducktown, Tennessee. The light beds are rich in mica and the dark beds are rich in quartz (the photograph is a negative print). The metamorphic bands are here quite thin (and therefore called cleavage), run up and down, are seen best in the lower half of the photograph, and consist mostly of mica and chlorite, in places emphasized by bands of associated opaque minerals (here seen as white stringers). The large white patches are pyrite [21]. Photograph and description generously provided by S. Sutton, Univ. of Cincinnati, Ohio.

Figure 9. Skarn from San Leone, Sardina [16]. The light bands, 1 to 2 mm thick, consist of andraditic garnet (calcium and ferric-iron bearing). The dark ones, 5 to 8 mm thick, consist of magnetite and quartz. Photograph and description generously provided by Bernard Guy of the School of Mines of Saint-Etienne, France. White rectangle is 1 cm long.

Figure 10. Orbicular diorite from Epoo, Finland. The concentric shells are alternatively richer in biotite (dark) and plagioclase (light). The radius of the orbicule is ∿ 10 cm.

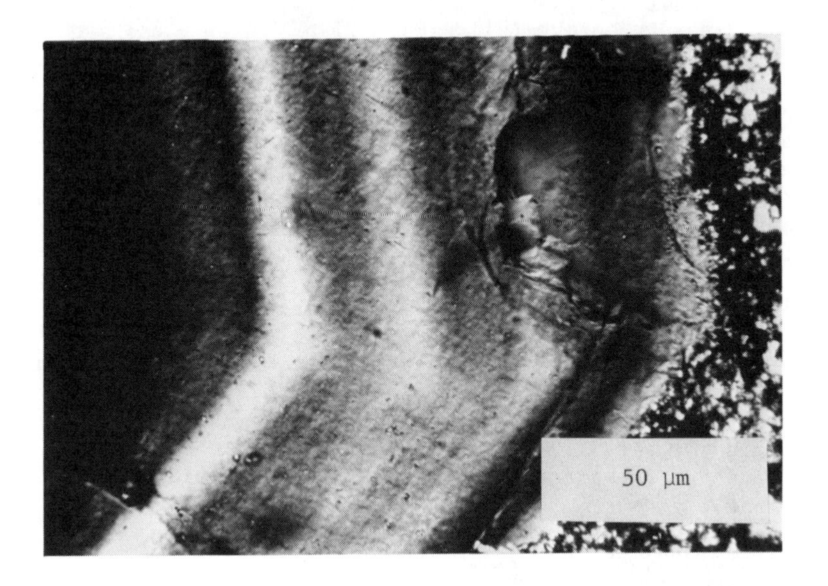

Figure 11. Plagioclase crystal approximatively 0.5 mm across displaying oscillatory zoning ; the concentric layers are alternatively richer (dark) and poorer (light) in anorthite. View between crossed polars.

REFERENCES

1. Hedges, E.S. and Myers, J.E. 1922, The problem of physico-chemical periodicity. New York, Longmans, 95 p.
2. Davis, E.F. 1918, Univ. California, Department of Geology Bull. 11, pp. 235-432.
3. Bailey, E.H., Irwin, W.P. and Jones, D.L. 1964, Calif. Division Mines & Geology Bull. 183, 177 p.
4. Hartman, R.J. and Dickey, R.M. 1932, Jour. Phys. Chem. 36, pp. 1129-1135.
5. Liesegang, R.E. 1913, Geologische diffusionen. Dresden and Leipzig, Steinkopff, 180 p.
6. Liesegang, R.E. 1915, Die Achate. Dresden and Liepzig, Steinkopff, 118 p. (Also, Chemie d. Erde 4, pp. 526-528.)
7. Knopf, A. 1908, U.S. Geol. Survey Bull. 358, 71 p.
8. Ostwald, W. 1925, Kolloid Z. 36, pp. 380-390.
9. Prager, S. 1956, Jour. Chem. Phys. 25, pp. 279-283.
10. Nicolis, G. and Prigogine, I. 1977, Self-organization in non-equilibrium systems. New York, Wiley, 491 p.
11. Ortoleva, P. 1978, In Eyring, H. (ed.), Theoretical Chemistry v. 4, New York, Acad. Press, pp. 235-286.
12. Ross, J. 1976, Berichte der Bunsengesellschaft physikalische chemie 80, pp. 1112-1125.
13. McBirney, A.R. and Noyes, R.M. 1979, J. Petrology 20, pp. 487-554.
14. Haase, C.S., Feinn, D., Chadam, J. and Ortoleva, P. 1980, Science 209, pp. 272-274.
15. Allègre, C.J., Provost, A. and Jaupart, C. 1981, Nature 294, pp. 223-228.
16. Guy, B. 1981, C.R. Acad. Sci. Paris 292, Série II, pp. 413-416.
17. Ortoleva, P., Merino, E. and Strickholm, P. 1982, Amer. J. Science 282, pp. 617-643.
18. Boudreau, A.E. 1982, Fine-scale layering in the Stillwater Complex, Montana. Univ. of Oregon, Master's thesis, 67 p.
19. Merino, E., Ortoleva, P. and Strickholm, P. 1983, Generation of evenly-spaced pressure-solution seams during late diagenesis: a kinetic theory. Contrib. Mineral. Petrology, in press.
20. Strickholm, P., Ortoleva, P. and Merino, E. 1983, The self-organization of differentiated layering. Submitted to J. Geophysical Research.
21. Sutton, S. 1983, Metamorphic differentiation in slates of the Ocoee Gorge, Tennessee. Ph.D. thesis, Univ. of Cincinnati (in progress).
21a. Hobson, R.A. 1930, Jour. Roy. Soc. Western Australia 16, pp. 57-70.
22. Frondel, C. 1978, American Mineralogist 63, pp. 17-27.
23. Frondel, C. 1982, American Mineralogist 67, pp. 1248-1257.
24. Milliken, K.L. 1979, J. Sediment. Petrology 49, pp. 245-256.

25. Jones, F.T. 1952, American Mineralogist 37, pp. 578-587.
26. Mel'nik, Y.P. 1982, Precambrian banded iron formations. Developments in Precambrian Geology 5, Amsterdam, Elsevier, 310 p.
27. James, H.L. and Sims, P.K. 1973, Econ. Geol. 68, pp. 913-914.
28. Trendall, A.F. 1972, Geol. Soc. Australia 19, pp. 287-311.
29. Grubb, P.L.C. 1971, Econ. Geology 66, pp. 281-292.
30. Feeney, R., Schmidt, S., Strickholm, P., Chadam, J. and Ortoleva, P. 1983, Periodic precipitation and coarsening waves: applications of the competitive particle growth model. Jour. Chem. Phys., in press.
31. Dunnington, H.V. 1967, Seventh World Petroleum Congress Proceedings (Mexico) 2, pp. 339-352.
32. Pettijohn, F.J. 1957, Sedimentary Rocks, New York, Harper and Row (2nd ed.), 718 p.
33. Stockdale, P.B. 1922, Indiana University Studies 9, 97 p.
34. Alvarez, W., Engelder, T. and Lowrie, W. 1976, Geology 4, pp. 698-701.
35. Bathurst, R.G.C. 1975, Carbonate sediments and their diagenesis, Elsevier Develop. Sedimentology 12 (2nd ed.), 677 p.
36. Tremolières, P. and Reulet, J. 1978, Rev. Inst. Francais du Pétrole 33, pp. 331-348.
37. Heald, M.T. 1955, J. Geology 63, pp. 101-114.
38. Fletcher, R.C. and Pollard, D.D. 1981, Geology 9, pp. 419-424.
39. Hudson, J.D. 1975, Geology 3, pp. 19-22.
40. Wong, P.K. and Oldershaw, A. 1981, J. Sediment. Petrology 51, pp. 507-520.
41. Hobbs, B.E., Means, W.D. and Williams, P.F. 1976, An outline of structural geology. New York, Wiley, 571 p.
42. Williams, P.F. 1972, Amer. J. Science 272, pp. 1-47.
43. Vidale, R. 1974, In Hoffman, A.W. and others (eds.), Geochemical transport and kinetics, Washington D.C., Carnegie Inst. Washington, pp. 273-286.
44. Turner, F.J. and Weiss, L.E. 1963, Structural analysis of metamorphic tectonites, New York, McGraw-Hill, 545 p.
45. Orville, P.M. 1969, Amer. J. Science 267, pp. 64-86.
46. Brady, J.B. 1978, Carnegie Inst. Washington Year Book 78, pp. 577-580.
47. Williams, H., Turner, F.J. and Gilbert, C.M. 1982, Petrography (2nd ed.), San Francisco, Freeman & Co., 626 p.
48. Burt, D. 1974, In Hoffman, A.W. and others (eds.), Geochemical transport and kinetics, Washington D.C., Carnegie Inst. Washington, pp. 287-293.
48a. Moore, C.H., Hettmer, J., Auchmuty, G.J., Ortoleva, P. and Merino, E. 1983, Mineral bands produced by redox-front propagation in sandstones: flow/reaction instabilities. (In preparation)
49. Feinn, D., Ortoleva, P., Scalf, W., Schmidt, S. and Wolff, M. 1978, Jour. Chem. Phys. 69, pp. 27-39.

50. Chen, C.F. and Turner, J.S. 1980, Jour. Geophys. Research 85, pp. 2573-2593.
51. Ohle, E.L. 1980, Econ. Geol. 75, pp. 161-172.
52. Fontboté, L. and Amstutz, G.C. 1982, In Amstutz, G.C. and others (eds.), One genesis--The state of the art, Heidelberg, Springer, pp. 83-91.
53. Fontboté, L. and Amstutz, G.C. 1980, Rev. Inst. Investigaciones Geológicas (Diputación Provincial, Univ. Barcelona) 34, pp. 293-310.
54. Beales, F.W. and Hardy, J.L. 1980, Soc. Econ. Paleont. Mineralogists, Special Publication no. 28, pp. 197-213.
55. Enz, R., Kudo, A.K. and Brookins, D.G. 1979, Geol. Soc. America Bull. II, 90, pp. 349-380 (also ibid. I, 91 (1980) pp. 246-247).
56. Thompson, T.B. and Giles, D.L. 1974, Geol. Soc. America Bull. 85, pp. 911-916 (and ibid. I, 91 (1980) pp. 245-246).
57. Lasaga, A.C. 1982, Amer. J. Science 282, pp. 1264-1288.
58. Maynard, J.B. 1983, Geochemistry of sedimentary ore deposits. New York, Springer, 305 p.
59. Smith, J.V. 1974, Feldspar minerals, New York, Springer, 690 p.
60. Ribbe, P.H., editor 1983, Feldspar mineralogy. Mineralogical Soc. America Short Course Notes vol. 2 (2nd ed.), in press.

ACKNOWLEDGEMENTS

It is a pleasure to thank Sally Sutton, Bernard Guy, Rafael Fenoy, Richard Hereford, and C. Stephen Haase for supplying some of their samples and photographs, and Rodney Hackler, Karen Walker, and George Ringer for their help.

MODELING NONLINEAR WAVE PROPAGATION AND PATTERN FORMATION AT GEOCHEMICAL FIRST ORDER PHASE TRANSITIONS

Peter J. Ortoleva

Departments of Chemistry and Geology
Indiana University, Bloomington, Indiana 47405 and
Geo-Chem research Assoc.,
700 North Walnut St., Bloomington, Indiana 47401

From among the many beautiful patterns of mineral distribution, a number have been modeled via equations of phase transition kinetics and transport. Mathematical models for metamorphic layering and stylolites will be presented and their predictions summarized. Nonlinear redox waves and deposition, and porosity instabilities in aquifers are also modeled and discussed.

I. OVERVIEW

The many beautiful examples of geochemical pattern formation presented by E. Merino (see these proceedings) underscore the breadth of phenomena in these systems and the interest in modeling them. Here we shall discuss the mathematical modeling of some of these phenomena and, in addition, the modeling of a number of nonlinear reaction-transport waves in porous media flow-reaction problems. All these phenomena have the common feature that they arise out of the coupling of the kinetics of first order phase transitions to transport. Interest in these phenomena is not new but dates past the beginning of this century - see [1] and [2] for early citations.

There are a number of routes to self organization in reaction transport systems. The types of linear instability resulting in the monotonic growth of perturbations has been discussed [3] and are summarized in Fig. 1. Shown are four cases of the dependence of stability eigenvalues for perturbations, of wave vector k, from the uniform state. The first two cases were distinguished in [4] where the extrinsic type of instability was introduced. In the intrinsic case patterns arise at a well defined wave vector

G. Nicolis and F. Baras (eds.), Chemical Instabilities, 329–340.

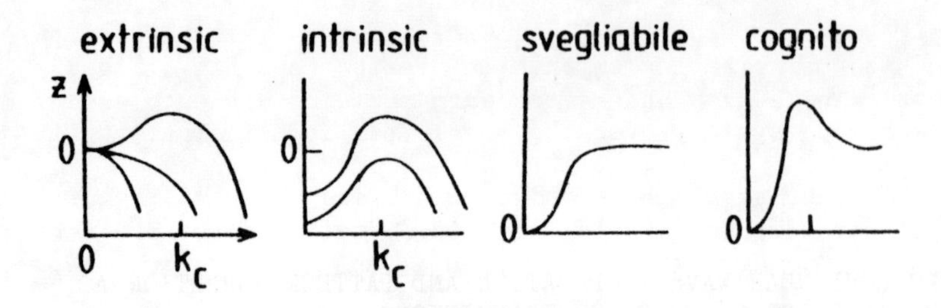

Figure 1.

as some system parameter passes through the bifurcation point. In these cases the pattern lenght is on the order of $2\pi/k_c$ (k_c being the wave vector of the perturbations that first become unstable) and hence on a scale inherent to the reaction-transport parameters. For the extrinsic case the first mode to become unstable is that of zero k, i.e. infinite wave lenght. Hence the bifurcation parameter must be increased further until the point where the eigenvalue passes the zero is at a k value such that $2\pi/k$ is on the order of magnitude of the dimensions of the system. Then patterns of this scale can emerge. Thus pattern formation is extrinsically length scaling systems, the pattern lenght is determined by the system size and not any inherent length scale in the reaction-transport parameters.

Of greatest relevance to the question of pattern formation in precipitating systems are the cases of svegliabile and cognito behavior. In both cases the stability eigenvalue approaches a constant as $k \to \infty$, i.e. as the perturbation wave length goes to zero. This zero wave length instability is due to the presence of variables (such as crystal size) which do not diffuse in these problems. Thus unlike the usual reaction-diffusion case where all eigenvalues are proportional to $-k^2$ as $k \to \infty$, there is a growing mode as $k \to \infty$. This is a reflection of the presence of single crystal instabilities in these systems such that a given crystal, favorably "disposed", may grow at the expense of the other crystals in the system. For the case of the svegliabile instability the $k \to 0$ modes are in fact the fastest growing. How then can such a system ever display banding on a macroscopic lenght scale ? It was shown for the Liesegang ring system (see the other article by the author in these proceedings) that a svegliabile system subject to a reactive boundary condition will have macroscopic (small k) modes excited much more strongly than the single particle (large k) modes. Hence such a system can be awoken to manifest its potentiality for macroscopic banding ("svegliabile" means "capable of being awoken" in Italian). It is found that once the macroscopic pattern perturbations are excited, the nonlinearities of the problem seem to lock in the linear patterns.

In the case of cognito (legal Italian for "well known") patterning, the perturbation exponential factor (see Fig. 1) has a maximum at finite k. Thus macroscopic patterns win out, even for the homogeneous system, over the single particle instability.

Both cases, svegliabile and cognito, are observed in our theory of metamorphic layering and stylolization to be presented here. Liesegang banding is discussed in the other article by the the author in these proceedings.

II. GENERAL REMARKS ON PRESSURE AND DEFECT DENSITY SOLUBILIZATION

The solubility or free energy of a grain depends on its state of stress and the density of crystal defects. These dependencies can provide the basis of a feedback loop that can destabilize the state of uniform texture of a rock under stress.

A grain in a rock under overall stress experiences stresses that depend on the distribution and mechanical properties of the surrounding grains. This stress, a function of the local environment, will then effect the grain's solubility (pressure solubilization) and hence we arrive at the "mean field" conjecture that for the purpose of a macroscopic theory there exists a functional relationship between a grain's molar free energy μ and the surrounding "texture", $\underset{\sim}{T}$ - i.e. the distribution of sizes, orientations and shapes of all mineral grains in the vicinity of the given grain. The dependence $\mu[\underset{\sim}{T}]$ is a functional one - μ depends not only on $\underset{\sim}{T}$ at the location of the grain but also on the distribution of $\underset{\sim}{T}$ in the vicinity of the grain. To set forth a theory of pressure solution in rocks we must determine the properties of the functional $\mu[\underset{\sim}{T}]$.

Similarly the rate of change of the defect density depends on the stresses experienced by a grain and, in the mean field theory, these are a functional of the texture.

One expects these functional relationships to be short range - i.e. they only depend on the immediate neighborhood of the grain of interest. We expect that the effect of the next "coordination layer" of grains only acts in a very average way via the averaged stress field of the surrounding visco-elastic average medium which, in turn, is related to the flow velocity $\vec{v}$ of the rock. All these couplings are clearly complicated and do not lend themselves to simple formulae for the functional relation between μ and $\underset{\sim}{T}$ and $\vec{v}$.

The Curie principle is a symmetry rule by which one can relate things of one tensorial type to another [5,6]. For example the rate of reaction in a gas cannot depend linearly on the gradient of temperature - if it did then the contribution of the gradient to the rate would change sign upon a change of the reference frame so that all coordinates change sign. Thus the Curie principle limits the functional form of linear laws. (Readers not interested in mathematical details may proceed to the last paragraph of this section.)

For example let us investigate the change in μ , $\delta\mu$, due to a change of texture $\delta\underset{\sim}{T}$ from uniformity and a change of flow $\delta\vec{v}$ from some flow $\vec{v}$ of the uniform rock. Then to linear order in $\delta\underset{\sim}{T}(\vec{r})$, $\delta\vec{v}(\vec{r})$ (repressing time for the moment) we have

$$\begin{aligned}\delta\mu(\vec{r}) &= \int d^3r' \, \underset{\sim}{K}(\vec{r},\vec{r}') * \delta\underset{\sim}{T}(\vec{r}') \\ &+ \int d^3r' \, \vec{Q}(\vec{r},\vec{r}') \cdot \delta\vec{v}(\vec{r}')\end{aligned} \tag{1}$$

where $\underset{\sim}{K}$ is a set of kernels and $\underset{\sim}{K} * \underset{\sim}{T}$ reflects a sum over all textural variables and $\vec{Q} \cdot \vec{v}$ is a simple vector dot product of $\vec{Q}$ with $\delta\vec{v}$. The integrals over $\vec{r}'$ reflect the nonlocality of the relationships - i.e. μ depends on the texture and velocity profiles and not only on the texture and velocity at the grain itself. To illustrate how the Curie principle can yield valuable information about $\delta\mu$ let us neglect the velocity effects for the moment, i.e. $\vec{Q} = \vec{0}$ and $\underset{\sim}{K}$ does not depend on the presence of the overall flow $\vec{v}$ of the uniform rock. Then if $\delta\mu$ really only depends on the behavior of $\delta\underset{\sim}{T}$ in the vicinity of the location of the grain at $\vec{r}$, then $\underset{\sim}{K}$ decays rapidly in $|\vec{r} - \vec{r}'|$ and hence we can expand the $\vec{r}'$ dependence about the values of quantities near $\vec{r}$, i.e. put $\delta\underset{\sim}{T}(\vec{r}') = \delta\underset{\sim}{T}(\vec{r}) + \vec{\nabla}\delta\underset{\sim}{T}(\vec{r})\cdot(\vec{r}' - \vec{r}) + \cdots$.
If $\delta\underset{\sim}{T}$ does not vary too rapidly in space we may truncate this gradient expansion to some low order. Substitution into (1) and carrying out the integrals over $\vec{r}'$ we obtain

$$\delta\mu = \underset{\sim}{A}*\delta\underset{\sim}{T} + \sum_{\alpha=1}^{3} \underset{\sim}{B}_\alpha * \frac{\partial\delta\underset{\sim}{T}}{\partial r_\alpha} + \sum_{\alpha,\beta=1}^{3} \underset{\sim}{C}_{\alpha\beta} * \frac{\partial^2\delta\underset{\sim}{T}}{\partial r_\alpha \partial r_\beta} + \cdots \tag{2}$$

where the $\underset{\sim}{A}$, $\underset{\sim}{B}_\beta$, and $\underset{\sim}{C}_{\alpha\beta}$ quantities are appropriate integrals or moments of $\underset{\sim}{K}$ over $\vec{r}'$ and $\partial/\partial r_\alpha$ represents a partial derivative with respect to the α-th spatial direction r_α. Since $\delta\mu$ for a grain at $\vec{r}$ is independent of an arbitrary rotation of coordinates about $\vec{r}$ (the "Curie principle") one can easily show [6] that

$$\underset{\sim}{B}_\alpha = 0 \ , \quad \underset{\sim}{C}_{\alpha\beta} = \underset{\sim}{C}_\alpha \delta_{\alpha\beta} \tag{3}$$

where $\delta_{\alpha\beta} = 0, \ \alpha \neq \beta \ ; \ = 1, \ \alpha = \beta$.
Hence we obtain, retaining up to second order derivatives only,

$$\delta\mu = \underset{\sim}{A}*\delta\underset{\sim}{T} + \sum_{\alpha=1}^{3} \underset{\sim}{C}_\alpha * \frac{\partial^2\underset{\sim}{T}}{\partial r_\alpha^2} \ . \tag{4}$$

If the rock is subject to overall isotropic stress then $\underset{\sim}{C}_\alpha$ is independent of α ; if there is only one special stress direction then the two of $\underset{\sim}{C}_\alpha$, for α the directions perpendicular to that special direction, are equal. With this it is seen that the Curie principle can be used to give rather general information on the functional relationship between grain free energy and textural variations, and similarly for velocity variations.

This technique was introduced in [6] and applied there and in [7] to the study of the stability to the formation of textural patterns of the uniform state of a rock under stress. This is clearly a rather powerful technique for analyzing these phenomena in random elastic or other media and must be investigated in greater detail to completely unravel all the effects contained in the mechano-chemical coupling that can lead to pattern formation in stressed rocks. In the next two sections we investigate the application of these concepts to metamorphic layering and stylolization.

III. STYLOLITES

(Before reading this section, please see the discussion by E. Merino on this subject in these proceedings.)

A simple model of stylolite formation based on pressure solution kinetics has been set forth in [8] . The intuitive basis of the model is that a grain in a stressed porous rock experiences a greater stress the greater local porosity. Thus pressure solution demands that the grains in a region of higher porosity, ϕ, are more soluble than in an adjacent lower ϕ region and hence the grains in the lower ϕ region will grow, by influx of solutes, at the expense of the grains in the higher ϕ region. Since grain dissolution increases ϕ we see that this is clearly a runaway effect : a local maximum in ϕ in a stressed rock tends to self amplify.

Consider the simplest situation of quartz where the solid is taken to dissolve into $SiO_2(aq)$. Let c be the concentration (moles per pore volume) of $SiO_2(aq)$. Then in the rock under stress, the equilibrium value of c for a grain in a region of porosity ϕ(= volume fraction of rock that is pore space) will, by the above argument, be a ϕ-dependent function, $c^{eq}(\phi)$. With this a simple model of stylolite formation is easy to set forth.

Let the grains in a monomineralic (quartz) rock have volume V and surface area proportional to $V^{2/3}$. Then we conjecture that the grain volume evolves according to

$$\frac{\partial V}{\partial t} = kV^{2/3} [c - c^{eq}(\phi)] \tag{5}$$

where k is a rate coefficient and $V(\vec{r},t)$ is the volume of grains in a small volume element about the point $\vec{r}$. If n is the number density of grains then we have

$$\phi + n V = 1. \tag{6}$$

Thus, assuming n to be constant in time and space, we have

$$\frac{\partial\phi}{\partial t} = - kn^{1/3} (1 - \phi)^{2/3} [c - c^{eq}(\phi)] \quad . \tag{7}$$

To complete the c, ϕ formulation, we write the reaction-diffusion equation for c,

$$\frac{\partial(\phi c)}{\partial t} = \vec{\nabla} \cdot [\phi D(\phi) \vec{\nabla} c] + \rho\frac{\partial\phi}{\partial t} \tag{8}$$

where ρ is the molar density. The last term on the rhs accounts for the fact that as the grains grow, c decreases ; $D(\phi)$ is the ϕ-dependent diffusion coefficient.

Stability analysis of the model equations (7,8) shows that since $\frac{dc^{eq}}{d\phi} > 0$ the uniform state of porosity $\bar{\phi}$, $\bar{c} = c^{eq}(\phi)$ is unstable. The pattern forming instability is of the svegliabile type. Numerical simulations show that a boundary interaction yields a regular array of dissolution seams whereas an initially microscopic random deviation is amplified into an array of widely separated seams that appear to dominate adjacent initially growing seams that get recemented by the dissolution products of the neighboring dominant seam. This cascade to longer length scales seems to be another rather general feature of svegliabile systems - i.e. similar results were found in the studies of [9] on Liesegang bands and other precipitate patterning.

The simple porosity feedback model neglects an important feature of sedimentary rock evolution - compaction. This occurs (for the well cemented rocks of interest here, and not for loose sediments) by another manifestation of pressure solution. Since grain-to-grain contacts are under higher stresses than free surfaces, the very local value of c^{eq} is generally higher on the facets perpendicular to (say) the overall applied compression than on the facets parallel to it. Hence grains flatten by dissolution on the highly stressed facets and reprecipitation on the facets generally parallel to the overall applied compression stress. Thus in a more complete model, stylolization instability occurs when the porosity feedback dominates the flattening effect that tends to decrease ϕ most strongly where it is highest. Such a model is presently under investigation by the author and coworkers.

IV. METAMORPHIC LAYERING

This phenomenon (see the report of Merino, these lectures) has been modeled using equations of pressure solution coupled to the transport of solute species along intergrain boundaries [6, 7]. The central approach was to consider any given crystal to evolve in an average environment specified by the local texture. Consider a crystal of mineral i of volume $L_i^3(\vec{r},t)$ and located in the vicinity of point $\vec{r}$; L_i^3 is taken to evolve according to

$$\frac{\partial L_i^3}{\partial t} = - \vec{v} \cdot \vec{\nabla} L_i^3 + G_i \quad . \tag{9}$$

The $\vec{v}$ term accounts for the fact that the overall deformation flow of velocity $\vec{v}$ will cause the observed value of L_i^3 at a given point to change as grains move by while the crystal growth rate is G_i. We model G_i via, for example,

$$G_A = k_A L_A^2 [XY - k_A(\underset{\sim}{T})] \tag{10}$$

for the reaction A = X + Y ; k_A is the rate coefficient, L_A^2 is proportional to the grain surface area and $k_A(\underset{\sim}{T})$ is the texture $\underset{\sim}{T}$ dependent equilibrium constant. It is the existence of such an equilibrium "constant" as a functional of texture, as for stylolites, where the equilibrium constant is a functional of porosity, that provides the feedback for instability. Its functional dependence can be examined via the Curie principle as discussed in Section II.

The number density n_i of i-crystals is a conserved variable in the absence of nucleation. Thus in this case

$$\frac{\partial n_i}{\partial t} = - \vec{\nabla} \cdot (n_i \vec{v}) \tag{11}$$

A more complex nucleation theory to account for the common occurence of this process in metamorphic conditions may be set up via a formalism similar to that outlined by the author in the article on precipitate patterning in these proceedings.

Since the pore space is negligible in metamorphic rocks, we have

$$\sum_{i=1}^{N} n_i L_i^3 = 1 \quad , \tag{12}$$

for the N-mineral system, $n_i L_i^3$ being the volume fraction occupied by i.

The solute species concentrations evolve according to reaction-transport equations. For the concentration X_α (moles/rock volume) of species α we obtain

$$\frac{\partial X_\alpha}{\partial t} = - \vec{\nabla} \cdot \vec{J}_\alpha + R_\alpha \tag{13}$$

where the flux $\vec{J}_\alpha$ is taken to be $- \phi D_\alpha \vec{\nabla}(X_\alpha/\phi) + X_\alpha \vec{v}$ and the reaction rate source R_α for species α is related to the G_i via

$$R_\alpha = \sum_{i=1}^{N} \nu_{\alpha i} n_i \rho_i G_i \tag{14}$$

through the stoichiometric coefficients $\nu_{\alpha i}$ and the molar density ρ_i of solid i. (Additional terms must be added to the rhs of (14) if reactions among solutes are important.)

The above equations with a phenomenology for the equilibrium constants $K_i(\underset{\sim}{T})$ provide a self consistent description in one dimensional problems since the condition (12) yields the one required equation to determine $\vec{v}$, a single number in one spatial dimension. It also is a complete theory for determining the linear stablity analysis since for each Fourier stability mode for the state at rest ($\vec{v} = \vec{0}$) the perturbation (proportional to $\exp(i\vec{k}\vec{r})$ for wave number $\vec{k}$) involves only the longitudinal (i.e. parallel to $\vec{k}$) component of the velocity.

Results of the stability theory of the above simple nucleation free model show that metamorphic pattern formation can arise via svegliabile or cognito behavior. Applied vectors, such as an applied pressure gradient in the pore fluid, resulting in a flow through, can orient banding by making banding perturbations of a preferred orientation most rapidly growing. The theory also shows that preferred band orientation can occur when the rock is subject to an overall shear flow. The concept of a self organization phase diagram was introduced for this problem ; this diagram in the space of the texture of the initially unbanded rock, shows which initial texture domains can lead to banding. (For details and citations see [6,7].)

Progress towards a complete three dimensional theory is underway via the analysis of velocity equations for the viscoelastic medium. Other areas under study are the development of preferred orientation of crystals with respect to bands [15], nucleation effects (along the lines developed in the chapter on Liesegang phenomena by the author, these proceedings) and the dependence of the grain solubility on the defect density [7].

V. POROSITY INSTABILITIES DURING REACTIVE PERCOLATION

Flow through in porous media of fluids that can react with the medium can lead to an interesting coupling between chemistry and flow that can yield instability and symmetry breaking. (See Ref. 10 and references cited.) It is easy to demonstrate. Suppose one takes a porous medium (a sugar cube) and imposes a flow across it of liquids (distilled water) that will react with and dissolve the medium (Sugar (solid) $\rightleftarrows$ Sugar(aq)). Then the exiting fluid will be in equilibrium with the solid for a sufficiently thick sample (producing saturated sugar-water). If the porosity at any one point in the medium is originally slightly higher than the initial average, then because of Darcy's Law, the flow-through will be faster there and hence the reactive fluids will further increase the porosity. This clearly leads to a runaway wherein the local porosity maxima increase faster and eventually holes form. These holes tend to elongate and take on the character of worm holes oriented roughly parallel to the overall flow. These worm holes themselves compete for the flow by a similar mechanism and the tendency is then to form a few large channels that dominate the flow. Thus we have the cascade porosity instability → worm holes → channels → caves.

A related problem that leads to more tractable mathematics is the case of a porous medium made of two or more minerals, one of which is insoluble in the imposed reactive flow. In this problem we do not have the complication of coupling the hydrodynamic equations in the holes and channels with the porous medium (percolation) equations within the undissolved matrix.

The simplest model for the transport and reaction of the assumed single solute species of concentration c is

$$\frac{\partial}{\partial t}(\phi c) = \vec{\nabla} \cdot \phi \, [D(\phi) \, \vec{\nabla} c + c \, K(\phi) \, \vec{\nabla} p] + \rho \frac{\partial \phi}{\partial t} \tag{15}$$

where ϕ is the porosity (volume fraction of medium that is pore space), ρ is the solid molar density, D is the ϕ-dependent diffusion coefficient, p is the pressure and $K(\phi)$ is the ϕ-dependent Darcy's constant that relates the fluid velocity $\vec{v}$ to the pressure gradient, i.e. $\vec{v} = - K(\phi) \, \vec{\nabla} p$.
Assuming the solvent is neither consumed or liberated in the dissolution process, is incompressible and its molar volume does not change appreciably upon introduction of the solute leads to an equation for the pressure,

$$\vec{\nabla} \cdot [\phi \, K(\phi) \, \vec{\nabla} p] = \frac{\partial \phi}{\partial t} .$$

To complete the theory we need an equation for the porosity. Let $V(\vec{r},t)$ be the volume of the soluble grains in the vicinity of point $\vec{r}$ at time t. Then if n is the number density of these grains we have

$$\phi_I + \phi + nV = 1 \tag{16}$$

stating that the volume fraction of inert material, ϕ_I, plus the porosity plus the volume fraction occupied by the soluble grains must account for all the medium. The dissolution kinetics of the soluble grains is expected to be proportional to their surface area (and hence $V^{2/3}$) and to the concentration difference between c and its equilibrium value c^{eq}. Putting these ideas together one finds

$$\frac{\partial\phi}{\partial t} = \frac{1}{\varepsilon}(1 - \phi - \phi_I)^{2/3}(c - c^{eq}) \tag{17}$$

where ε is a constant. Note that the reaction stops when either the fluid is satured ($c = c^{eq}$) or all the soluble mineral is removed, $\phi = 1 - \phi_I$.

The above ϕ, c, p model contains the wash-out instability of th soluble mineral grains. There is a planar wash-out, wherein all variables depend only on the spatial direction z along which the flow takes place. For the infinite medium with a specified value of $\partial p/\partial z$ as $z \to -\infty$, one can obtain an essentially exact solution for a plane wave wash-out advancing at constant velocity. This planar solution is found to be unstable to nonplanar perturbations, indicating the symmetry breaking instability leading to the analogue of worm holes in the present problem with an insoluble, background matrix.

The above unstable wash-out problem may be cast as a moving boundary problem when the dissolution kinetics are fast, i.e. as $\varepsilon \to 0$ in equation (17). In the simplest case where the diffusion of c is not important relative to flow we obtain the following. The surface at which the fast dissolution takes place is denoted $S(\vec{r},t)=0$. The domain on which the soluble component has been removed is denoted $S < 0$ and the univaded region is $S > 0$. Then S is found to satisfy the equation

$$\frac{\partial S}{\partial t} + u\,|\vec{\nabla}S| = 0, \tag{18}$$

where u is the normal advancement speed of the interface. Since the porosity is constant in the two domains $S \gtrless 0$, we have

$$\nabla^2 p = 0. \tag{19}$$

Matched asymptotic expansion techniques yield a jump condition on the normal derivative of p across the interface and a value for u - see [10] for details.

The analysis is now reduced to that of a moving boundary problem. Exact planar and nonplanar solution can be attained and their stability may be adressed via shape stability analysis [11,12].

VI. PROPAGATING REDOX FRONTS

When waters that can react with a rock are forced to percolate through such a rock, zones of chemical transformation move through the system along the flow. If the rock is oxygen poor, containing reducing materials such as pyrite or carbonaceous matter and the waters are oxygen rich (meteoric waters from above) the transition zone is called a redox front. Since many metal ions are relatively soluble under oxidizing conditions and form relatively insoluble minerals under reducing conditions, the redox interface can provide a trap that collects metal ions in the incoming waters or liberated as the front moves through the rock. One example of such a phenomenon is a roll type deposit, economically interesting because of their ability to trap uranium. Other examples include copper and gold deposits. See Ref. [13] for citations to the redox deposit literature. We have examined te possibility of reaction-transport instability of various redox front phenomena. Since perhaps the most interesting phenomenon here is that of the banding of deposited minerals [12,13], we defer most of the discussion to the other article by the author in these proceedings. Also very interesting possibilities exist when one combines these redox front banding instabilities and deposition phenomena with the washout instability of equations (15-19). Indeed scalloping is observed in uranium roll type deposits.

VII. CONCLUSIONS

We believe that the study of instability and pattern formation in geochemical systems is at a very exciting juncture. Many phenomena that had been viewed as anomalous curiousities are now, it appears, to be viewed under a unified theory of instability to macroscopic pattern formation in reaction-transport systems undergoing first order phase transitions. While it cannot be said that our and other modeling efforts are conclusive -- much laboratory work is needed to verify this -- it seems that our unified approach is justified and we look forward to rapid growth of interest in this area in the next few years. While having many common features with the purely chemical self organization phenomena, elegantly discussed in Ref. [15], the phase transition-transport patterning phenomena have unique characterization of their own.

REFERENCES

1. Liesegang, R.E., 1913, "Geologische Diffusionen", Steinkopf, Dresden and Leipzig
2. Hedges, E. and Myers, J.E., 1926, "The Problem of Physico-Chemical Periodicity", Longmans-Green, New York
3. Chadam, J., Merino, E. and Ortoleva, P., Macroscopic Self Organization at Geological and Other First Order Phase Transitions, Science (submitted for publication)
4. Nitzan, A., Ortoleva, P. and Ross, J., 1974, J. Chem. Phys. 60, pp. 3134
5. de Groot, S. R. and Mazur, P., 1963, "Nonequilibrium Thermodynamics", North Holland Pub. Co., New York
6. Merino, E., Ortoleva, P. and Strickholm, P., 1982, Amer. J. Sci. 282, pp. 617
7. ibid, The Self Organization of Metamorphic Layering, J. Geophys. Res. (submitted for publication)
8. ibid, A Kinetic Theory of Stylolite Generation and Spacing, Contrib. to Min. and Pet. (to appear)
9. Feeney, R., Schmidt, S.L., Strickholm, P., Chadam, J. and Ortoleva, P., 1983, J. Chem. Phys. 78, pp. 1293
10. Chadam, J., Goncalves, J., Hagstrom, S., Hettmer, J., Hoff, D., Larter, R., Merino, E., Ortoleva, P. and Sen, A., Porosity Instability at Reactive Percolation, Physica D (submitted for publication)
11. Chadam, J. and Ortoleva, P., in "Nonlinear Phenomena in Chemistry", Hlavacek, V., ed.
12. Moore, C., Auchmuty, G., Chadam, J., Hettmer, J., Larter, R., Merino, E., Ortoleva, P., Ripley, E. and Sen, A., 1983, in Proc. of 4th Inter. on Water-Rock Interactions, Misasa, Japan
13. Merino, E., Moore, C., Ripley, E., Auchmuty, G., Chadam, J., Hettmer, J. and Ortoleva, P., Kinetic Modeling of Redox Roll Fronts and Their Instabilities (submitted for publication)
14. Ortoleva, P., Auchmuty, G., Chadam, J., Hettmer, J., Merino, E., Moore, C., Ripley, E., Redox Front Propagation and Banding Modalities, Physica D (submitted for publication)
15. Nicolis, G. and Prigogine I., 1977, "Self Organization in Nonequilibrium Systems", Wiley, New York

CHEMICAL INSTABILITIES AND "SHOCKS" IN A NON-LINEAR CONVECTION PROBLEM ISSUED FROM GEOLOGY

B. GUY , F. CONRAD , M. COURNIL and F. KALAYDJIAN

Ecole des Mines, 42023 St Etienne cedex, France

The chemical transformation of a rock by a pervading convecting fluid in disequilibrium with it is modelled by a non-linear partial differential equation of hyperbolic type. Starting from a continuous concentration profile, the solution may become discontinuous within a finite time in some cases. We show that this is due to the instability, in the sense of the second principle of thermodynamics, of a range of the possible concentrations with respect to this dynamical phenomenon.
The simple model exposed gives a good account for the characteristic pattern of this phenomenon, which is the self spatial organization of the concentrations in different domains separated by strong gradients.

INTRODUCTION, AIM OF THE PAPER

The convection of fluids through rocks is a general phenomenon in geology, and it occurs as well at the surface of the earth as at depth. This may induce migrations of chemical components and chemical reactions ; one conspicuous feature that may be thus produced is the possible appearance of spatial heterogeneities in the form of domains of different concentrations for the chemical components and separated by strong gradients, or discontinuities. The rocks showing such features may have several meters to hundreds of meters of extension. The problems of chemical transformations in rocks have been studied by various authors and for different aspects. We will here contribute to a thermodynamical study of the simplest model accounting for the appearance and the maintenance with time of discontinuities. The foundations of this model were given by Korzhinskii [10] ; cf. also [3], [4], [9].

G. Nicolis and F. Baras (eds.), Chemical Instabilities, 341–348.

NOTATIONS

We will consider here only one chemical component c, migrating along one spatial dimension, the x axis. The rock will be considered as divided into a solid part and an interstitial fluid part filling its pores. The porosity, i.e. the volumic ratio pore/pore + solid will be called p. We will call c_s and c_f the volumic concentrations of the component in the solid and fluid parts of the rock respectively. At last the velocity of the pervading fluid will be called v and the diffusion coefficient (if necessary) of the component in the fluid phase will be D, taken as constant ; p and v may be functions of space and time. No diffusion in the solid part of the rock will be considered.

BASIC ASSUMPTIONS OF THE MODEL : CHEMICAL LOCAL EQUILIBRIUM, ABSENCE OF DIFFUSION

We know that fluid velocities in geological processes are very slow and, by comparison, we will suppose that the chemical kinetics are rather fast (this is particularly true in the case of rocks formed at temperatures higher than 300°C, such as skarns [6], [8]). For those two reasons, we will consider that, at given x and t, there is a chemical equilibrium between the solid and fluid portions of the rock, expressed by a law : $c_f = f(c_s)$ Eq.(1). This law is called isotherm and may be given by experimental data for mineral-fluid equilibria. The major point is that it is generally non linear.
Another simplification will be to consider that the diffusion is negligible. We indeed are interested by a scale (meters to kilometers) where the main way to move components is convection, not diffusion.

MASS BALANCE AND BASIC EQUATION

The conservation of the chemical component in a section of the rock inclued between any abscissa x_1 and x_2 reads :

$$\frac{d}{dt}\int_{x_1}^{x_2} [pc_f(x,t) + (1-p)c_s(x,t)]\, dx = pvc_f(x_1,t) - pvc_f(x_2,t) \qquad (2)$$

where the flux of the chemical component is due only to convection. We will consider here the particular case where p and v are constant. With Eq.(1), the derivation of (2) with respect to x leads to :

$$(1-p)\,\frac{\partial c_s}{\partial t} + p\,\frac{\partial f(c_s)}{\partial t} + pv\,\frac{\partial f(c_s)}{\partial x} = 0 \qquad (3)$$

what we want to solve by giving for instance a profile $c_s(x)$ at $t = 0$.

A GEOMETRICAL SOLUTION

Eq.(3) may be rearranged in $\partial c/\partial t + avg'\partial c/\partial x = 0$, where $g' = f'/1 + af'$, with $a = p/1-p$. This is a non-linear first order partial differential equation of hyperbolic type. A semi-explicit analytic solution is possible and leads to a geometrical representation of the solution : this is the ruled surface defined in the (c,x,t) space by horizontal (parallel to the plane x,t) straight lines called the characteristics and figuring the propagation of the initial condition $c_s(x,0)$ at a velocity $dx/dt = avg'(c_s)$ which is equal to $pvf'(c_s)$ for p small and f' bounded. This construction is done on figure 1a in the case of a smooth decreasing initial profile connecting two domains with different concentrations, and with a non-linear isotherm such as on figure 1b. We see here that the initial profile is steepened and the surface is folded : in the plane (x,t) a triple solution for c appears within the domain indicated on figure 1c and limited by the cusp curve :

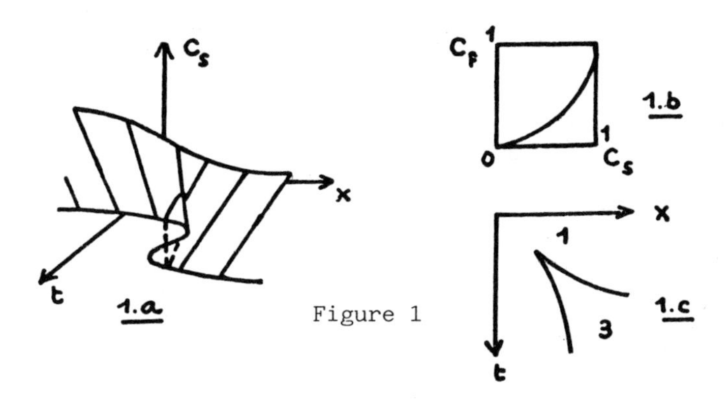

Figure 1

CHOICE OF A DISCONTINUOUS PROFILE

Before the surface is folded, the phenomenon made appear infinite gradients and we may think that it is because we have neglected diffusion. To avoid the physically meaningless triple solution and keeping our simple model, we may fit in the triple valued profile a discontinuity figuring a sharp change of concentration (fig.1a). If we choose to keep an equation such as (2) with chemical equilibrium between c_s and c_f, this gives a constraint on the way to obtain the discontinuous solution (equality of the matter limited by the curve $c(x)$, [17]).
At the appearance of the shock, the phenomenon shows a bifurcation.

We may notice here that time and/or space are here the bifurcation parameters. We could also say that we have here a non-equilibrium phase transition figured by a cusp catastrophe [15].

WEAK SOLUTIONS

Another approach is also possible : we can indeed consider the problem in a mathematical framework suitable for the discontinuous solutions : we will look for a solution c(x,t) in the sense of distributions ("weak solutions"), which must be, on the basis of chemical equilibrium, directly that given in the preceding section. It can be shown that, in the new framework, the solution of the equation is no longer unique. Mathematicians use to add a so-called "entropy condition" that makes the selection of the solution with physical meaning [11]. In the following sections, we will see the correspondence between the approach of mathematicians and that we can follow in thermodynamics. As may seem obvious for a phenomenon that involves strong gradients, the diffusion will be considered as the perturbative phenomenon used for the choice of the "physical" solution [1].

ENTROPY BALANCE (THERMODYNAMICS)

We will examine this question in a particular case, in order to explain our approach. The most complete case may be examined in the future. We will consider first that there is no variation of the volumes of the solid and fluid phases. We will also assume that the exchange reaction between the solid and fluid phase is athermic and that there is no heat flux. In absence of other energy terms, we can thus set the problem at constant energy and volume. The chemical equilibrium between the solid and fluid phase is then expressed in the form $dS_f/dc_f = dS_s/dc_s$, where S_s and S_f are the volumic entropies of the solid and the fluid respectively because the chemical potentials are here $\mu = -T(dS/dc)_{e,v}$.
In the complete case where there is diffusion, the entropy balance may be written according to :

$$p\frac{\partial S_f}{\partial t} + (1-p)\frac{\partial S_s}{\partial t} + pv\frac{\partial S_f}{\partial x} = pD\frac{dS_f}{dc_f}\frac{\partial^2 c_f}{\partial x^2} \qquad (4)$$

with an entropy flux composed of convection and diffusion (all terms rearranged, after [5]). The second principle is expressed by

$$p\frac{\partial S_f}{\partial t} + (1-p)\frac{\partial S_s}{\partial t} + pv\frac{\partial S_f}{\partial x} - pD\frac{\partial^2 S_f}{\partial x^2} \geq 0 \qquad (5)$$

THE ENTROPY CONDITION (IN THE SENSE OF MATHEMATICIANS)

If we involve a fickian diffusion of the component in the fluid phase Eq. (3) is replaced by :

$$p\,\frac{\partial c_f}{\partial t} + (1-p)\,\frac{\partial c_s}{\partial t} + pv\frac{\partial c_f}{\partial x} - pD\frac{\partial^2 c_f}{\partial x^2} = 0 \qquad (6)$$

Let us take an arbitrary function, regular and concave $H_f(c_f)$, representing an entropy. Let us multiply Eq. (6) by $H_f(c_f) = dH_f/dc_f$; we then have :

$$p\,\frac{\partial H_f}{\partial t} + (1-p)\frac{dH_f}{dc_s}\,\frac{\partial c_s}{\partial t} + pv\frac{\partial H_f}{\partial x} = pD\frac{\partial^2 H_f}{\partial x^2} - pD\frac{d^2 H_f}{dc_f^2}\,\left(\frac{\partial c_f}{\partial x}\right)^2 \qquad (7)$$

As H_f is concave, the last term of the second member is of constant sign (negative) and we have the inequality :

$$p\frac{\partial H_f}{\partial t} + (1-p)\frac{dH_f}{dc_f}\,\frac{\partial c_s}{\partial t} + pv\frac{\partial H_f}{\partial x} - pD\frac{\partial^2 H_f}{\partial x^2} \geq 0 \qquad (8)$$

Let us now see what happens when D goes to zero : the term $pD\partial^2 H_f/\partial x^2$ goes to zero in the sense of distributions and Eq. (8) leads to :

$$p\,\frac{\partial H_f}{\partial t} + (1-p)\frac{dH_f}{dc_f}\,\frac{\partial c_s}{\partial t} + pv\frac{\partial H_f}{\partial x} \geq 0 \qquad (9)$$

that must be fulfilled whatever H_f concave. The identification with the above relations (4) and (5) leads to identifying H_f to S_f and to imagine a new function H_s, homologous of S_s and that verifies $dH_s/dc_s = dH_f/dc_f$. With the mentionned hypotheses, this condition is actually fulfilled and expresses the local chemical equilibrium. We have soforth shown the correspondence between the two approaches.

ENTROPY AND SHOCKS

The preceding results allow us to discuss the question of the entropy bound to the shocks. In the entropy condition (9) obtained after (7), let us notice that, in absence of diffusion, the first member is equal to zero when there is no shock. The inequality has interest only for shocks : it then appears infinite gradients $\partial c/\partial x$ and, though D goes to zero, we do not know the convergence of the last term in Eq. (7). The only thing we know is its sign.

As for the other term, it goes to zero, so the inequality. This is due to the choice of considering the problem at a scale where strong variations are taken as discontinuities and to "ignore" what happens in the shock. We can say that we have here an example where irreversibility and quantification are connected [2], [14] - the quantification here lies in the concentration discontinuity-.

SUITE OF COMPOSITIONS

The entropy condition (9) also leads to another interesting result : the concave functions H_f are indeed generated by the functions $-|c-k| + cste$ and, in the case of a shock Eq. (9) leads to the condition :

$$\frac{pv(c_f^+ - c_f^-)}{p(c_f^+ - c_f^-) + (1-p)(c_s^+ - c_s^-)} \geq \frac{pv(c_f - c_f^-)}{p(c_f - c_f^-) + (1-p)(c_s - c_s^-)} \quad (10)$$

whatever c_f and c_s within the intervals (c_f^+, c_f^-) and (c_s^+, c_s^-) where the superscripts + and - indicate the two sides of a shock. In the simple case of an arrival of fluid of constant composition in a rock of also constant composition, this condition allows to determine the suite of the concentrations, with possible discontinuities, by the consideration of the concavity of the isotherm between the points figuring the starting rock and the source fluid. In the general case, at a given time, this condition also involves the initial and boundary conditions. This condition may be rather simply understood in terms of velocities of the different concentrations : the faster velocities are in advance on the slower ones (cf. supra). It can be shown that the velocity of a shock is a generalization of that of a concentration. Eq. (10) expresses in the same time a local condition of stability (something like a second derivative) and a global one (instability of a part of the suite of concentrations and possible apperance of discontinuities).

NUMERICAL SIMULATIONS

A condition such as (9) is compulsory for the numerical simulations in order to get the appropriate discontinuous solutions. We have done them by the Godunov method [12] and we present here on figure 2 the evolution of the concentration suite for the initial profile indicated (with an arbitrary time scale) and the isotherm of fig. 1b. An initial inhomogeneity has been taken. The isotherm shows small velocities for low concentrations. The greatest velocites are for the highest concentrations : so the inhomogeneity is here washed out as seen on the figure. The final shock is between 0 and 1 as the whole isotherm is unstable (cf. Eq. (10)).

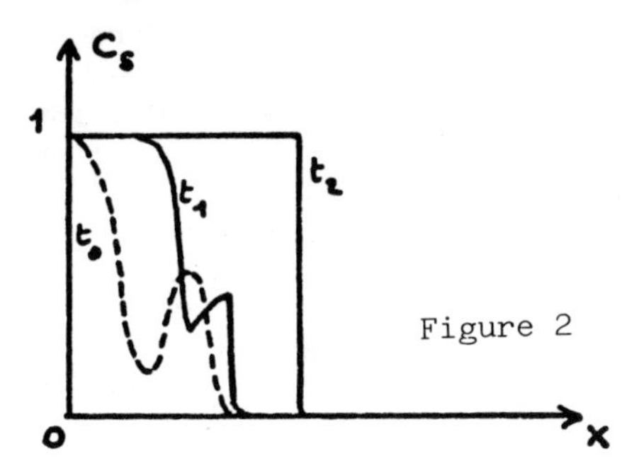

Figure 2

CONCLUSIONS AND ADDITIONAL COMMENTS

According to the original intuitions and works of Korzhinskii [10], we can explain the spatial and temporal organization created by "infiltration metasomatism", in the form of the existence of zones separated by sharp fronts (at a given place a series of mineralogical changes may be seen), as produced in the course of a single phenomenon, and not by the action of different events following in time. The choice of discontinuities and the coherence of the whole suite of compositions (cf the law of the progression on the velocities of Korzhinskii) is ruled in our model by the second principle of thermodynamics, which expresses the possible instability of a part of the concentrations ; the organization is occuring after a bifurcation and through an entropy production. The strong gradients are simulated by appropriate schemes based on the entropy condition. This geological phenomenon is similar to the solitons in another field. It may also exhibit oscillating patterns [7] that are under study (cf also [13]).

ACKNOWLEDGEMENTS

We wish to thank MM. Fer, Fargue and Valour [16] for helpful discussions. Part of this work has been supported by CNRS (ATP Geochimie). I (B.G.) would like at last to thank Prof. G. Nicolis and I. Prigogine for inviting me at the workshop on Chemical Instabilities and stimulating me to write this paper.

REFERENCES

1. Conrad F., Cournil M. et Guy B. (1983), C.R. Acad Sc. Paris, in press.
2. Fer F. (1977), L'irréversibilité, fondement de la stabilité

du monde physique, Gauthiers Villars, 135 p.
3. Fletcher R.C. and Hofmann A.W. (1974), Carnegie Inst. Wash. pub. 634, pp. 243-259.
4. Fonteilles M. (1978), Bull. Minéral., 101, 166-194.
5. Glansdorff P. et Prigogine I. (1971), Structure, stabilité et fluctuations, Masson, Paris, 288 p.
6. Guy B. (1979), Thèse Ing. Doct., Ecole des Mines, Paris, 270 p.
7. Guy B. (1981), C.R. Acad. Sc., Paris, 292, II, pp. 413-416.
8. Guy B. (1981), Springer Verlag, Synergetics, vol. 12, p. 263.
9. Hofmann A. (1972), Am. J. Sc., 272, pp. 69-80.
10. Korzhinskii D.S. (1970), Theory of metasomatic zoning, Clarendon Press, Oxford, 162 p.
11. Lax P.D. (1971), E.M. Zarantonello ed., Acad. Press, pp. 603-634.
12. Le Roux (1974), Thèse, Rennes.
13. Merino E. (1983), Workshop on Chemical Instabilities, Austin.
14.Prigogine I.(1980), Physique, temps et devenir, Masson, 275 p.
15. Thom R. (1977), Stabilité structurelle et morphogénèse, Interéditions, Paris, 351 p.
16. Valour B. (1983), Equipe Anal. Num. Lyon St Etienne, 17, 34 p.
17. Whitham G.B. (1974), Linear and nonlinear waves, J. Wiley.

PART V

MATERIALS SCIENCE

DYNAMICAL PROCESSES DURING SOLIDIFICATION

J.H. Bilgram and P. Böni

Laboratorium für Festkörperphysik
Eidgenössische Technische Hochschule
CH-8093 Zürich / Switzerland

A diffuse layer is observed at the surface of a crystal which is growing into the melt. The building up of this layer and the dynamics of entropy fluctuations have been studied by quasi elastic light scattering. Water and salol have been used as test substances. In a stationary non equilibrium steady state the layer is several thousand lattice constants thick. The diffusion constant, which determines the dynamics of the entropy fluctuations in this layer, is $D_i \approx 3 \cdot 10^{-8}$ cm^2/s. It is isotropic in space. During freezing fluctuations in order can not be separated from fluctuation of heat. Therefore we interpret D_i as J. Frenkel's constant of "structure diffusion".

1. INTRODUCTION

The freezing transition is a first order phase transition.
Two processes are necessary for solidification:
i.) ordering of the molecules (atoms) at the crystal surface and
ii.) transport of the latent heat away from the solid-liquid interface.

Therefore the dynamics of freezing can be studied only under non equilibrium conditions, when the interface is driven continuously into the melt.
Dynamical Processes can be studied at three scales:
i.) at macroscopic scale
ii.) at the scale of fluctuations
iii.) at atomic scale.

G. Nicolis and F. Baras (eds.), Chemical Instabilities, 351–366.

1.1. Dynamics on Macroscopic Scale

There exist many data where the solidification rates have been measured as a function of the supercooling ΔT at the solid-liquid interface. ΔT is difficult to measure without disturbing the system. Best reproducible data have been obtained in experiments where crystals grow freely into supercooled melt. In this case it is not necessary to measure the supercooling at the interface. Rates of free growth have been measured as a function of (T_m-T_∞) for ice [1], succinonitrile [2] and for salol [3,4]. T_m is the melting temperature of the crystal and T_∞ is the temperature of the melt far away from the crystal. Ice and succinonitrile grow in a dendritic shape, whereas salol forms many facetts. Therefore ice and succinonitrile seem to be an ideal couple for a comparison of theories of dendritic growth with the experiment. Unfortunately both substances grow with the same rate at a given supercooling:

$$v_{ice}\,(T_m-T_\infty) = v_{succinonitrile}\,(T_m-T_\infty)$$

Hence these two substances are not a good couple to check a scaling law for dendritic solidification, as it has been done by Langer et al. [5].

Salol forms many facetts and therefore a growth behavior is expected, which cannot be compared with dendritic growth. (For a detailed discussion of growth mechanisms see [6]). For high supercooling Cahn et al. [7] have predicted for facetting crystal surfaces a transition from lateral to a normal growth mechanism. After this transition - at high supercooling - growth can be compared with dendritic growth. Such a transition has been observed in the free growth experiments of salol where the dependence of growth rate on (T_m-T_∞) changes from an exponential law at small supercooling to a power law at high supercooling. Therefore salol can be used as a third substance to study free growth of crystals into supercooled melt. In section 4 it will be shown how the dynamical processes at the solid-liquid interface influence the growth rates of crystals.

1.2. Dynamics on the Scale of Fluctuations

Light scattering experiments are well known in the study of critical phenomena, where an increasing density-density correlation lenght and a slowing down of the decay rate of fluctuations are observed. In difference to critical phenomena the freezing transition can be studied neither at equilibrium conditions nor in an extended volume.

First dynamic light scattering experiments at the solid-liquid interface of a growing crystal have been performed at the ice water interface [8]. In these experiments the scattering vector has been oriented in the interface plane. With such a scattering geometry no information about the extension of the interface region into the liquid or into the solid can be obtained. Experiments, where the scattering vector has been directed out of the interface plane, give evidence, that there is a region at the interface of a growing crystal, which has properties different from the ones of ice and water. In section two experiments will be discussed, which give information about the properties of this interface region.

These experiments have been performed with H_2O and D_2O. No isotope effect has been found as it would have been expected if hydrogen bonding would be important. Experiments performed with salol [9] led to results which can be compared with those of water by means of a simple scaling law. Therefore we assume that the effects, we observe in light scattering experiments, are typical for the melting transition, or perhaps for experiments where great gradients in the chemical potential occur.

1.3. Dynamics on atomic/molecular scale

There are simulations of crystal growth using molecular dynamics technique. Starting with a configuration of a crystal and a supercooled liquid, the approach to equilibrium of the whole configuration has been studied by Landman et al.[10]. A layering is found in the liquid, which is followed by an intra layer ordering.

Growth rates of crystals from a Lenard-Jones liquid have been calculated in a molecular dynamics study by Broughton et al.[11]. It is found, that in difference to ordinary liquids no thermal activated steps are involved in this crystallization process.

The speed of molecular growth steps crossing the solid-liquid interface has been measured by Doppler velocimetry [9]. This is the only experiment where such measurements have been done in a real system where the crystal grows into its melt.

2. DYNAMIC LIGHT SCATTERING

The intensity of light, scattered by density fluctuations in an ordinary liquid, is proportional to the isothermal compressibility χ_T. Two types of density fluctuations can be distinguished: propagating ones, they give rise to Brillouin scattering, and non propagating ones, which lead to a broadening of the central line

(Rayleigh scattering). The intensity of the Rayleigh line relative to the intensity of the Brillouin lines is given by the Landau-Placzek ratio

$$\frac{I_R}{2I_B} = \frac{C_p - C_v}{C_v} = \frac{\chi_T - \chi_S}{\chi_S} \tag{1}$$

where C_p, C_v and χ_T, χ_S are the specific heats and the compressibilities at constant pressure, volume, temperature and entropy respectively. Water has a density maximum at $4^{\circ}C$. As the thermal expansion vanishes there $C_p = C_v$. This is the reason why Rayleigh scattering is very small in water close to the melting point. This property makes water to a unique substance for the study of dynamical processes at the solid-liquid interface: all the light, which is scattered quasielastically from the interface region, contains information about ordering processes at the solid-liquid interface.

For light scattering experiments at the solid-liquid interface it is very important to reduce background scattering from impurities in the melt or from grain boundaries in the crystal. The experiments described below have been performed in situ during zone refining. Details on the experimental set up have been published elsewhere [12].

At conditions close to equilibrium the ordering process during freezing takes place in a region which is comparable to molecular scale and the interface is sharp. At conditions far from equilibrium there is a transition from this sharp to a diffuse interface. This transition and the properties of the diffuse interface will be discussed in the following.

2.1. The Transient State

The power spectrum of the light, scattered from the ice-water interface at conditions close to equilibrium, contains only Raman and Brillouin contributions. There is no Rayleigh scattering. At a critical growth rate $v_{crit.}$ an onset of quasi elastic light scattering is observed. The intensity of the scattered light increases continuously with time and reaches a constant level after about one hour. In fig. 1 a sequence of power spectra is shown as obtained by analysing the light which is scattered out of a laser beam illuminating the solid-liquid interface from the liquid side in grazing incidence. A Fabry-Pérot interferometer is used with a free spectral range of 7.3 GHz. The frequency shift 5.10 GHz ± 0.6 % agrees with the value measured by Teixeira et al. [13]

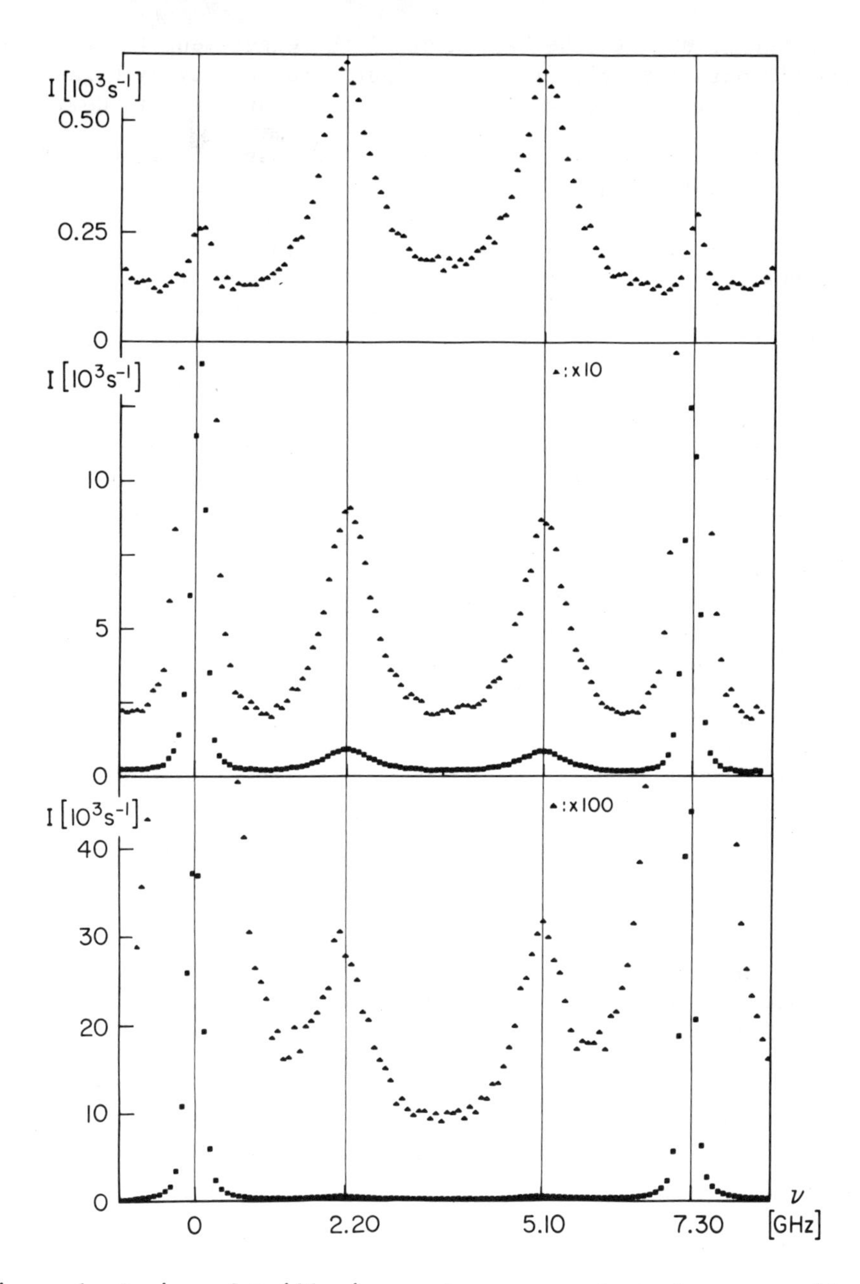

Figure 1. Series of Brillouin spectra measured at the water side of the interface during the onset of scattering. The magnification of the spectra (triangles) from the top to the bottom is one, ten and one hundred (scattering angle = 90°).

for water at $0^{o}C$. At the beginning of the experiment (fig. 1 a) the Brillouin intensity is high compared to the Rayleigh intensity. In the steady state (fig. 1c) the Rayleigh intensity is dominant.

The critical growth velocity for the onset of light scattering at the solid-liquid interface depends on the thermal conditions at the interface. Thermal gradients G_i in the ice have been measured during the growth of the crystal just below its surface by freezing a thermocouple into the ice. The thermal flow through the liquid can be calculated using G_i and heat balance at the interface.

The freezing of a thermocouple into the crystal disturbes the crystal perfection. As crystals with low perfection do not scatter light reproducible at the solid-liquid interface, two experiments are necessary to determine the thermal conditions during the onset of scattering. In a first experiment a perfect crystal is used and $v_{crit.}$ is measured when the onset of scattering is observed. In a second experiment another crystal is used with the thermocouple placed in the zone. The same experimental conditions are established as in the first experiment. Now the temperature is measured when the thermocouple crosses the interface. In fig. 2 the critical growth velocity is plotted as a function of the thermal gradient in the ice. Within the range $2\ K/cm < G_i < 4.5\ K/cm$, $v_{crit.}$ is proportional to G_i. For high gradients and growth rates $v_{crit.}$ increases slower , than it is expected by extrapolation from low values.

The dashed line in fig. 2 corresponds to zero heat flow in the liquid close to the interface. This line is defined by $v = G_i K_i / L$ where v, K_i, and L are the growth rate of the ice, the thermal conductivity in the ice and the latent heat respectively. In all our experiments there is a heat flow from the water to the interface. In our zone refining apparatus we cannot supercool the melt. Hence there is always a heat flow from the melt to the interface. The interface is always stable and a Mullins-Sekerka instability has never been observed. The region A is not accessible in our experiment. In region B we always observe onset of light scattering. In region C light scattering is only observed if the crystal has grown at conditions of region B first.

Simultaneous measurements of the intensity and the linewidth of the scattered light show, that the linewidth decreases during the increase of the intensity. Up to now we are not sure, which are the good parameters to characterize the system in this transient state. From several experiments we know that the growth velocity or the time after the onset of scattering are no good parameters. The scattering power of the interface region seems to be

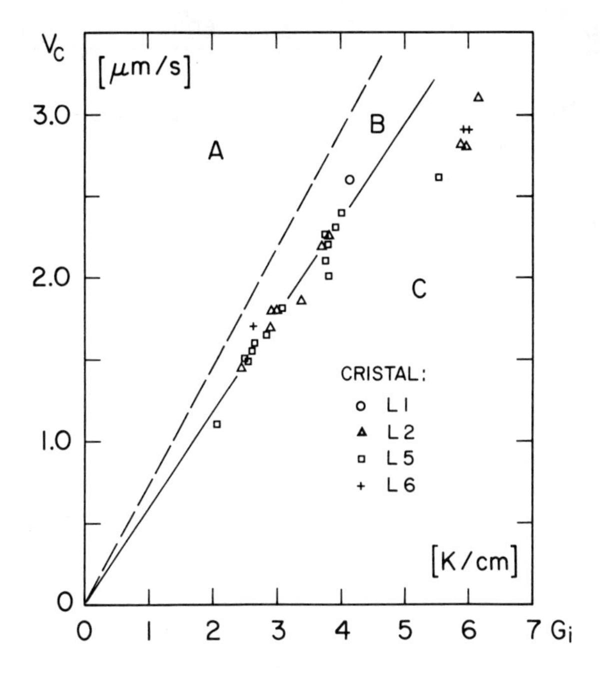

Figure 2. Critical growth velocity versus temperature gradient in the ice, just beneath the interface. The dashed line is calculated by $v = G_i K_i / L$

a better parameter. In fig. 3 the line width measured during the transient state is plotted as a function of the scattering power. It can be fitted by a power law:

$$\Gamma = f \left(\frac{I_c}{1000 \text{ counts/s}}\right)^{-\alpha} \tag{2}$$

with the fitting parameters $\alpha = 0.23 \pm 1\%$ and $f = 2750$ rad/s $\pm$ 1%. I_c is the intensity of the quasielastically scattered light. Contributions from Brillouin scattering and elastic background scattering, which are also detected by the photomultiplier, have been subtracted. There is a weak dependence of α and f on the thermal conditions at the interface. In fig. 4 α and f are plotted as a function of the thermal gradient in the ice. For high gradients a tendency for an increase of f is observed. Fig. 4 shows also the scatter of the experimental data from run to run. For all our experiments the parameter α is within the limits $\alpha = 0.25 \pm 20\%$.

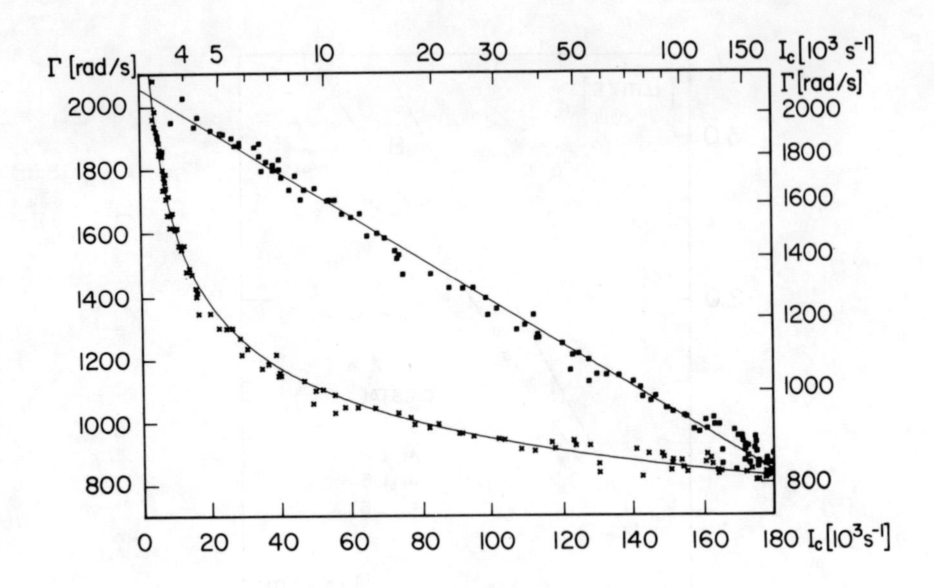

Figure 3. Linewidth of the scattered light versus the intensity in a linear measure (crosses) and in a log-log representation (squares). The solid lines are obtained by linear regression.

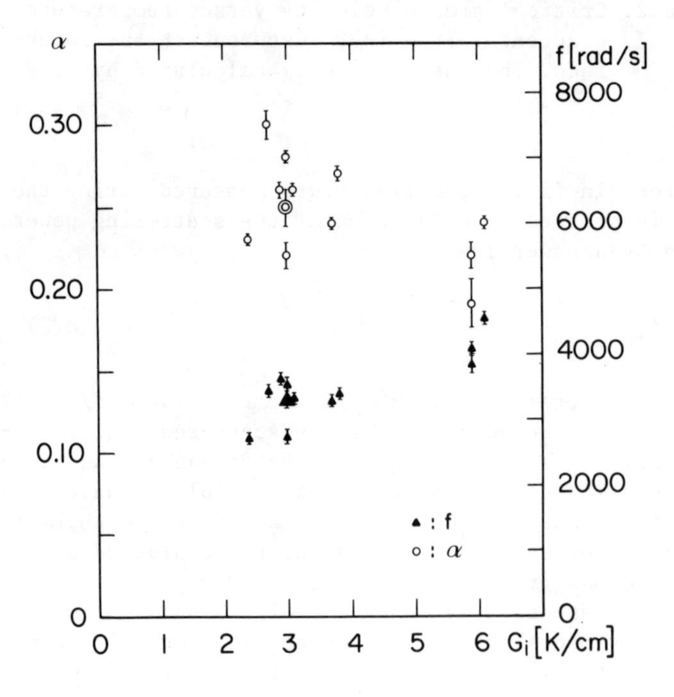

Figure 4. Dependence of the parameters α (circles) and f (triangles) from equation (2) on the temperature gradient in the ice. The emphasized points represent data measured during a run where the intensity was decreasing and the linewidth increasing.

2.2. The Far From Equilibrium Steady State

We obtain information about the nature of the dynamical process at the solid-liquid interface from the statistical properties of the photon counts of the photomultiplier, which detects the light, which is scattered in the interface region. The experimental observations and the conclusions, which can be drawn about the scattering inhomogeneities, are:

1.) The autocorrelation function of the photoncounts can be fitted by one single exponential, which is characterized by a decay time τ. The decay rate $1/\tau = \Gamma$ is the linewidth of the quasi-elastically scattered light.
   - *There is one relaxation process.*

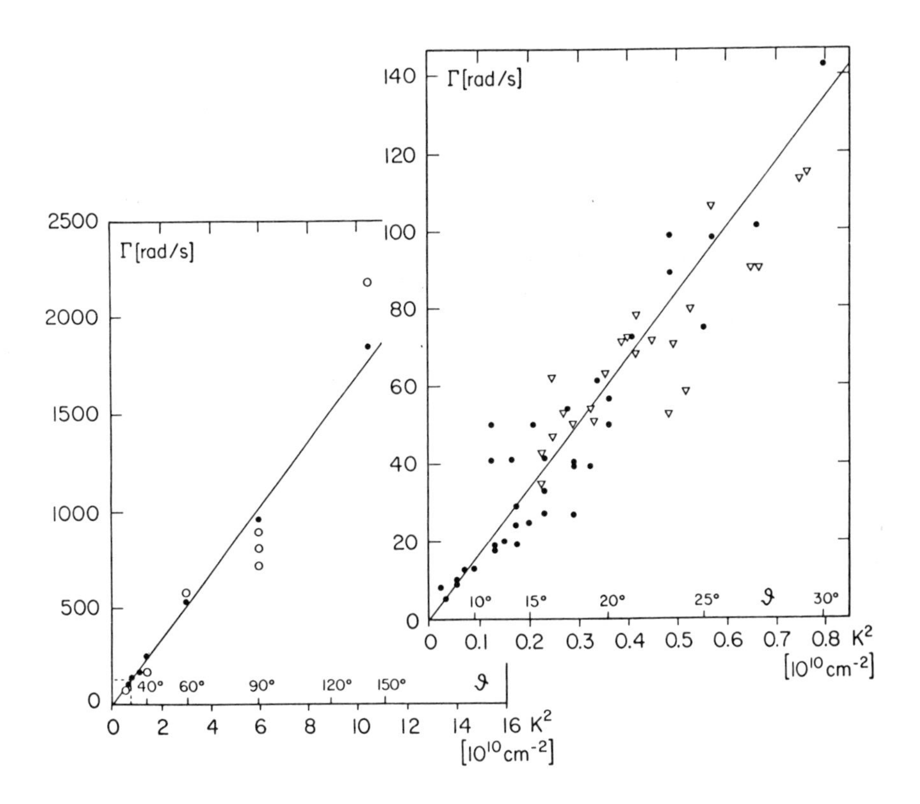

Figure 5. Linewidth of the light scattered with k parallel (open circles and full circles) and k perpendicular (triangles) to the solid-liquid interface. The insert is expanded by a factor of twenty (growth rate = 1.35 μm/s).

2.) The linewidth of the scattered light is proportional to the square of the scattering vector k
- *The growth and decay of the density fluctuations which scatter light can be described by a diffusion law:*

$$\Gamma = D_i \, k^2 \tag{3}$$

D_i is the diffusion constant at the interface. A typical value for D_i is $3 \cdot 10^{-8}$ cm^2/s.

3.) D_i does not depend on the orientation of the scattering vector relative to the interface. D_i depends only on the magnitude of k. (Fig. 5)
- *The diffusion process is isotropic in space.*

4.) By illuminating the interface from the ice side the light is first refracted and then scattered.
- *Light scattering takes place at the liquid side of the interface.*

5.) The intensity distribution in space is like the one of a radiating Hertzian dipole.
- *The dimensions of the scattering inhomogeneities are small compared to the wavelength of visible light.*

6.) The linewidths measured simultaneously for two different scattering vectors vary in the same sense with time. They always lead to an identical D_i.
- *Changes in the linewidth originate in a change of the scattering medium and not in the statistics of the autocorrelation function.*

7.) Light can be detected which is scattered quasielastically at a scattering angle $\theta = 15^o$ when the scattering vector is oriented perpendicular to the interface. (Fig. 5)
- *There is a Fourier component of the density modulation in front of the crystal surface with a spatial wavelength of 1.4 μm. This is a lower limit for the thickness of the scattering region.*

8.) The proportionality between Γ and k^2 is independent of the wavelength of the light.

9.) The intensity of the scattered light as a function of the angle of incidence ψ_o has a maximum at the critical angle of total reflexion $I_c(\psi_o > \psi_{crit}) \geq 4 \cdot I_c(\psi_o \ll \psi_{crit})$. I_c does not vanish for grazing incidence (points in fig. 6).

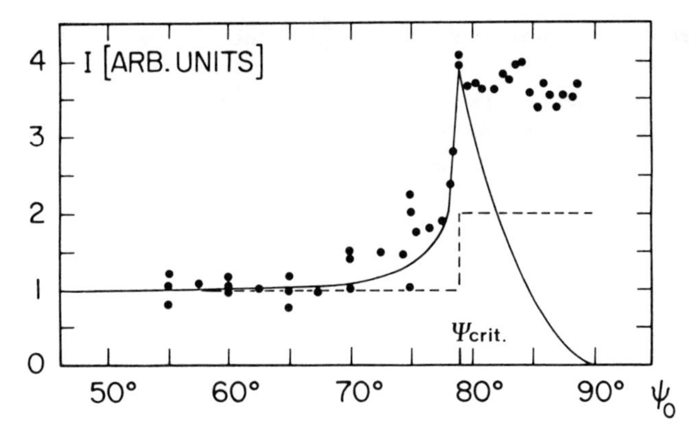

Figure 6. Dependence of the scattered intensity on the angle of incidence (growth rate = 1.3 μm/s). The solid line is the calculation for a rough interface. The dashed line is the calculation for a sharp, plane interface with suspended particles in front of it.

3. DISCUSSION

The questions to be discussed in this paragraph are: what does scatter light at the solid-liquid interface and what is performing a random walk process, which leads to the diffusion law observed in the experiment. Three models have been developed:

Model A) There is a corrugated interface.

Model B) There are particles immersed in the melt in front of the crystal.

Model C) There is a homogeneous phase inbetween solid and liquid with properties different from those of the melt and the crystal.

If the interface is illuminated in grazing incidence and the scattering plane is in the interface plane, then these models can not be distinguished by quasielastic light scattering. In all three models the dynamics can be described by a diffusion law: $\Gamma = D_i k^2$.

Model A) The corrugations grow and decay again. Using the Gibbs-Thomson equation one can calculate [8]

$$D_i = \mu S/\gamma \qquad (4)$$

where μ is the dynamical constant of crystal growth, S the melting entropy and γ the solid-liquid interfacial energy.

Model B) D_i is the diffusion constant of particles.

Model C) Fluctuations in the intermediate phase are slowed down entropy fluctuations. D_i is the constant of "structure diffusion".

These three models can be distinguished if linewidth and intensity of the scattered light are measured at scattering vectors which are oriented out of the interface plane.

Model A) According to observation 5) the amplitudes of the corrugations are small compared with their spatial wavelength. For small amplitudes the intensity of the scattered light as a function of the angle of incidence can be calculated [14] (fig. 6 drawn line). Under the condition that the amplitude of the corrugations is small compared to the wavelength of the scattered light, all the changes in the density distribution take place in the interface plane. Therefore only the projection of the scattering vector onto the interface plane should be effective in the determination of D_i. This is in contradiction to the observations 3) and 7). Therefore from intensity as well as linewidth measurements we have to conclude that Model A) is not compatible with our experiments.

Model B) Particles in front of the interface lead to an isotropic diffusion process. The intensity of the scattered light does not depend on the angle of incidence ψ_o except at the critical angle of total reflexion ψ_{crit}. For low angles of incidence ($\psi_o < \psi_{crit}$) the incident laser beam crosses the sampling volume once. For $\psi_o > \psi_{crit.}$ the incident beam is totally reflected and crosses the scattering volume twice. Therefore an increase of the scattered intensity by a factor 2 is expected at $\psi_o = \psi_{crit.}$ (dashed line in fig. 6). It is different from the experiment.

Model C) The model is depicted in fig. 7. Following assumptions are made: According to observation 7) the thickness of the layer

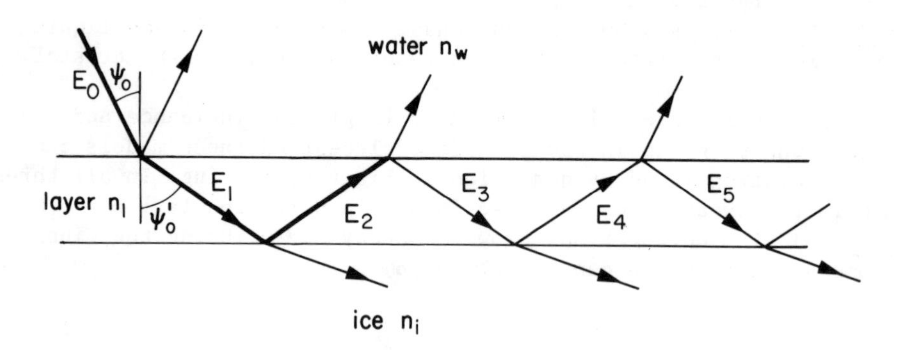

Figure 7. The model of the interface layer. ψ_o: angle of incidence, ψ_o': angle of refraction within the layer, E_i: electric field amplitudes, n_x (x = w, l, i): refraction indices of water, layer and ice.

is large compared to the wavelength of the light. Inside of the layer the light is scattered by entropy fluctuations. The scattering intensities arising from different parts of the zigzag path of the primary beam inside the layer are additive (observation 5). The interfaces water-layer and layer-ice are sharp. For the calculation of the coefficients of transmission and reflexion we used the Fresnel formula to obtain the electric field inside of the layer as a function of ψ_o for different indices of refraction in the layer n_1. For $n_1 = 1.333$ we obtain the best agreement of the calculated dependence of the intensity inside of the layer on ψ_o with the measured scattering intensity. In this simple model no mechanism is included for coupling of light into the layer at grazing incidence. To do this a gradient in n_1 might be introduced. For illustration: $n_1 = 1.333$ corresponds to the index of refraction of water at -8.5°C.

4. PROPERTIES OF THE LAYER

The index of refraction of the material in the layer can be estimated to be 1.333. The total intensity of the quasi elastically scattered light is proportional to the isothermal compressibility. From intensity measurements we estimate it to be about a factor 500 higher than the compressibility of water at 0°C. A diffusion constant, which describes the dynamics of the entropy fluctuations, has been determined to be about $3 \cdot 10^{-8}$ cm²/s. This value has to be compared with the thermal diffusivity in water $\alpha \approx 10^{-3}$ cm²/s or the constant for self diffusion in water $D_{H_2O} \approx 10^{-5}$ cm²/s.

The crystal is growing into the layer. Freezing is accompanied by a change in entropy and structure. Therefore we assume that the freezing rate is determined by D_i. For the calculation of the freezing rate we use the Wilson-Frenkel equation [15]:

$$v = \frac{D}{\ell} \frac{L}{T_E} \frac{1}{k_B} \frac{\Delta T}{T_E} \tag{5}$$

Here D is the rate determining diffusion constant - we use D_i. ℓ is the thickness of the boundary between the layer and the crystal - we use $\ell = \gamma/L$. This quotient is independent of substance [16], the numerical value is a few Angstrom. L/T_E is the melting entropy and k_B is the Boltzmann constant. $\frac{L}{T_E k_B} = 2.65$ for the water-ice transition. For the transition layer-ice the entropy change is somewhat smaller, we use $S/k_B \approx 1$.

Thus we obtain for the freezing rate

$$v = D_i \frac{L}{\gamma} \frac{\Delta T}{T_E} \qquad (6)$$

v depends only on D_i and $\Delta T/T_E$. L/γ is material independent. With $v/\Delta T = \mu$ we obtain

$$D_i = \mu \cdot S/\gamma \qquad (7)$$

This is the same expression which has been obtained assuming a corrugated interface and using Gibbs-Thomson equation (eq. 4). In table 1 the values calculated from eq. 7 are compared with the D_i values as determined from the light scattering experiment for ice and salol. μ, S and γ have been determined experimentally in different laboratories. The agreement of the results is striking. It confirms or assumption, that D_i is a quantity, which plays an important role in the dynamics of solidification. This is supported by the observation that the dendritic growth of crystals also can be described by D_i [4,23].

		ICE	SALOL
S	J/cm^3K	1.1 [17]	0.324 [26]
γ	J/cm^2	$2.9 \cdot 10^{-6}$ [18]	$2 \cdot 10^{-6}$ [9]
$\mu = v/\Delta T$	$cm/s \cdot K$	10^{-2} [19,20]	$2.3 \cdot 10^{-4}$ [22] 10^{-4} [9]
D_i from eq. 7	cm^2/s	$2.6 \cdot 10^{-8}$	$1.5 \cdot 10^{-9}$ $6.5 \cdot 10^{-10}$
light scattering experiment D_i	cm^2/s	$3 \cdot 10^{-8}$	10^{-9} [9]

Table 1. Comparison of measured and calculated values of the constant of "structure diffusion" D_i for ice and salol.

The Gibbs-Thomson equation and the Wilson-Frenkel equation are both based on free energy considerations and both lead to $D_i = \mu \cdot S/\gamma$. Probably this combination of thermodynamic quantities is a means for book keeping of the free energies which are involvolved in the freezing process.

During the freezing process ordering of the molecules and heat transport take place. As mentioned by Frenkel [24] it is difficult to separate these two processes. Therefore he used the expression "structure diffusion". Probably this is the best word to describe the quantity D_i, which we measure in our experiment. The question is raised: why is "structure diffusion" so slow ?

All attempts to give an answer to this question have to take into account, that we perform our experiments at non equilibrium conditions. The increase of the scattering intensity can be correlated to an increase of the correlation length of the density fluctuations. We also observe a slowing down of the dynamics of these fluctuations. Such a behavior is well known in critical phenomena. For non equilibrium experiments such a behavior is expected for fluctuations in a gradient in the chemical potential or close to an instability point. The detection of these long range fluctuations during a first order phase transition might contribute to a better understanding of homogeneous nucleation in condensed phases. Experiments, where nucleation in liquid-liquid decomposition [25] or nucleation in supercooled melt [26, 27] have been studied, lead to discrepancies between theory and experiment which cannot be explained by simple modifications of the classical nucleation theory.

ACKNOWLEDGEMENTS

This work is supported by the Swiss National Science Foundation.

REFERENCES

1. Sekerka, A.F. 1976, in: "Proc. of the Darken Conference" R.A. Fisher Ed., pp. 301-327
2. Huang, S.-C. and Glicksman, M.E. 1981, Acta Metall 29, pp. 701-715
3. Schneider, H. 1982, Diploma Thesis ETH Zürich
4. Bilgram, J.H. 1982, Naturwissenschaften 69, pp. 472-478
5. Langer, J.S., Sekerka, R.F. and Fujioka, T. 1978, J. Crystal Growth 44, pp. 414-418
6. Bilgram, J.H. 1982, in "Nonlinear Phenomena at Phase Transitions and Instabilities", T. Riste Ed. Plenum Press, New York, pp. 343-370
7. Cahn, J.W., Hillig, W.B. and Sears, G.W. 1964, Acta Metall. 12, pp. 1421-1439

8. Bilgram, J.H., Güttinger, H. and Känzig, W. 1978, Phys.Rev. Lett. 40, pp. 1394-1397
9. Dürig, U. and Bilgram, J.H. 1982, in "Nonlinear Phenomena at Phase Transitions and Instabilities", T. Riste Ed., Plenum Press, New York, pp. 371-378
10. Landman, U., Cleveland, Ch.L. and Brown, Ch.S. 1980, Phys.Rev.Lett. 45, pp. 2032-2035
11. Broughton, J.Q., Gilmer, G.H. and Jackson K.A. 1982, Phys.Rev.Lett. 49, pp. 1496-1500
12. Güttinger, H., Bilgram, J.H. and Känzig, W. 1979, J.Phys. Chem. Solids 40, pp. 55-66
13. Teixeira, J. and Leblond, J. 1978, J. Physique Lett. 39, pp. L 83 - L 85
14. Bilgram, J.H. and Böni, P. 1980 in: "Light Scattering in Liquids and Macromolecular Solutions, Ed.: V. Degiorgio, M. Corti and M. Giglio, Plenum Press, New York, pp. 203-213
15. Jackson, K.A. 1975 in "Treatise on Solid State Chemistry, Vol. 5, Changes of State" Ed.: Hannay H.B., Plenum Press, New York, pp. 233-282
16. Holloman, J.H. and Turnbull, D. 1953 in: "Progr. in Metal Physics, Vol. 4" Ed.: Chalmers, B., Pergamon Press, London pp. 333-388
17. Hobbs, P.V. 1974, "Ice Physics" Clarendon Press, Oxford
18. Hardy, S.C. 1977, Philos. Mag. 35, pp. 471-484
19. James, D.W. and Sekerka, R.F. 1967, J.Crystal Growth 1, pp. 67-72
20. Hillig, W.B. 1958, in: "Growth and Perfection of Crystals" Ed.: Doremus,R.H., Roberts, B.W. and Turnbull, D., Wiley, New York, pp. 350-360
21. Pollatschek, H. 1929, Z.phys.Chem. A 142, pp. 289-300
22. de Leeuw den Bouter, J.A. and Heertjes, P.M. 1971, J.Crystal Growth 5, pp. 19-25
23. Bilgram, J.H. 1983, Ann. New York Acad.Sci., 404
24. Frenkel, J. 1946, "Kinetic Theory of Liquids" Oxford, The Clarendon Press.
25. Heady, R.B. and Cahn, J.W. 1973, J.Chem.Phys. 58, pp.896-910
26. Dorsey, N.E. 1948, Trans.Amer.Philos.Soc. 38, pp. 245-328
27. Brown, A.C., Unger, U. and Klein, W. 1982, Z.Phys.B. - Condensed Matter 48, pp.1-4

INSTABILITIES IN THE PULSED LASER IRRADIATION INDUCED MELTING AND SOLIDIFICATION

E.Rimini and S.U.Campisano

Istituto Dipartimentale di Fisica - 57 Corso Italia
I95129 Catania - Italy

Irradiation of materials with pulsed laser or electron beam of nanosecond duration causes a deposition of a large amount of energy in short times and into the near surface region of the sample. As a result of the irradiation the surface layer can melt. Melting and solidification occur at extremely high velocities: 100 m/sec for the melt-in front propagation and 10 m/sec for the resolidification front. It is possible to by pass solid state thermally activated processes and then to transform an amorphous layer into a strong undercooled liquid. During solidification the high velocity and then the large undercooling causes in some cases the formation of an amorphous phase being the interface unstable. In the presence of dissolved impurities the melting and solidification process modifies the impurity distribution. Supersaturated alloys are formed in semiconductors and in metals. The segregation coefficient increases with the solid-liquid interface velocity and again at high velocities interfacial instabilities arise and formation of cell structures will result.

INTRODUCTION

Pulsed laser or electron beam offer a unique method to deposit a large amount of energy into the first few microns of the irradiated material so rapidly that the bulk temperature is not affected. This method was applied initially to anneal the damage in ion implanted semiconductors [1] and to activate electrically the

G. Nicolis and F. Baras (eds.), Chemical Instabilities, 367–383.

implanted species. Since this initial motivation many applications not only limited to the semiconductor technology have been considered.

In the meantime a large effort [2] has been paid to the investigation of the laser light interaction with materials at extremely high photon fluxes. The coupling of the laser beam with the target involves the generation of free carriers and the energy transfer from the excited carriers to the lattice. As a result of the irradiation the sample is heated and by a suitable energy density value of the impinging beam the near surface can be melted.

Extremely high heating and cooling rates can be reached so that melting and solidification occur in a novel regime at high velocities. According to the purity of the material, metastable phases, new phases are formed. Before to discuss some phenomena which might be of interest to the chemical instabilities subject, we consider briefly the experimental conditions [3] of irradiation. Nd:YAG or glass (λ=1.06 μm), ruby (λ=0.69 μm) and excimer (λ=0.342 μm) lasers are mainly used. The pulse duration is either in the nanosecond or in the picosecond regime, and the energy density amounts to 1 J/cm^2. Under suitable condition the surface is melted to a depth of several hundredth nanometers. During solidification liquid phase epitaxial regrowth can occur from the underlying substrate. The velocity of the resolidification front depends upon the rate at which the latent heat of crystallization can be extracted from the solidifying interface. In these cases the thermal gradients range between 10^6-10^7 K/cm with concomitant quenching rates of 10^9-10^{10}K/sec. The solidification velocity is of the order of 1-10^2 m/sec, to be compared with 10^{-5} m/sec, typical growth rate of conventional liquid phase crystal growth as in the Czochralski process.

The basic mechanism responsible for instance of the annealing of ion implanted semiconductors is shown schematically in Fig.1. A single crystal substrate is overlaid with an amorphous layer and is irradiated with a pulsed electron or laser beam. The absorbed energy is converted into heat and at a particular energy density value the near surface is melted. If all the amorphous layer is liquid the subsequent solidification occurs on a single crystal substrate and a liquid phase epitaxy will result.

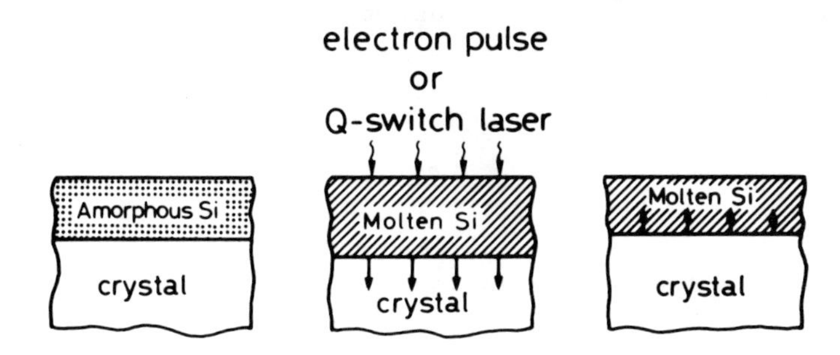

Fig.1. Crystallization of amorphous Si on single crystal substrate by pulsed laser or electron beam irradiation.

If the thickness of the molten layer is lower than the amorphous thickness a polylayer will be obtained. This example justifies also the tendence to indicate under "laser annealing" a large number of phenomena associated to pulsed laser irradiation although not directly related to the removal of damage.

In the following we will describe some experiments performed in the field of melting and solidifications at high rates. In particular we will consider the transition from the amorphous to the liquid phase [4,5], by passing crystallization, and the amorphous formation by irradiation of Si crystal [6]. The solidification front becomes unstable at high velocities, and amorphous nucleation is initiated at the interface breakdown points.

Impurities dissolved in the liquid can influence and are influenced by the solidification process [7]. Trapping, dependence of the interface segregation on the velocity during resolidification, instabilities of the planar interface and formation of cell structures represent another area of interest to the meeting and some relevant experiments will be described. At this stage we prefer, in view also of our background to point out mainly, the experimental results and a simple thermal treatment [8].

THERMAL TREATMENT AND RESOLIDIFICATION VELOCITY

A simple description of the heating effects induced by laser irradiation can be done on the assumption of an "instantaneous" ($\leq 10^{-11}$sec) conversion of the photon energy into heat which then propagates inside the sample according to the heat equation prediction. In the one dimensional problem and for a light incident along the z axis of a target with composition homogeneous in the X-Y plane the heat diffusion equation becomes

$$\frac{\partial T}{\partial t} = \frac{\alpha}{\rho C_p} I(z,t) + \frac{\partial}{\rho C_p \partial z}(K\frac{\partial T}{\partial z}) \quad (1)$$

where I(z,t) is the power density of the laser light at depth z and time t, ρ, C_p, K and α are the density, specific heat, thermal conductivity and absorption coefficient of the sample. In a homogeneous medium $I(z,t)=I_o(t)(I-R)\exp(-\alpha z)$, being $I_o(t)$ the temporal power output from the laser and R the target reflectivity. The variations of optical and thermal parameters with temperature and structure of the irradiated material, absorption and release of latent heat during melting and solidification must be taken into account in Eq.1 and it must be solved [9] numerically.

Several informations are obtained by the solution of Eq.1 with the appropriate boundary conditions: temperature distribution inside the sample and time dependence, melt propagation and penetration, velocity of the resolidification front. As an example the dynamic of the melt penetration and its dependence on the energy density is shown in Fig.2 for 30 nsec ruby laser pulse ir-

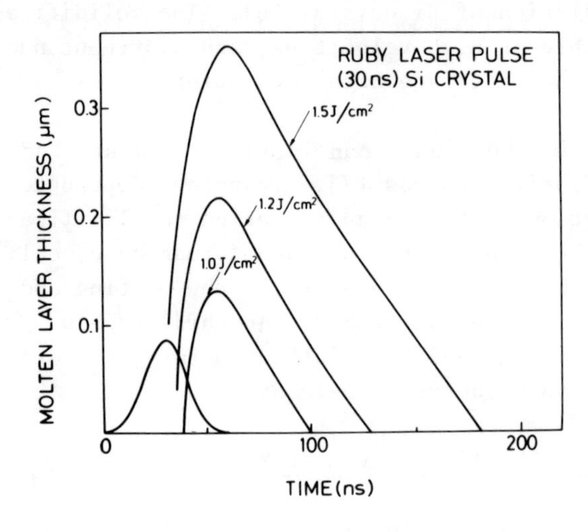

Fig.2

Kinetics of melt front in Si crystals irradiated with 30 ns ruby laser pulses of different energy densities.

radiation of Si single crystal. The threshold for surface melting is $\sim$0.8 J/cm^2. The thickness of the molten layer increases with a typical rate of 0.6 μm/J/cm^2. The melting occurs usually during irradiation, solidification instead takes in a time interval of about ten times the pulse duration.

In a pure substance solidification is governed entirely by heat flow. The rate of solidification at any point along the liquid-solid interface is determined by how rapidly the latent heat generated at that point can be conducted into the bulk of the sample. In addition solidification requires a certain degree of undercooling. The data of Fig.2 were obtained neglecting undercooling effects and temperature gradient in the liquid. The estimated resolidification velocity given by the slope of the curves of Fig.2 is of about few meters per second and exceeds of several orders of magnitude the conventional growth rate. Measurements during irradiation by transient electrical conductance support [10] nicely the velocity values estimated on the basis of the simple thermal model. In these experiments one takes advantage of the large increase ($\sim$30 times)of Si conductance upon melting.

In a typical experiment the sample is irradiated with a pulsed laser and for each pulse the electrical conductivity is measured as a function of time. Simultaneous measurements of the reflectivity of a probe laser light check the melting occurrence, its starting and its duration. A typical set of measurements is shown in Fig.3. The time dependent voltage v(t) is reported for a variety of laser energy densities in the lower part of the figure. The signal observed at energy densities below the melt threshold, 0.8 J/cm^2, is due to the photoconductivity.

The reflectance signal, in the upper part of Fig.3, indicates the onset of the high reflectivity phase and its duration. Comparison of experimental data with calculation is shown in Fig.4, where the voltage data are converted into melt depth through a detailed analysis. The solid-liquid interface velocity as determined by the slopes of the curves agrees quite well with the estimated values.

The use of Si on sapphire because of the extremely short lifetime of the photogenerated carriers has allowed measurements [11] and determination of the melt-in front propagation. In the case of ultraviolet Q-switched pulse of 10 nsec duration, velocities of 100 m/sec were determined [12].

FAST HEATING AND COOLING

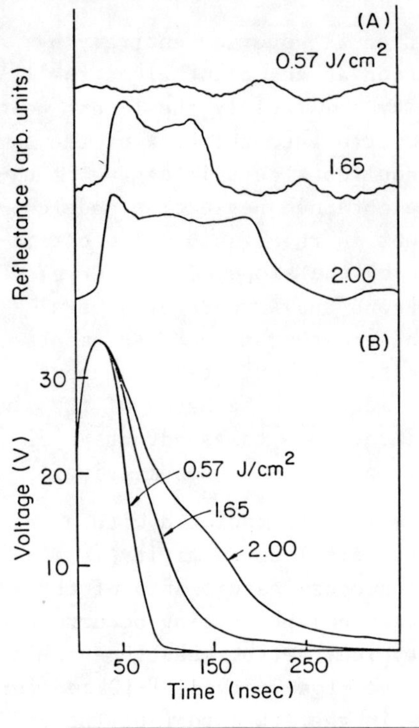

Fig.3. Reflectance of a He-Ne probe vs time for three different energy densities of ruby laser pulses (a). Electrical conductance, measured by the voltage across the scope load resistor, as a function of time (b).

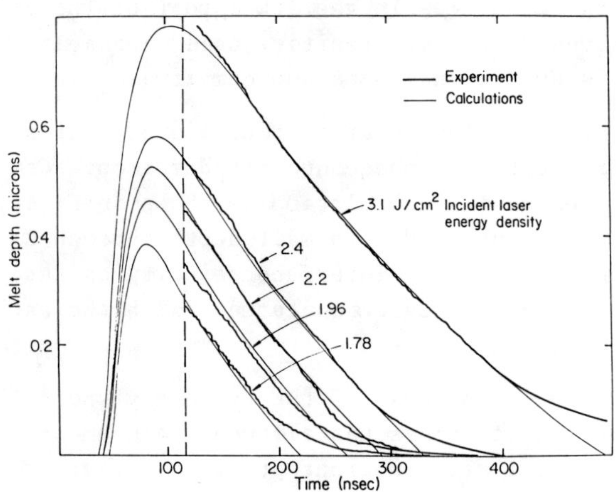

Fig.4. Experimental (heavy lines) and computer calculated (light lines) melt depths as a function of time for several energy densities of laser pulses.

As an application of the unique possibilities offered by pulsed heating methods we consider the melting of an amorphous layer and the single crystal to amorphous transition in irradiated Si samples.

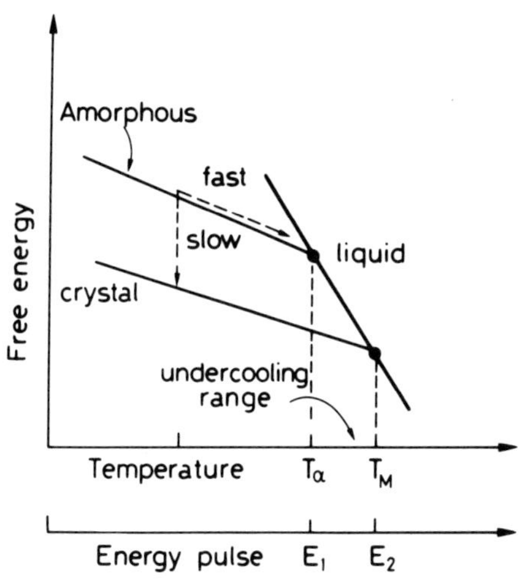

Fig.5. Free energy diagram of amorphous, crystalline and dense metallic liquid silicon. T_{α} and T_m are the melting points of the amorphous and of the crystalline phase respectively. The molten layer is undercooled for the E_1-E_2 energy density range.

The free energy of the amorphous phase is higher than that of the crystalline one. A diagram for silicon is shown [13] in Fig.5. The intersection of the free-energy curves of crystalline and liquid silicon is by definition the equilibrium melting point T_m. Assuming that the phase transition from the tethraedral fourfold coordinated amorphous phase to the close packed [11-12] coordinated metallic liquid is first order, the melting point, T_{α}, of the amorphous is lower than T_m. Calculations based on the measured change of enthalpy give few hundredth degrees as depression.

The usual heating procedure does not allow the observation of the melting point depression because the unstable amorphous phase will crystallize before T_{α} can be reached. Pulsed irradiation can avoid crystallization, by passing solid state processes thermally activated. Not only fast heating but also suitable thermal gradients are required. Electron beam [4] or laser irradiation [5] on the backside of the sample produce similar temperature distributions characterized by a small gradient.

Experiments [4-5] indicate that the amorphous layer melts at temperature lower than T_m. The liquid is then in a regime of strong undercooling and a layer with a double structure results: nucleation of crystalline grains at the sample surface and solification at high velocities on the single crystal substrate. The estimate melting temperature and enthalpy of the amorphous silicon are T_{α}=1200K and $\Delta H_{\alpha-\ell}$=1250 J/gr to be compared with T_m=1700K and $\Delta H_{c-\ell}$=1790 J/gr respectively. Heat of crystallization of amorphous Si obtained by ion implantation was measured [14] recently by differential scanning method. The obtained $\Delta H_{\alpha-\ell}$ value was 11.3±0.8 and the estimated T_α was below T_m.

The resulting structure of an amorphous layer irradiated by front laser (a) and back or front electron beam (b) is illustrated in Fig.6. In the case of front laser irradiation the sample surface reaches T_α at an energy density E_1. With increasing the ener-

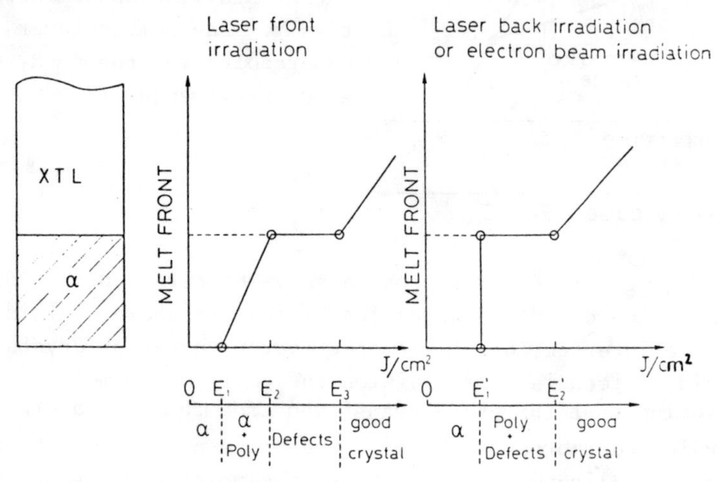

Fig.6. Melt front penetration for front and back irradiation as a function of the pulse energy density. The resulting structures are shown in the lower part.

gy density the molten region at T_α extends deeper and at E_2 all the amorphous layer is liquid. In the range E_2-E_3 the absorbed energy increases the temperature of the liquid. Normal liquid at $T>T_m$ near the surface and undercooled liquid at $T<T_m$ near the interface coexist.

The E_2 value represents the energy density threshold for the amorphous-to-single crystal transition but the crystalline quality of the regrown layer is very poor. Nucleation and growth of defects easily occur from the undercooled liquid in contact with the single crystal substrate. Only at energy values higher than E_3 the initial amorphous layer is above T_m, with part of the underlying single crystal substrate molten.

In back laser or electron beam irradiation the amorphous layer starts to melt at an energy density value $E_1^{\prime}$. All the layer, nearly, becomes liquid simultaneously for the low thermal gradient. Up to E_2' value the underlying single crystal is still solid. In the E_1'-E_2' range the molten layer is undercooled, nucleation and fast crystallization occur easily. Above E_2' a good crystalline epitaxial regrowth results.

If the free-energy curve of the liquid silicon can be followed from right to left, the molten layer will be found in a condition of strong undercooling. This amorphous phase [6] [12] can only nucleate when the temperature of the undercooled melt falls below the equilibrium melting point of α-Si. Experiments should be performed using particular conditions in such a way that the thermal gradients allow a fast propagation velocity of the solid-liquid interface. The corresponding undercooling should be enough to reach the equilibrium amorphous melting point.

Amorphous layers have been obtained using either picosecond pulses of 0.53μm [15] or nanosecond pulses of 0.347μm [11]. The thickness of the layers depends on the substrate orientation, it is much higher for (111) than for (100) orientation. The amorphous-crystalline interface, as shown by transmission electron microscopy, is undulating with an amplitude of about 30% of the average thickness. The measured liquid-solid velocity for the amorphous formation is 15 m/sec.

The fact that not all molten layer becomes amorphous would suggest that there are strong kinetic limitations on the nucleation of the amorphous phase. Probably nucleation is initiated by interface instabilities at high velocities, which needs some time to form, alternatively the delay might be attributed to the time required to undercool the liquid below the transition temperature. In any case the solidification process becomes unstable and the front advances into a metastable phase, i.e. into an undercooled melt.

Enhancement of the amorphous thickness was obtained [16] in Bi-doped Si irradiated with picosecond pulses. The results are shown in Fig.7: the amorphous thickness is about 50% higher than in the undoped crystal. The influence of the impurities, also at concentrations of 1 at%, in reducing the barrier for amorphous nucleation seems then relevant. For most metallic elements amorphous phase formation cannot be achieved without alloying additions.

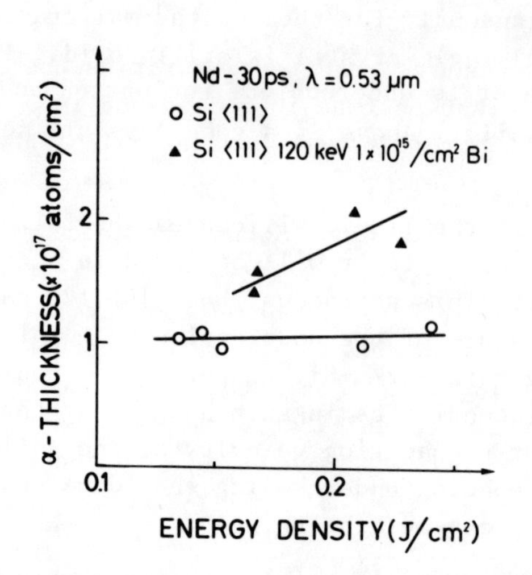

Fig.7. Thickness of amorphous Si obtained by picosecond irradiation of undoped and Bi-doped Si single crystals.

IMPURITY REDISTRIBUTION AND CELL STRUCTURE

Irradiation of sample containing solute in dilute concentration is also used [7] to study the behaviour of alloy at these high velocities. The formation of a liquid layer modifies strongly the dopant distribution. The diffusion coefficient of most impurities in liquid Si is of the order of 10^{-4} cm^2/sec. The surface layer remains molten for a time of 10^{-7}sec so that the broadening of a narrow impurity profile amounts to 40-100nm that can be easily measured by several techniques.

The dopant depth profile is changed by the movement of the liquid-solid interface. Each impurity is characterized by an equilibrium distribution coefficient K_o, ratio of solubilities in solid and liquid phase respectively. In Si this quantity is usually less than one for all the impurities. The solid front during resolidification rejects then impurities in the liquid and the process is described in terms of a normal freezing.

If equilibrium parameters are used to describe the laser induced redistribution, a very large fraction of the dissolved impurities should be rejected at the surface. For example K_o for Te in Si is 4×10^{-6}, so that only a negligible fraction of impurities should be retained inside the sample. Instead a large amount of impurities in great excess of the maximum solid solubility limit is found after irradiation, in substitutional lattice sites. The supersaturated alloys formed by laser irradiation indicate clearly that the solidification process occurs far from thermodynamic equilibrium conditions.

To account for the large amount of retained impurities the interfacial segregation coefficient must be increased and in a comparison of experimental depth profile with a model calculation it can be used as a fitting parameter.

The impurity profile in the solidified layer cannot be calculated analytically and a numerical [9] procedure must be used. In the model the impurity is allowed to diffuse in the liquid layer according to the concentration gradient and the known diffusivities. The advancing solid front rejects impurities into the liquid according to the fitting segregation coefficient parameter. An example is shown in Fig.8. The concentration profile in the so-

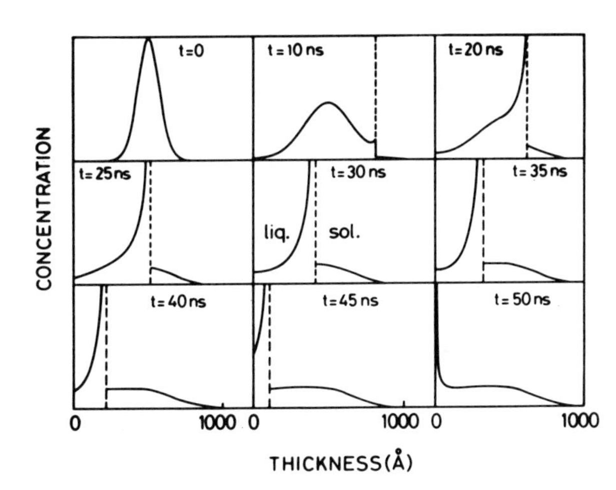

Fig.8. Computed impurity profiles during solidification. The liquid-solid interface moves to the left at 2.0 m/s. The diffusion coefficient of the impurity in the liquid was assumed $10^{-4}cm^2/s$ and the segregation coefficient K=0.1.

lid and in the liquid is reported at several times of the solidification process. The continuous rejection of impurities at the interface will result in surface accumulation during the freezing of the outer surface layer.

Experimental profiles in the case of Te implanted Si samples irradiated at different substrate temperatures with 1.5 J/cm^2-30 nm ruby laser pulses are shown in Fig.9. The concentrations as measured by channeling effect technique in combination with MeV He^+ Rutherford backscattering, indicate that the higher surface accumulation is obtained at 600K substrate temperature. The difference in the profiles is caused by the change of the solidification velocity with the substrate temperature.

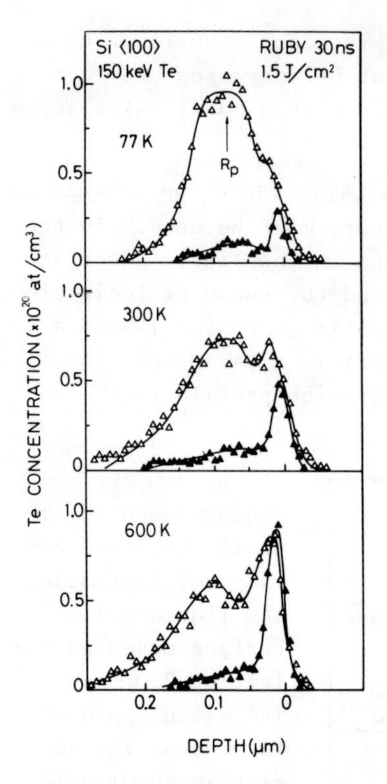

Fig.9. Concentration profiles of Te-implanted into <100> Si and irradiated with 1.5 J/cm^2, 30 ns ruby laser pulses at different substrate temperatures. (Δ) random incidence, (▲) aligned incidence.

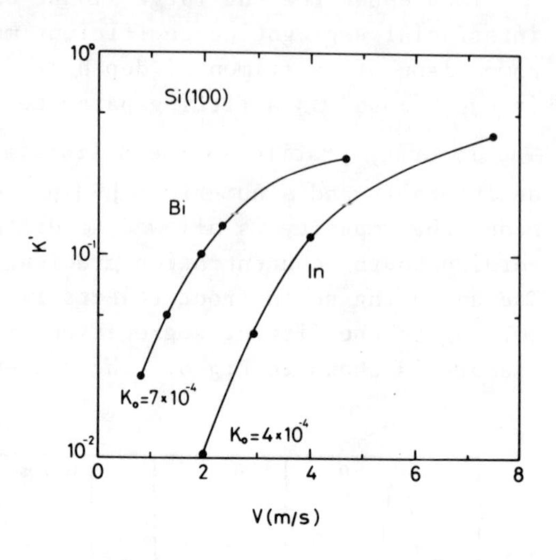

Fig.10. Measured segregation coefficients of Bi and In implanted into <100> Si substrates as a function of the liquid-solid interface velocity.

The major result of these experiments is the unique relation between the K' value needed to fit the surface accumulation and the solid-liquid interface velocity independent of the way it was obtained. Typical (17) K'-v trends are shown in Fig.10 for Bi and In-implanted Si systems in the 0.8-8.0 m/sec range. K' increases quite abruptly (note the log scale in the ordinate axis) and saturates at a value lower than 1.

Several models have been considered to describe non equilibrium dopant segregation at a fast moving liquid interface. A common feature of all these treatments is the diffusion coefficient D_i of the impurity at the interface region. It has been shown [18] that a reasonable estimate of D_i is given by $(D_s D_\ell)^{\frac{1}{2}}$ being D_s e D_ℓ the diffusion coefficients in the solid and in the liquid respectively. It is possible on this hypothesis, to predict if a given impurity will be trapped or rejected to the surface. If $D_i/\lambda \geq v_1\ \lambda$ interface few monolayers thick, the impurity is faster than the ℓ-s interface and it will accumulate in the liquid. The final profile will be characterized by an almost complete surface accumulation. If $D_i/\lambda \leq v$ the impurity will be trapped by the advancing solid front.

During the solidification process the impurities rejected in the liquid are allowed to diffuse. However due to the competition between growth and diffusion they accumulate in a layer of thickness D_ℓ/v, called "coherence length" (see the upper left part of Fig.11). The melting point of a dilute alloys depends on the

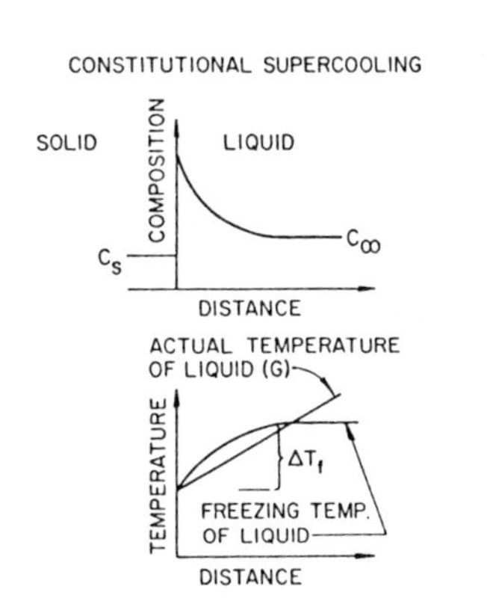

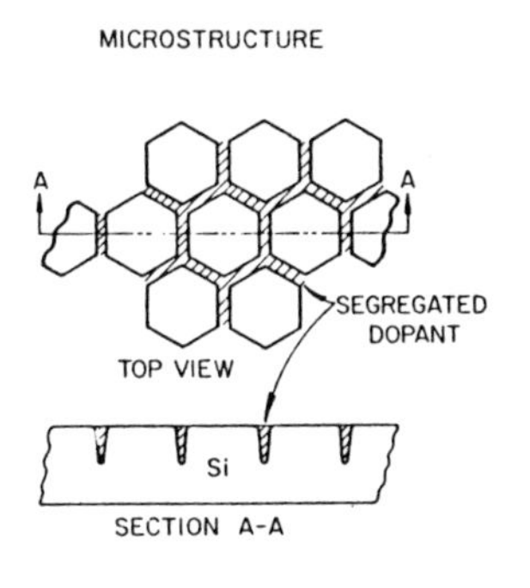

Fig.11. Illustration of the constitutional supercooling process: depth concentration and the corresponding temperature profile are shown in the left part, the microstructure is shown in the right hand part.

impurity concentration according to the liquidus line of the phase diagram. In the liquid layer close to the interface the solidification temperature changes according to the concentration profile. The resulting gradient must be compared with the real temperature gradient created by the heat flux (see lower left part of Fig.11).

The solidification temperature can then be smaller or larger than the real temperature. In the first case the molten is stable and a flat interface can propagate. In the second case a liquid layer of thickness comparable with the coherence length is constitutionally supercooled and the interface becomes unstable. It can be easily shown that the condition [19] for the propagation of a flat interface is

$$G_\ell > \frac{mC_s(i)v}{D_\ell} \left(\frac{1-K'}{K'}\right) \qquad (2)$$

The term on the left is the temperature gradient in the liquid and that on the right is the temperature gradient obtained from the phase diagram taking into account the concentration dependence of the solidification temperature. The function $D_\ell G_\ell/mC_s(i) \cdot K/1-K$ is reported in Fig.12 versus v. The step shape of this cur-

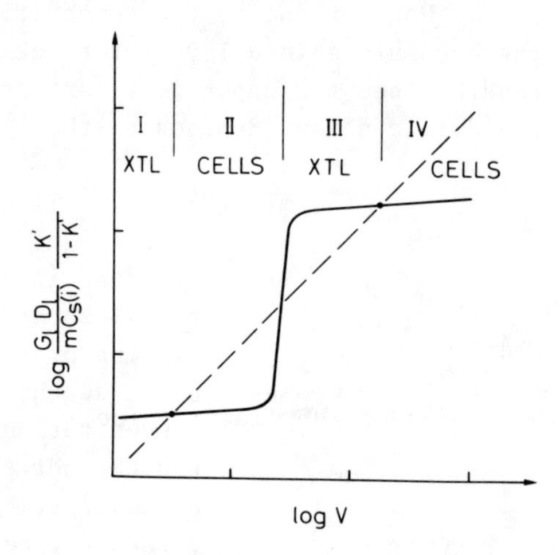

Fig.12. Schematic representation of the four different velocity regions alternating the conditions for the growth of a good single crystal to those for the occurrence of constitutional supercooling. The velocity dependence of K' is taken into account.

ve is mainly determined by the velocity dependence of K', if it saturates at large velocities to a value less than unity. This step like curve intersects the slope 1 line in three points determining four regions. In region I the rejected impurities have enough time to diffuse in the liquid giving rise to a small solidification temperature gradient. A flat interface can propagate then. Increasing the solidification velocity the rejected impurities form a narrow region with large concentration gradient. The conditions for constitutional supercooling are reached and a cell structure is developping as schematically shown in the right part of Fig.11. The transition from region I to region II occurs at low velocities and has been investigated in details in classical metallurgical works. The cell size is of of the order of the coherence length i.e. $\sim 10^{-4}$-10^{-3} cm. A further increase in the velocity changes drastically the interfacial segregation coefficient and a lower amount of impurities is rejected in the liquid. In region III then a flat interface can propagate again and, due to the large value of K' a large amount of impurities is incorporated in the growing solid phase. If the velocity is increased to extremely high values and K'<1 the concentration gradient starts to build up again in a narrower coherence length. Constitutional supercooling can then occurs again and the size of the cell structure will be rather small. Transmission electron observation [20-21] in In or Bi implanted Si indicated a cell size of the order of 10-100 nm as shown in Fig.13 for Bi-implanted

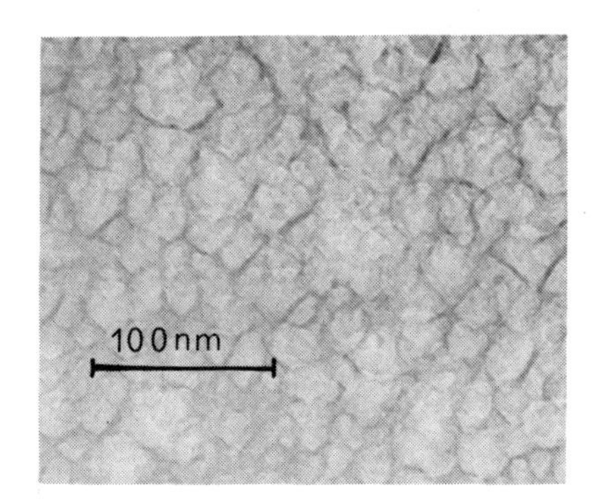

Fig.13. Transmission electron micrograph showing the cell structure in a Bi-implanted Si sample irradiated with 1.5 J/cm^2-30nsec ruby laser at 77 K substrate temperature. (Courtesy of A.G.Cullis).

Si irradiated at 77K with 1.5 J/cm^2-30ns ruby laser pulse.

A continuous increase in the solidification velocity of a liquid diluted alloy gives rise thus to an oscillating behaviour in the interface stability. It is interesting also to note that the width of the four regions depends upon dopant concentration. An increase in dopant concentration lowers the step-like function thus decreasing the maximum velocity at which a flat interface can propagate [22]

CONCLUSION

In this paper we have tried to describe briefly some relevant and new experimental data obtained by pulsed irradiation of materials. In the case of semiconductors, mainly Si, it has been possible to take advantage of the extremely high quenching rates, thermal gradients, and high velocity of the melting and solidification front to study crystal growth in a novel regime. Instabilities arise both in pure and doped semiconductors. In the first case amorphous Si structure is obtained at solidification velocity in excess of 15 m/sec. In doped material the interface instability causes, according to the segregation coefficient value, cell formation.

Most of the work has been done in silicon, in a comparison [23] with metals one must note that usually interface velocities are higher due to the greater thermal conductivity. Supersaturated alloys have been obtained by ion implantation in combination with electron or laser beam pulsed heating. In some cases the coupling between thermal and matter transport, i.e. the Soret effect, has been evidenced [24]. Precipitation in the liquid phase has been shown in Sb-implanted Al system together with the measurement of submicrosecond nucleation times. Amorphous phase formation requires usually alloys.

REFERENCES

1. Poate,J.M. and Mayer,J.W. editors "Laser Annealing of Semiconductors", 1982, Academic Press, N.Y.
2. Poate,J.M.,Foti,G.,and Jacobson,D.,C.,editors "Surface Modification and alloying", 1983, Plenum Press, N.Y.
3. Rimini,E., 1982, Physica Sripta T1, pp.56-61
4. Baeri,P.,Foti,G.,Poate,J.M. and Cullis,A.G., 1980, Phys. Rev. Lett. 45, pp. 2036-2039
5. Baeri,P.,Campisano,S.U.,Grimaldi,M.G., and Rimini,E., 1982, J. Appl. Phys. 53, pp. 8730-8733
6. Cullis,A.G.,Webber,H.C.,Chew,N.G.,Poate,J.M. and Baeri,P., 1982, Phys. Rev. Lett. 49, pp.219-221
7. Campinoso,S.U., 1983, Appl. Phys. A30, pp.195-211
8. Baeri,P.,Campinoso,S.U.,Foti,G., and Rimini,E., 1979, J. Appl. Phys. 50, pp.788-796
9. Baeri,P., and Campinoso,S.U., in ref.1 (Chap.4)
10. Galvin,G.J.,Thompson,M.O.,Mayer,J.W.,Peercy,P.S., Hammond, R.B., and Paulter,N., 1983, Phys. Rev. B27, pp.1079-1086
11. Thompson,M.O.,Galvin,G.J.,Mayer,J.W.,Peercy,P.S., and Hammond,R.B., 1983, Appl. Phys. Lett. 42,pp. 445-447
12. Thompson,M.O.,Mayer,J.W.,Cullis,A.G.,Webber,H.C.,Chew,N.G., Poate,J.M. and Jacobson,D.C., 1983, Phys. Rev. Let. 50, pp. 896-899
13. Turnbull,D., and Spaepen,F. in ref.1 (Chap.2)
14. Poate,J.M. in Nucl. Instr. Meth., 1983 (to be published)
15. Yen,R.,Liu,J.M.Kurz,H., and Bloembergen,N., 1982, Appl. Phys. A27, pp.153-162
16. Campisano,S.U.,Baeri,P.,Zhang,J.P.,Rimini,E. and Malvezzi, A.M., 1983, Appl. Phys. Lett., in press
17. Baeri,P.,Foti,G.,Poate,J.M.,Campisano,S.U., and Cullis,A.G., 1981, Appl. Phys. Lett. 38, pp. 800-802
18. Jindall,B.K., and Tiller,W.A., 1968, J. Chem. Phys. 49, pp. 4632-4639

19. Mullins,W.W. and Sekerka,R.F., 1964, J. Appl. Phys. 35, pp. 444-452
20. Cullis,A.G.,Hurle,D.T.J.,Webber,H.C.,Chew,N.G.,Poate,J.M., Baeri,P., and Foti G., 1981, Appl. Phys. Lett. 38, pp. 642-644
21. Narayan,J., 1981, J. Appl. Phys. 52, pp. 1289-1297
22. Lauger,J.S., 1980, Rev. Mod. Phys. 52, pp. 1-28
23. Picraux,S.T., and Follstaedt,D.M. in ref.2 (Chap.II)
24. Miotello,A. and Dona delle Rose L.F., 1981, Phys. Lett. A 91 pp. 153-156.

SPATIAL STRUCTURES IN NONEQUILIBRIUM SYSTEMS

G. Dewel*, P. Borckmans* and D. Walgraef*

Université Libre de Bruxelles, Chimie-Physique II,
C.P. 231, Campus Plaine, Bd. du Triomphe,
B - 1050 Bruxelles (Belgique).

We want to present in this note some topics in the study of stationary twodimensional nonequilibrium structures. Two experiments which have stimulated a reniewed interest in the theory of spatial patterns are described. The problem of pattern selection both in isotropic and anisotropic systems is reviewed. The dynamics of the phase is also presented.

1. INTRODUCTION : EXPERIMENTAL ASPECTS

It is now well known that many systems undergo an instability leading to stationary periodic structures when they are driven sufficiently far from thermal equilibrium [1]. Classical examples are provided by the Rayleigh-Benard (R.B) or the Marangoni instabilities in isotropic fluids.

It has long been predicted that the coupling between chemistry and hydrodynamics could give rise to stationary chemical waves [1 ,2]. Unfortunately experimental examples of such structures are scarce. However after the pioneering work of Möckel [3] a series of photochemical patterns have been experimentally characterized[4, 5]. Typically in these experiments a shallow solution (Petri dish) containing a chromogenic (irreversible) or a photochromic (reversible) compound is irradiated with U.V. or visible light. First the absorption layer, at the illuminated surface, colors uniformly but thereafter breaks down into inhomogeneous regions which then invade the bulk of the solution.

The mechanism of this patterning was however far from clear. In that context Micheau and al. have shown that there exists a li-

G. Nicolis and F. Baras (eds.), Chemical Instabilities, 385–400.

Fig I :

Photochemical structure : Mercury dithizonate irradiated with visible light.

near relationship between the average wavelength of the photochemical structures and the depth of the layer.

Then they have clearly demonstrated the importance of evaporation in the generation of these structures. Indeed on closure of the reactor the shapes fade away but reform on opening. Furthermore, using the schlieren technique they obtained photographs of striations (prepatterns) [6] existing in the fluid prior to irradiation. The existence of these vermiculated convective rolls was decisively proven by depositing ink on the bottom of the dish.

As a consequence in this series of experiments the photochemical reaction serves only to reveal preexisting convective motions which are induced by evaporative cooling [7]. On the other hand since similar prepatterns have been characterized in the pure solvents reported by Avnir and al., we think that the same phenomenon is responsible for the onset of structures in these open systems too. Similarly, any surface reaction (for instance with oxygen) yielding a coloured product will provide a way to visualize bulk motions as the coloured product is then advected on the convective rolls. Such a mechanism is probably also responsible for the onset of most of the stationary mosaic patterns observed in shallow layers of chemical systems [8].

Many dissipative structures have also been experimentally studied in anisotropic media : the convective structures in liquid crystal films [9], the superlattices of voids in irradiated metals

[10], the long range periodic precipatation in illuminated solid solutions [11]. Liquid crystals present a rich variety of hydrodynamic instabilities (R - B, electrohydrodynamic instability and the instability of a nematic subjected to an elliptical shear (N.S.E.S)) [9]. These convective structures can be observed directly thanks to the large birefringence of the nematics and moreover large systems (200 rolls ; 10^4 squares) can easily be prepared ; they really provide "nonequilibrium crystals". The price to be paid is the complexity of the corresponding hydrodynamic equations describing the coupling between the velocity, the temperature and the director fields. As an example let us briefly describe the instability of a (N.S.E.S) because these experiments performed by Guyon and coworkers have furnished a lot of informations on the "elasticity" of nonequilibrium structures [12, 13]. A nematic is sandwiched between two rectangular plates with the long molecules perpendicular to the boundary surfaces. It is subjected to an elliptical motion by imposing the following displacements respectively to the lower and upper plates

$$X = X_o \cos \omega t$$

$$Y = Y_o \sin \omega t$$

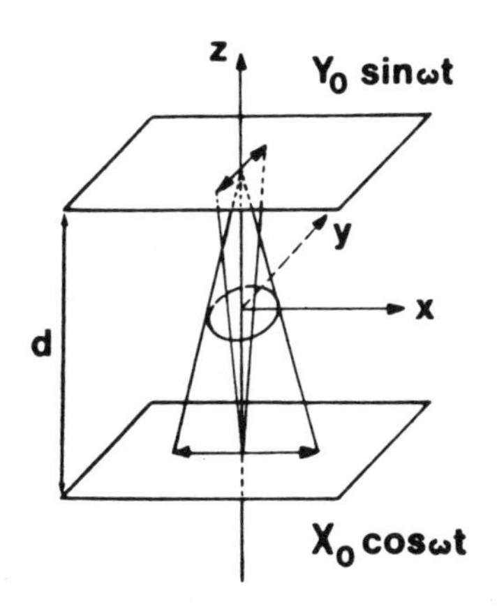

Fig. II :

Experimental set up in the case of a nematic subjected to an elliptical shear (After E. Guazzelli and E. Guyon J. Phys. 43 (1982) 985).

The parameter controlling the instabilities is $N = X_o Y_o \omega / D$ where D is the diffusivity for the nematic orientation. The experiments are performed for increasing values of N keeping the ellipticity $E = X_o / Y_o$ constant. At low shear rates, the director moves at frequency ω on an elliptic cone the axis of which is perpendicular to the plates but the system is homogeneous in a horizontal plane.

- As the shear exceeds a certain threshold N_1, a roll pattern of wavelength Λ ($\Lambda \simeq 2d$) develops. The orientation of these rolls is fixed with respect to the axis of the elliptical shear (0_x, 0_y).

- Above a higher value $N_2 > N_1$ a static square pattern is obtained in a reversible and continuous fashion. In certain instances, hexagons can also appear. When E = 1 (degenerate case) one has a direct transition to the square pattern, rolls only appear then as transient structures

- By further increasing the shear, the progressive disorganization of this twodimensional structure takes place. It is characterized by the proliferation of structural defects such as dislocations or grain boundaries. This situation presents analogies with recent models of melting of twodimensional equilibrium solids by unbinding of structural defects.

2. PATTERN SELECTION

The constitutive equations of motion generally admit a trivial solution which is invariant under a continuous transformation group but as the control parameter b exceeds a critical value b_c this solution loses stability and new solutions bifurcate which are only invariant under a subgroup of the original group. The conventional stability analysis around the trivial solution determines the value of b_c and the corresponding critical wavelength Λ_c. In fact, in isotropic unbounded systems the pattern selection involves three different aspects. First, one encounters the problem of wavelength selection. For b slightly above b_c, the trivial solution is linearly unstable for a continuous family of nonequivalent wavelengths, yet a unique wavelength which does not correspond to the maximal linear growth is selected in some real experiments [4 , 5]. Secondly in unbounded media there is an orientational degeneracy associated to the unrestrained choice for the directions of the critical wave vectors. This degeneracy can only be removed in isotropic systems by taking into account the effects of lateral walls. Finally, for dimensions higher than one, to a given wavelength there corresponds a large variety of possible planforms (rolls, squares, triangles, hexagons in 2d). It has been shown that the nonlinear terms induce a selection among these various structures already in the weakly nonlinear regime [16]. We now illustrate this property on a simple variational model.

2.1. Isotropic systems

Close to the bifurcation point, one can generally derive a contracted description of the dynamics only in terms of the order parameter which is associated to the unstable mode. The corresponding equation of motion in many cases takes a form similar to the time-dependent Ginzburg-Landau equation (T.D.G.L.) familiar from the theory of equilibrium phase transition :

$$\partial_t \sigma_q = \omega_q \sigma_q - \sum_{q_1}{}' v(q_1,q)\, \sigma_{q_1} \sigma_{-q_1+q} - \sum_{q_1}\sum_{q_2}{}' u(q_1,q_2,q)\, \sigma_{q_1}\sigma_{q_2}\sigma_{-q_1-q_2+q} \qquad \text{(II.1)}$$

with

$$\omega_q = \varepsilon - (q^2-q_c^2)^2 \quad ; \quad \varepsilon = b - b_c / b_c \qquad \text{(II.2)}$$

In (II.1), the summation is restricted to wavenumbers close to q_c. In the case of the R.B instability, the quadratic term is due to non-Boussinesq effects (e.g. temperature dependent transport coefficients) a similar term always appears in the case of the Marangoni or the Turing instability. When these equations have a gradient structure and this property is sometimes satisfied near the bifurcation point, one can define a Lyapunov functional that decreases in any dynamics. In that case if one describes local fluctuations by a gaussian white noise term then the (T.D.G.L.) equation can be written as :

$$\partial_t \sigma_q = -\Gamma \frac{\partial V}{\partial \sigma_{-q}} + \eta_q \qquad \text{(II.3)}$$

$$\langle \eta_q(t)\, \eta_{q'}(t') \rangle = 2\Gamma\, \delta(t-t')\, \delta_{q,-q'}$$

When the nonlinear coupling terms do not depend on the angles between the interacting wave vectors, the Lyapounov functional takes the simple form (Brazovskii's model) :

$$V = \frac{1}{\Gamma}\Big[\sum{}' \omega_q |\sigma_q|^2 + \frac{v}{3!} \sum_{q_1 q_2}\sum{}' \sigma_{q_1}\sigma_{q_2}\sigma_{-q_1-q_2} + \frac{u}{4!} \sum_{q_1 q_2 q_3}\sum\sum{}' \sigma_{q_1}\sigma_{q_2}\sigma_{q_3}\sigma_{-q_1-q_2-q_3} \Big] \qquad \text{(II.4)}$$

This functional then plays the role of a generalized potential far from equilibrium. Similar functionals have been derived in the case of the Rayleigh-Benard instability [17], the Turing instability [18] or the hydrodynamic instabilities in nematic liquid crystals [19].

Each pattern can be characterized by m pairs of wavevectors (q_i, $-q_i$). For the sake of simplicity we consider explicitly in the following only the structures which minimize the potential (II.4) ($|q_i| = q_c$). In space variables the corresponding order parameter becomes :

$$\sigma(r) = 2 \sum_{i=1}^{m} a_i \cos q_i \cdot r \qquad \text{(II.5)}$$

Two classes must be considered :

A) The structures described by m independent pairs. In that case the quadratic terms in the equations of motion do not contribute. They appear supercritically though a second order like phase transition ; we get indeed from the stationary condition of equations (II.1) :

$$a_m = \begin{cases} 0 & b < b_c \\ \left[\dfrac{2(b-b_c)}{(2m-1)u} \right]^{1/2} & b > b_c \end{cases} \qquad \text{(II.6)}$$

B) The structures the wave vectors of which satisfy the triangular condition

$$q_1 + q_2 + q_3 = 0$$

$$\text{(II.7)}$$

$$\sigma_3(r) = 2a_3 \left[\cos q_c x + \cos \frac{q_c}{2}(x+\sqrt{3}y) + \cos \frac{q_c}{2}(x-\sqrt{3}y)\right]$$

Depending on the sign of the cubic term v in (II.1), the maxima respectively define a triangular ($v < 0$) or a honeycomb lattice ($v > 0$). These patterns correspond to the Benard l-hexagons (upward motion in the centre of the hexagons) or g-hexagons (downward motion in the centre) in non Boussinesq fluids [20]. These structures appear subcritically through a first order like transition. The amplitude jumps indeed discontinuously to a finite value. Such a behavior has been experimentally verified by Berge and coworkers in water near the 4°C anomaly [21] and by Pantaloni and coworkers in the case of the Marangoni instability in silicone oil [22].

The physically possible patterns correspond to the fraction of all the stationary solutions which is stable with respect to arbitrary disturbances of infinitesimal amplitude. In the case of the Brazovskii's model (II.1), all the solutions of the first class are unstable with the exception of the case m = 1 corresponding to a stationary wave periodic in one direction (rolls). In this case the stability analysis strongly restrict the manifold of possible solutions ; only two structures remain : rolls and hexagons (cf. Fig. III)

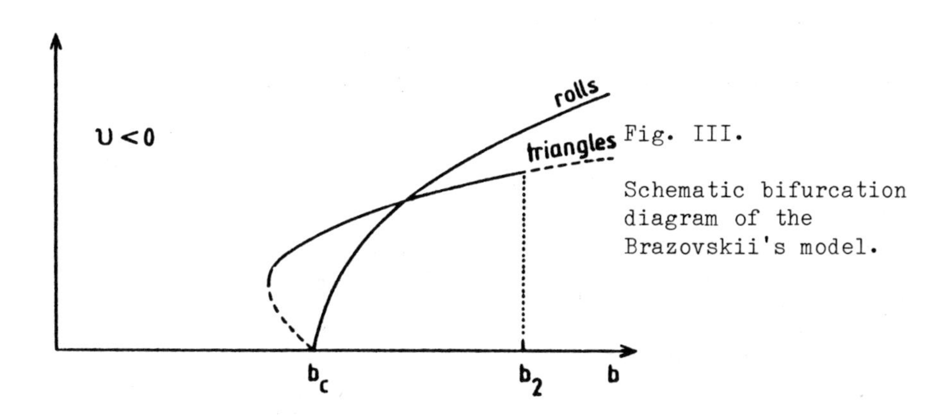

Fig. III.

Schematic bifurcation diagram of the Brazovskii's model.

The same principle of selection through stability can still be applied to more complex problems where a variational formulation is not possible any more.

In Fig. III, there is a range of parameters (from b_c to $b_2 = b_c + 8\, b_c\, v^2/u$) where rolls and hexagons coexist. In this regime, the potential can be used to calculate the relative stability of the coexisting patterns and determine the value of the control parameter at which one structure becomes more stable than the other. Such transitions between rolls and hexagons have been experimentally studied and hysteresis effects have been detected at this transition [21,23].

The bifurcation diagram of Brazovskii's model (u, v = constant) is unfortunately not universal. Indeed the angular dependepence of the coupling terms (u, v in Eq. III.1) can sometimes play an important role in the pattern selection. To illustrate this point we consider now the case of the variational model introduced by Sivashinsky [24] to describe planforms of the buoyancy driven instability in nearly insulated layers. The model is defined by the following equation for the order parameter $\sigma(r, t)$.

$$\partial_t \sigma = [\,\varepsilon-(\nabla^2+q_c^2)^2]\,\sigma + 3[\,(\frac{\partial\sigma}{\partial x})^2(\frac{\partial^2\sigma}{\partial x^2}) + (\frac{\partial\sigma}{\partial y})^2(\frac{\partial^2\sigma}{\partial y^2})\,]$$

$$+ (\frac{\partial\sigma}{\partial x})^2(\frac{\partial^2\sigma}{\partial y^2}) + (\frac{\partial\sigma}{\partial y})^2(\frac{\partial^2\sigma}{\partial x^2})$$

$$+ 4(\frac{\partial^2\sigma}{\partial x\partial y})(\frac{\partial\sigma}{\partial x})(\frac{\partial\sigma}{\partial y})\,. \tag{II.8}$$

In this case the only stable structure in the class A defined above is the square pattern. The amplitude of the squares can in general be determined by the following stationary condition :

$$\varepsilon a_i - g_D a_i |a_i|^2 - g_{ND} a_i |a_j|^2 = 0 \tag{II.9}$$

(i and j correspond to the two orthogonal directions of the structure). From Eq. (II.8) we get $g_D = 3\, q_c^4$ and $g_{ND} = 2\, q_c^4$ and the squares of amplitude $a_i = a_j = (\,\varepsilon\,/5\, q_c^4)^{1/2}$ appear through a second order like transition (exchange of stability with the homogeneous state) whereas the rolls are now unstable. More generally the squares will be selected whenever in Eq. (II.9) we have the inequality $g_D > g_{ND}$ as a result of the dependence of the nonlinear terms on the angles between the interacting wavevectors (here q_i and q_j). In Sivashinsky's model, hexagonal structures are marginally stable and could appear in the supercritical region. This situation presents also analogies with the experimental results found in the case of the (R.B) instability in homeotropic nematics (H = 0) [25] and with the case of a (N.S.E.S.) when the ellipticity is equal to one. In both cases one has a direct (reversible) transition to a square structure but hexagons are often met with the squares in the convective geometry. In these situations the hexagons can be stabilized by non-Boussinesq effects resulting for instance from the rapid variation of elastic and viscous coefficients with temperature.

2.2. Anisotropic systems

In anisotropic systems there is an intrinsic mechanism which raises the orientational degeneracy by inducing preferred directions for the wavevectors characterizing the structures. We illustrate this property on the rather academic problem of the Turing instability in the Brusselator model in a 2D uniaxial medium (the principal axis is parallel to O_x). The kinetic equations for the concentrations of the intermediate species α and β can be written

$$\partial_t \alpha = A - (B+1)\alpha + \alpha^2\beta + D_\perp^\alpha \nabla^2\alpha + D_a^\alpha \frac{\partial^2\alpha}{\partial x^2}$$

$$\partial_t \beta = B\alpha - \alpha^2\beta + D_\perp^\beta \nabla^2\beta + D_a^\beta \frac{\partial^2\beta}{\partial x^2} \tag{II.10}$$

where A and B are kept constant and B is the control parameter

$D_\perp^i$ represents the diffusion coefficient in the direction perpendicular to the principal axis ($\perp O_x$)
$D_a^i = D_\parallel^i - D_\perp^i$ is a measure of the anisotropy of the corresponding diffusion current.

This model can be considered as a caricature to describe recombination processes between excitations (e.g. phonons, vacancies, interstitials) in irradiated condensed matter systems [10,11]. As in the isotropic case [1] this model displays a symmetry breaking instability. Indeed the homogeneous solution $\alpha_0 = A$, $\beta_0 = B/A$ becomes unstable for $B > B_c = (1 + A\eta)^2 < 1 + A^2$ where

$$\eta(\phi) = \left(\frac{D_\perp^\alpha + D_\parallel^\alpha \cos^2\phi}{D_\perp^\beta + D_\parallel^\beta \cos^2\phi}\right)^{1/2}$$

against inhomogeneous fluctuations of wavelength q_c (ϕ) :

$$q_c^2(\phi) = A[(D_\perp^\alpha + D_\parallel^\alpha \cos^2\phi)(D_\perp^\beta + D_\parallel^\beta \cos^2\phi)]^{-1/2} \tag{II.11}$$

and making an angle ϕ with the principal axis. The prefered orientation ϕ_0 is obtained by minimizing B_c (ϕ). One finds that

if $D_\perp^\alpha / D_\perp^\beta < (D_\perp^\alpha + D_a^\alpha) / (D_\perp^\beta + D_a^\beta)$ then $\phi_0 = \pi/2$ and

the axis of the rolls is parallel to the principal axis whereas

when $D_\perp^\alpha / D_\perp^\beta > (D_\perp^\alpha + D_a^\alpha) / (D_\perp^\beta + D_a^\beta)$ then $\phi_0 = 0$ and the axis

of the rolls is perpendicular to the principal axis. In this simple example, the anisotropy in the transport coefficients induces preferred directions for the critical wavevector (easy axis) and this selection appears already in the linear analysis. In these anisotropic systems it is still possible to derive (T.D.G.L.) equations but now the frequency ω_q contains a symmetry breaking term. In the example discussed above the frequency takes the form (at the lowest order) :

$$\omega_q = \frac{b - b_c}{b_c} - (q^2 - q_c^2)^2 - \hat{A}q^2 \sin^2\phi \tag{II.12}$$

When $\hat{A} > 0$ this expression corresponds to an easy axis parallel to the principal axis ($\phi_O = 0$). A similar expression can be derived in the case of the (R.B.) instability in liquid metals in presence of a horizontal magnetic field which tends to favor longitudinal rolls i.e. rolls having their axis parallel to the field [26]. In that case ϕ is the angle between the wavevector and a direction perpendicular to the magnetic field.

In some experiments it is possible to change the sign of the anisotropic term. For instance in the case of the instability of a (N.S.E.S.) a linear analysis of the transition to the roll structure leads to the following expression for the threshold N in powers of (X_O/d) and (Y_O/d) [27] :

$$\begin{aligned} N &= N_o - \tilde{A}\left[\left(\frac{X_o}{d}\right)^2 + \left(\frac{Y_o}{d}\right)^2\right] \\ &\quad - \tilde{B}\left[\left(\frac{X_o}{d}\right)^2 - \left(\frac{Y_o}{d}\right)^2\right]\left(\tilde{\beta}\cos 2\Psi + \tilde{\gamma}\frac{N}{N_o}\sin 2\Psi\right) \end{aligned} \qquad (II.13)$$

where N_O is the threshold for thick samples, $\tilde{A}$, $\tilde{B}$, $\tilde{\beta}$ and $\tilde{\gamma}$ are parameters which can be expressed in terms of the properties of the system. Ψ is the angle between the normal to the rolls and the O_x axis defined in Section I. Clearly the ellipticity E measures the importance of the anisotropy of the shear ; when E = 1, the angular dependence disappears in Eq. (II.13) and one recovers a degenerate case. When E < 1, the selected orientation is $\Psi_1 = 1/2 \arctan(\tilde{\gamma}/\tilde{\beta}) + \pi/2$ whereas for E > 1, one finds $\Psi_2 = 1/2 \arctan(\tilde{\gamma}/\tilde{\beta})$. The ellipticity entirely fixes the orientation of the roll structure. Experiments [28] clearly show a jump of 90° near the value E = 1 (see fig.IV).

Anisotropies can also play an important role in the selection of the possible structures. Let us consider a system where in absence of anisotropy the angular dependence of the coupling term tends to select a square structure (cf. eq. II.8, 9). On the other hand we have seen that the introduction of an anisotropic effect induces roll structures making a well defined angle. The amplitudes of the rolls and squares of such a system can be determined by the following equations (cf. Eq. 9)

$$\begin{aligned} \varepsilon a_1 - a_1 \left[g_D|a_1|^2 + g_{ND}|a_2|^2 \right] &= 0 \\ (\varepsilon - \hat{A})a_2 - a_2 \left[g_D|a_2|^2 + g_{ND}|a_1|^2 \right] &= 0 \end{aligned} \qquad (II.14)$$

The anisotropy $\hat{A}$ tends to align the critical wavevector parallel to the direction "1". We take $g_D > g_{ND}$ so that when $\hat{A} = 0$

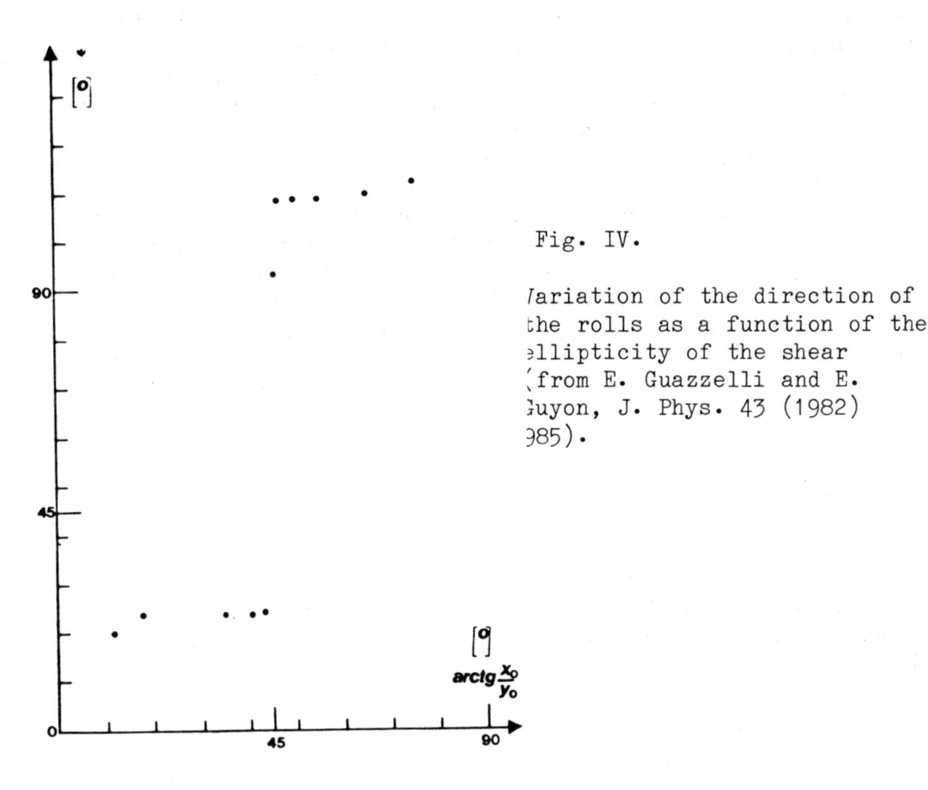

Fig. IV.

Variation of the direction of the rolls as a function of the ellipticity of the shear (from E. Guazzelli and E. Guyon, J. Phys. 43 (1982) 985).

the square pattern is selected. A stability analysis shows that rolls induced by the anisotropy become unstable for $\varepsilon > \tilde{\varepsilon} = g_D \hat{A} / (g_D - g_{ND})$ where one has a direct transition to a square structure. In this model there is thus a competition betwen anisotropy and nonlinear effects. First, rolls induced by the anisotropy abe selected but as the bifurcation parameter is increased, the nonlinearities become more important and favor the onset of a square structure. Here also this model presents similarities with the experimental results obtained in the case of (R.B.) in nematics heated from above in presence of a horizontal magnetic field or the instability of a (N.S.E.S.) when the ellipticity of the shear is different from 1 $(\hat{A} \equiv E)$ [12].

3. PHASE DYNAMICS

In confined convective geometries or in well stirred chemical reactors the transition to turbulence can be described in terms of few interacting modes ; a number of scenarios have been presented at this conference. On the other hand in large boxes or in distributed chemical systems one is faced with a problem involving a large number of degrees of freedom. They are associated to small local changes in the position and orientation of the structures. The corresponding dynamics may in some situations well be described in terms of a phase variable (vector) which obeys a diffusion equation [29]. To derive this equation one must first obtain a particular periodic stationary solution U_0 to the nonlinear problem $\partial_t U = N(U)$. If $\phi(r, t)$ is slowly varying in space and time, $U_0(r + \phi)$ describes a long wavelength distortion of this structure and one thus looks for solutions under the form

$$U = U_o + \phi . \nabla U_o + U^1 + U^2 \tag{III.1}$$

(U^1 and U^2 being formally of order $\nabla\phi$, $\nabla(\nabla\phi)$) and performs a phase gradient expansion. The solvability condition at the second order in the gradient can be written as a diffusion equation :

$$\partial_t \phi^i = \sum_k \sum_l D_j^{ikl} \frac{\partial^2 \phi^j}{\partial x_k \partial x_l} \tag{III.2}$$

More precisely in the case of a layered structure periodic along the O_x direction, the phase is a scalar which obeys the following equation

$$\partial_t \phi = D_{\|} \frac{\partial^2 \phi}{\partial x^2} + D_{\perp} \frac{\partial^2 \phi}{\partial y^2} \tag{III.3}$$

For polygonal structures the phase is a two dimensional vector. For instance in the case of a hexagonal structure, the phase diffusion equation takes the form

$$\partial_t \phi = D_{\perp}^h \nabla^2 \phi + (D_{\|}^h - D_{\perp}^h) \text{ grad div } \phi \tag{III.4}$$

It is interesting to compare with the basic equation of motion for the displacement field u of an isotropic solid : [30]

$$\rho \frac{\partial^2 u}{\partial t^2} = \mu \nabla^2 u + (\lambda+\mu) \text{ grad div } u \qquad \text{(III.5)}$$

where λ and μ are the Lamé coefficients. In the stationary case, the diffusion coefficients $D_{\parallel}^{h}$ and $D_{\perp}^{h}$ of the hexagonal structure play a role similar to the elasticity coefficients of a solid. However in the nonequilibrium case the phase obeys a generalized diffusion equation whereas the newtonian equation of motion for the displacement field admit elastic waves as solutions.

The phase diffusion coefficients have been experimentally obtained in the case of roll patterns for two different systems : the (R.B.) instability in silicone oil [31] and the instability of a N.S.E.S. [13].

From the lowest order (T.D.G.L.) equation one obtains the following expressions for the diffusion coefficients of a layered structure (isotropic case)

$$D_{\parallel} = \frac{\xi^2}{\tau} \left[\frac{\varepsilon - 3\xi^2 Q^2}{\varepsilon - \xi^2 Q^2} \right] \qquad \text{(III.6)}$$

$$D_{\perp} = \frac{\xi^2}{\tau} \frac{Q}{q_c} \qquad \text{(III.7)}$$

where $Q = q - q_c$ (q is the wavenumber of the structure) ξ is the characteristic length for horizontal coordinates and τ is the vertical thermal diffusion time. By studying the relaxation of thermally printed modulation of the pattern Croquette and al. [31] obtained the following value $D = 1.7\ 10^{-3} + 0.5\ \text{cm}^2/\text{s}$ which is in good agreement with the theoretical result (IV.6) $D_{\parallel} \simeq \xi^2/\tau \simeq 2.2\ 10^{-3}\ \text{cm}^2/\text{s}$. They have also experimentally verified expression (III.7). According to (III.6,7) near threshold, they find $D_{\perp} \ll D_{\parallel}$.

All the wavelengths which are amplified for $b > b_c$ do not correspond to physically realisable patterns. The curve $D_{\parallel}(q,\varepsilon) = 0$ and $D_{\perp}(q,\varepsilon) = 0$ delineate the region where the structure (rolls) is stable against inhomogeneous fluctuations. From Eq. (III.6,7), the system becomes unstable with respect to compression and dilatation of the rolls for $Q^2 > \varepsilon / 3\xi^2$ (Eckhaus instability) whereas for $Q < 0$, it is unstable against wavy distor-

tions of the layers (zig-zag instability). At this order, stable convection is thus limited in the range $0 < Q < \varepsilon^{1/2} / \sqrt{3}\xi$
The zig-zag distortion is clearly the most dangerous mode near threshold. Moreover at the lowest order the selected wave number sits at the margin of the zig-zag instability. This property has also been verified at higher orders for some systems [32] and for simplified models [29]. It has been proposed as a general principle for wavelength selection [33]. However in the case of the (R.B) instability higher order corrections (beyond $\varepsilon^{3/2}$) introduces in the phase diffusion equation a singular drift term resulting from large scale flows [34]. At low Prandtl number this singular contribution becomes important and modify the Manneville-Pomeau selection principle ($D_\perp = 0$).

In anisotropic systems, the transverse diffusion coefficient contains a contribution associated to the term which raises the orientational degeneracy. $D_\perp$ thus remains positive and finite even near threshold contrary to the isotropic case. For instance in the example defined in Eq. (II.12) one finds

$$D_\perp = \frac{1}{\tau}\left[\ \xi^2 \frac{Q}{q_c} + \hat{A}\ \right] \qquad \text{(III.8)}$$

(q_i parallel to the easy axis)
This property has been verified experimentally in the case of the instability of a (N.S.E.S.) by measuring the static deformations induced by an isolated dislocation [13]. From Eq. (III.8) the borderline for the zig-zag instability ($Q_{zz} = -A\, q_c/\ \xi^2$) is now shifted with respect to the corresponding isotropic case.

Describing the local fluctuations by a gaussian white noise term it is possible to derive a potential for the long wavelength phase fluctuations. In the case of hexagonal structures, it takes for instance the simple form :

$$V = \frac{1}{\Gamma} \int dr\left[\sum_i \sum_j D_\perp^h\, \Phi_{ij}^2 + \sum_i \left(D_\parallel^h - 2D_\perp^h \right) \Phi_{ii}^2 \right] \qquad \text{(III.9)}$$

where $\Phi_{ij} = \frac{1}{2}\left[\frac{\partial \phi^i}{\partial x_j} + \frac{\partial \phi^j}{\partial x_i} \right]$

Here again the analogy with the hamiltonian of an elastic solid is striking. As a result it is interesting to apply to the study of the disorganization of nonequilibrium structures [35] the methods which have been derived recently to explain the melting of 2D solids [36]. For increasing external constraints, isolated dislocations and grain boundaries may appear destroying the translational quasi long range order.

Acknowledgments.

We thank Professors I. Prigogine and G. Nicolis for their interest in this work. We are also gratiful to Dr. J.C. Micheaux and M. Gimenez for fruitful discussions and for the photograph of Fig. I.

REFERENCES

1. Glansdorff P., Prigogine I, 1971 : "Thermodynamics of Structure, Stability and Fluctuations" New York, Wiley-Interscience.
2. Turing A.M. 1952, Phil. Trans. Roy. Soc. Lond. B237 pp. 37-72
3. Möckel P. 1977, Naturwissenschaften 64 p. 224.
4. Kagan M., Levi A., Avnir D;, 1982 Naturwissenschaften 69 pp. 548-549.
5. Gimenez M., Micheau J.C. 1983, Naturwissenschaften 70 pp. 90-91.
6. Micheau J.C., Gimenez M., Borckmans P., Dewel G., 1983, Nature 305, pp. 43-45.
7. Berg J.C., Acrivos A and Boudart M., 1966 Adv. Chem. Eng. 6 pp. 61-123.
8. Orban M. 1980 : J. Am. Chem. Soc. 102 pp. 4311-14.
9. Dubois-Violette E., Durant G., Guyon E., Manneville P. and Pieranski P. 1978 in "Liquid Crystals" Liebert Ed. Academic Press pp. 147-208.
10. Krishan K. 1980 Nature 287 pp. 420-21.
11. Martin G. 1983 Phys. Rev. Lett. 50 pp. 250-52.
12. Dreyfus J.M., Guyon E. 1981, J. Physique 42 pp. 283-92.
13. Guazzelli E.,Guyon E., Wesfreid J.E. 1981 in "Symmetries and broken symmetries in condensed matter physics" N. Boccara ed. IDSET Paris pp. 455-61.
14. Koschmieder L., 1975 Adv. Chem. Phys. 32 pp. 109-33.
15. Langer J.S. Rev. Mod. Phys. 1980, 52 pp. 1-28.
16. Schluter A., Lortz D. and Busse F. 1965, 23 pp.129-144.
17. Swift J.,Hohenberg P. 1977, Phys. Rev. A15 pp. 315-28.
18. Walgraef D.,Dewel G., Borckmans P. 1980, Phys. Rev. A21, pp. 397-404.
19. Manneville P. 1978, J. Physique 39 pp. 911-25.
20. Busse F. 1978, Rep. Prog. Phys. 41 pp. 1929-67.
21. Dubois M., Bergé P., Wesfreid 1978, J. Physique 39 pp. 1253-57.
22. Pantaloni J., Bailleux R.,Salan J., Verlarde M.G. 1979, J. New-Equilib. Thermodyn. pp. 201-17.
23. Stengel K.C., Oliver D.S. and Boocker J.R., J. Fluid Mech. 120, pp.411-31 (1982).
24. Gertsberg V.L., Sivashinsky G.I. 1981, Prog. Theor. Phys. pp. 1219-29.

25. Pieranski P., Dubois-Violette E., Guyon E. 1973, Phys. Rev. Lett. 30 pp. 736-39.
26. Tabeling P. 1982, J. Physique 43 pp. 1295-1303.
27. Sadik J., Rothen F., Bertgen W. and Dubois-Violette E. 1981, J. de Physique 42 pp. 915-28.
28. Guazzelli E. and Guyon E. 1982, J. Physique 43 pp. 985-989.
29. Pomeau Y. and Manneville P. 1987, J. Phys. Lett. 40 pp. L609-12.
30. Landau L.D. and Lifshitz E.M. "Theory of Elasticity" Pergamon, New York 1970.
31. Croquette V. and Schosseler F. 1982, J. de Physique 43, pp. 1183-91.
32. Langer J.S., Müller-Krumbhaar M. 1983, Phys. Rev. A27 pp. 499-514.
33. Kramer L., Ben-Jacob E., Brand H. and Cross M.C. 1982, Phys. Rev. Lett. 49 pp. 1891-94.
34. Cross M.C. 1983, Phys. Rev. 27A pp. 490-98.
35. Walgraef D.,Dewel G., Borckmans P. 1982, Z. Physik B48, pp. 167-73.
36. Kosterlitz J.M. and Thouless J.D., 1973, J. Phys.C. : Solid State Phys. pp. 1181-1203.

FORMATION OF CHEMICAL AND MECHANICAL STRUCTURES DURING THE OXIDATION OF METALS AND ALLOYS

J.M. Chaix, G. Bertrand, A. Sanfeld.*

LA. 23, Université de Dijon, BP 138, F21004 Dijon Cédex.
* Chimie Physique II, ULB - B1050 Bruxelles.

A very peculiar morphology of the corrosion scale produced during the oxidation of metals and alloys is observed in some cases ; the scale is made of the stacking of a great number of identical layers. We show that these multilayered scales exhibit the main characteristics of far from equilibrium patterns. The formation itself is analyzed from a theoretical point of view. Chemical patterns can be explained by oscillations between two states in the kinetics of the propagating reaction front. In each of these states a definite compound is produced, so the relaxation oscillations give rise to the observed multilayered scale. Mechanical patterns involve a periodic fracturing process at the interface between the growing oxide and the metal.

INTRODUCTION

The reactivity of solids deals with their transformation due to chemical reactions. Generally it is concerned with the solid-gas reactions, in which a new solid phase is produced at the expense of the initial material. Thus the oxidation of a metal, the reduction of an oxide and the thermal decomposition of a carbonate or a hydrate come under such type of reactions [1, 2].

At present, all the results of studies in this field cannot be interpreted if the knowledge of the systems under far from equilibrium conditions is neglected. Although stability studies and bifurcation theories, when applied to solid systems give new example of dissipative structures, these structures show certain specific characters. We shall illustrate this proposal by means of studies on oxidation of metals and alloys which produce a very interesting ordering of the corrosion scale.

G. Nicolis and F. Baras (eds.), Chemical Instabilities, 401–413.

1. MORPHOLOGY OF CORROSION SCALES

Normally, the corrosion scale is either macroscopically homogeneous (mono - or polyphased) or duplex, formed by two underscales of different oxides which grow simultaneously [1-3]. More interesting examples are those where the scale is built up by the repetition of layers parallel to the initial metal surface. We observed such patterns through morphological and chemical analyses of the cross-sections of the oxidized samples. The oxidation was done on small polycristalline metallic slabs under fixed conditions of temperature and pressure of a pure gas (see Figure 1).

1.1 Sulfidation of Fe-22Cr-5.5Al alloy by H_2S [4]

In the case of the sulfidation of this alloy, alternating repetition of compact and microcrystallized layers is observed. The chemical (X ray energy dispersive spectrometry) and structural (X ray diffraction) analyses show that compact layers are rich in iron and contain mostly pyrrothite (FeS sulfide) whereas the microcrystallized layers are poor in iron and contain daubreelite ($Fe(Cr, Al)_2 S_4$ spinel). Very near the metal, a thin zone of variable composition and thickness (about several micrometers) is noticed. It can be verified that the layers are formed successively from this internal zone. An external zone of sulfide FeS develops during the reaction (see Figure 2a).
The observation of this pattern depends on the experimental conditions. (i) The layer thickness increases regularly from 5 to 40μm when the temperature is increased from 580°C to 740°C. At around 740°C the scale morphology changes brutally : the stratification disappears whereas a duplex scale arises. (ii) The layer thickness also varies with the composition of the alloy [5]. If the concentration of chromium and aluminum is too small, the pattern is not observed. (iii) A clear tendency to the disappearance of the layers is noticed, when the sample is kept under reaction conditions for a long time (Figure 2b-c).

1.2 Oxidation of titanium in oxygen [6](see Figure 2d-e-f)

In the case of the oxidation of titanium, the very regular repetition of the void bands between the rutile (oxide TiO_2) layers is observed. The layer thickness decreases when the temperature is increased from 700°C to 950°C, whereas the void bands evolve inversely. It does not seem to depend on the initial sample thickness ; however the layers are not formed when the sample is too thin. Finally, if the oxide is kept under the reaction conditions, the bands may disappear through recrystallization ; the layers bind together and the scale becomes homogeneous and compact.

1.3 Conclusion of the experimental part

The two types of self-organization described above result obviously

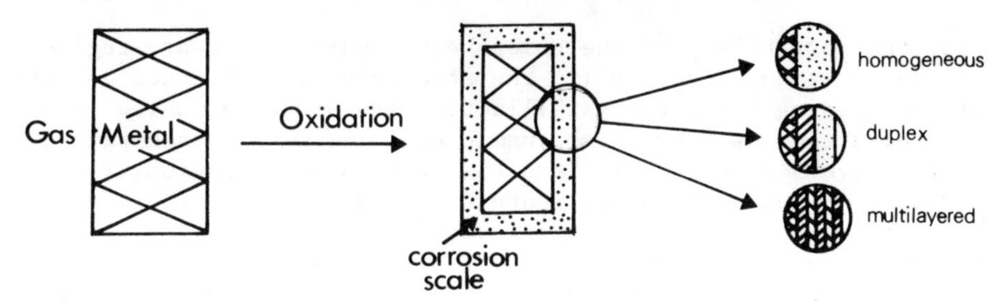

Figure 1. Schematic section showing the growth of the corrosion scales on a metallic slab.

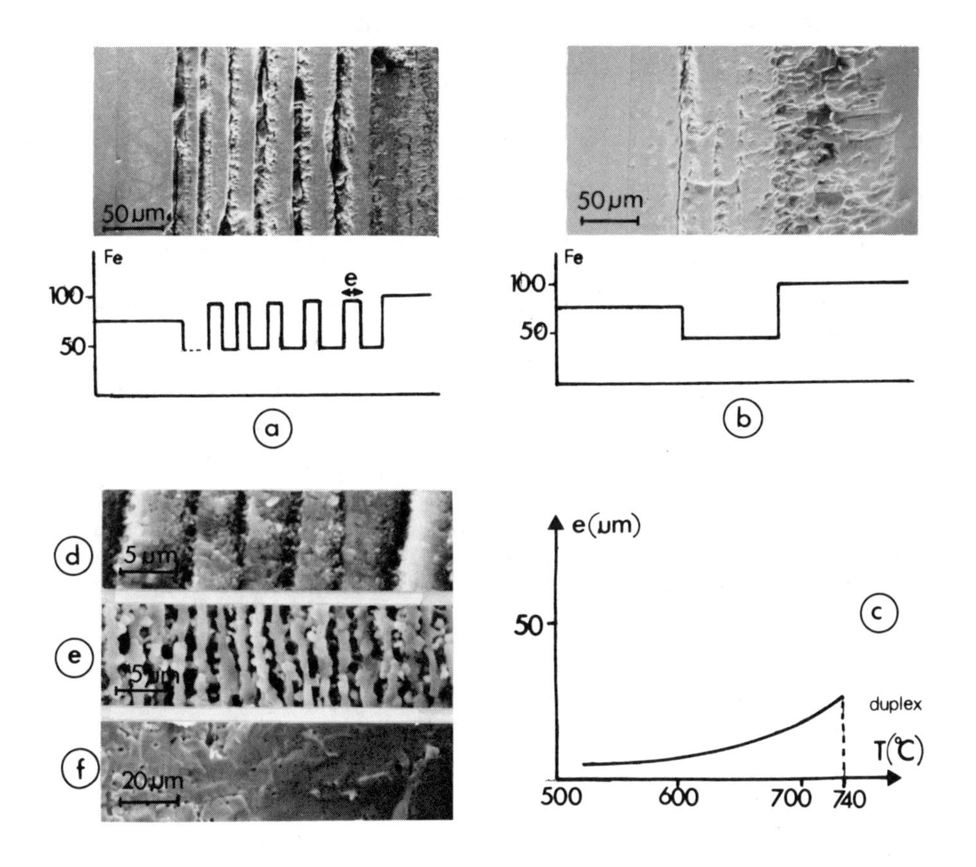

Figure 2. Morphology of the corrosion scales
- Sulfidation of a Fe-22Cr-5.5Al alloy (a) 580°C (b) 900°C
(c) evolution of the layer thickness vs temperature
- oxidation of titanium (d) 840°C (e) 965°C (f) recrystallized oxide.

from the instability of the homogeneous quasi-stationary regime of the reaction. Actually (i) the stratification is a non-equilibrium ordering - the recrystallization and sintering which lead the solid towards its thermodynamic equilibrium are accompanied by the layers disappearance and the scale homogenization. (ii) Different bifurcation parameters are established : reaction temperature, alloy composition, initial sample thickness.
Morphologically speaking, the layering is a spatial symmetry breaking inside the corrosion scale. However the reaction interface propagates continuously into the metal leaving behind the produced oxide. Due to this, the observed pattern can be spatial consequence of a time oscillating phenomenon at the interface.

2. THEORETICAL ANALYSES OF THE ORDERING MECHANISMS

Three complementary analyses are proposed below to describe the specific results arising in stability studies applied to examples of the reactivity of solids. How can the following affect the onset of the instability, the corrosion scale thickening, the mechanism of the structural transformation of the solid, the elastic mechanical effects at the solid-solid interface where stresses develop ?

2.1 Instabilities during the growth of a corrosion scale on a pure metal [7]

The scale is supposed to be growing out of a cationic diffusion process coupled with an exothermic and thermally activated reaction at the external interface. The analysis of this model should show the following results (see Figure 3 and Table I).

A tristationary behaviour appears in a certain temperature range. It is also revealed by the variation of the oxidized scale thickness. Independent of the stability study, the S-shaped curve gives the possibility of a hard transition between the two quasi-stationary states during the growth of the scale (see Figure 4).

The linear stability analysis with respect to homogeneous perturbations at the surface, shows that the scale thickness is a bifurcation parameter. Thus the quasi stationary state is stable up to a bifurcation point resulting from the destabilizing effects of the interfacial processes ; beyond a certain thickness the quasi stationary regime becomes stable again because of the stabilizing effects of the bulk diffusion.

A numerical calculation evidences the non stationary behaviour. Sustained oscillations of the surface values, of the limit-cycle type, propagate in the scale by getting damped. Consequently the reaction rate oscillates and the kinetics of growth of the oxidized

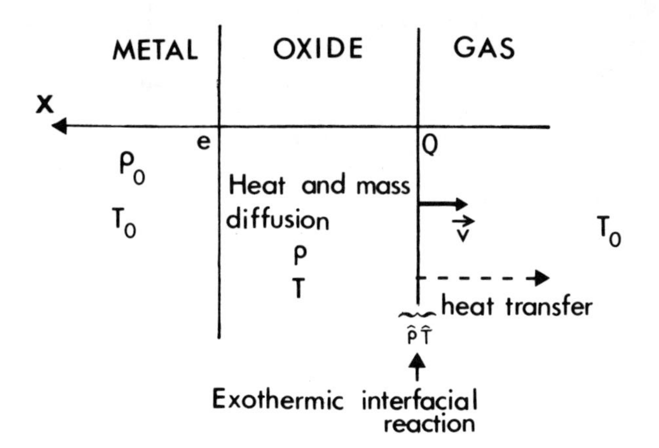

Figure 3.

Table I. Equations describing the growth of a protective scale on a pure metal by cation diffusion and surface reaction. (Symbols are listed at the end of the paper).

scale bulk $0<x<e$

$$\frac{\partial \rho}{\partial t} = D \frac{\partial^2 \rho}{\partial x^2} + v \frac{\partial \rho}{\partial x}$$

$$\frac{\partial T}{\partial t} = D_T \frac{\partial^2 T}{\partial x^2} + v \frac{\partial T}{\partial x}$$

external interface $x = 0$

$$\frac{\partial \hat{\rho}}{\partial t} = - k\hat{\rho} e^{-\frac{E}{R\hat{T}}} + D \left.\frac{\partial \rho}{\partial x}\right|_{x=0} + v\rho(o)$$

$$\hat{C}_p \frac{\partial \hat{T}}{\partial t} = Qk\hat{\rho} e^{-\frac{E}{R\hat{T}}} + \lambda \left.\frac{\partial T}{\partial x}\right|_{x=0} + v\, C_p\, T(o) + h(T_o - \hat{T})$$

boundary conditions

$$x = e \begin{cases} \rho(e) = \rho_o \\ T(e) = T_o \end{cases} \qquad x = 0 \begin{cases} \rho(o) = \beta\hat{\rho} \\ T(o) = \hat{T} \end{cases}$$

scale growth

$$\frac{de}{dt} = - v = \frac{1}{\rho_{ox}} k\, \hat{\rho}\, e^{-\frac{E}{R\hat{T}}}$$

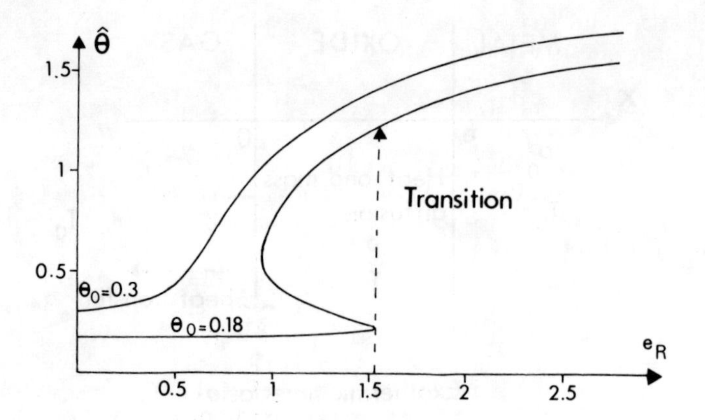

Figure 4. Steady-states multiplicity, function of the oxide thickness e_R (Θ_o : fixed temperature).

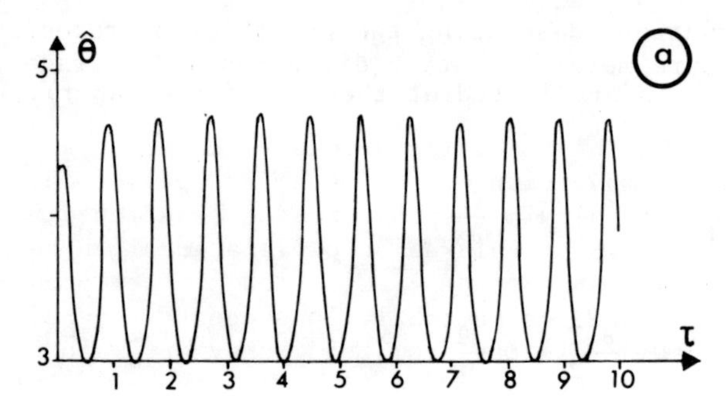

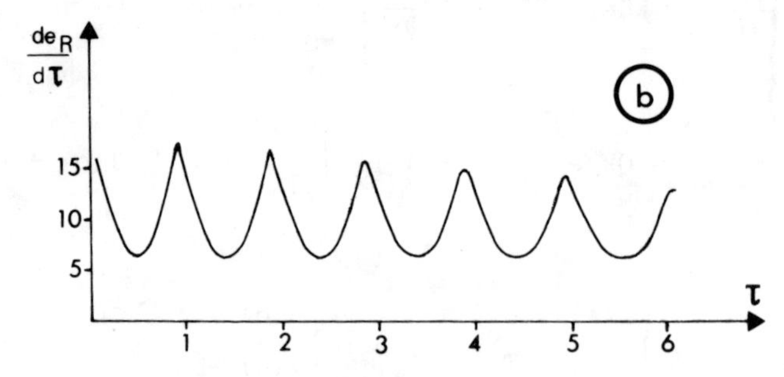

Figure 5. Numerical simulations in the range of unsteady solutions

(a) Surface temperature oscillations versus time.

(b) reaction rate oscillation versus time.

scale is made up of successive branches as has been observed sometimes [8, 9] (see Figure 5).

2.2 Multilayered chemical patterns

A purely chemical model can predict patterns as observed in sulfidation and oxidation of some alloys. Even though this model is applied to one specific example, it can be extended to other similar systems by changing the stoichiometry of the solid structures.

The diffusion of the ionic species inside the scale and the interfacial reactions are the basic phenomena involved in the balance equations and continuity conditions. Two solid structures can be formed competitively at the interface I_2 (see Figure 6) ; the mechanism of these structural transformations is modelized from the examination of the structural blocks, which can combine in different stacking-up giving both the structures. In addition the slight differences between the unit parameters can induce a cooperativity in the interfacial processes, which can be treated according to a temkin type mechanism.

All these considerations taken together will give rise to a bistable behaviour of the interfacial stationary state and to a variable (oscillating) composition of the metallic species in the zone I, acting as a reservoir in response to the reaction-diffusion processes. So interfacial relaxation oscillations can be observed between the two stable states [10], each of them producing a definite compound (FeS or $FeCr_2S_4$). Consequently the alternating spatial profile originates from this behaviour, since the corrosion scale grows due to the interfacial reaction (see Figure 7).

Numerical simulations not only confirm these predictions but also help to obtain the essential characters observed by selection of realistic values of the parameters, these characters being oscillatory formation of the layers, the multilayered morphology and the bifurcation leading to the duplex scale (see Figure 8).

2.3 Periodical ruptures of the metal-oxide interface

The periodical breaking observed in the example of the oxidation of titanium suggests the essential intervention of the mechanical processes during some kinetics of scale growth, stresses being created at the solid-solid interface. This intervention may be thought to be provoked by means of an instability, which interrupts the quasi-stationary growth when a critical oxide scale thickness is attained. The lengthwise rupture of the metal-oxide interface gives back the uncovered metal surface and thus the phenomena of growth and rupture get repeated periodically. Even if this analysis is not yet formulated in a general and solvable model, two simplified studies explain the fracturing in terms of a mechano-che-

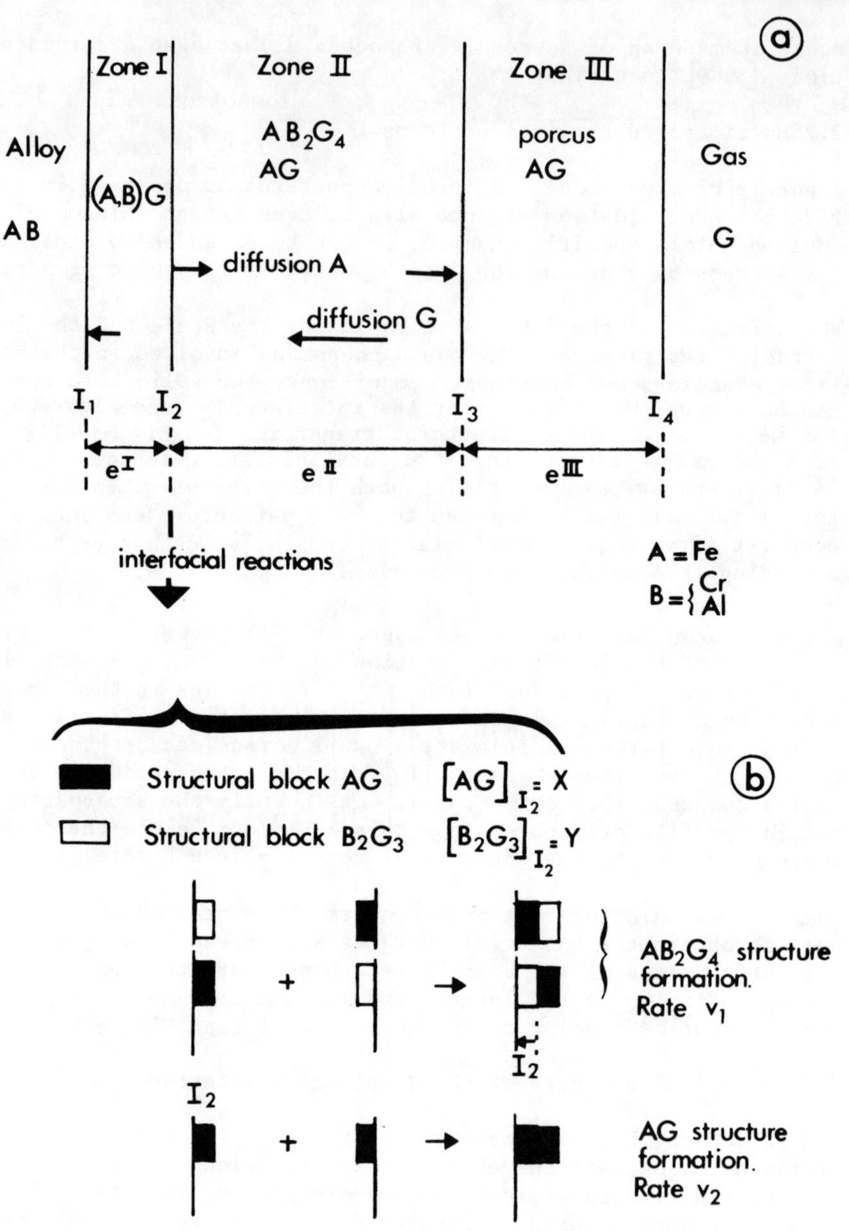

Figure 6. The formation of the multilayered oxidized scale
(a) the different zones as showed by the experiments

(b) the interfacial structural transformation.

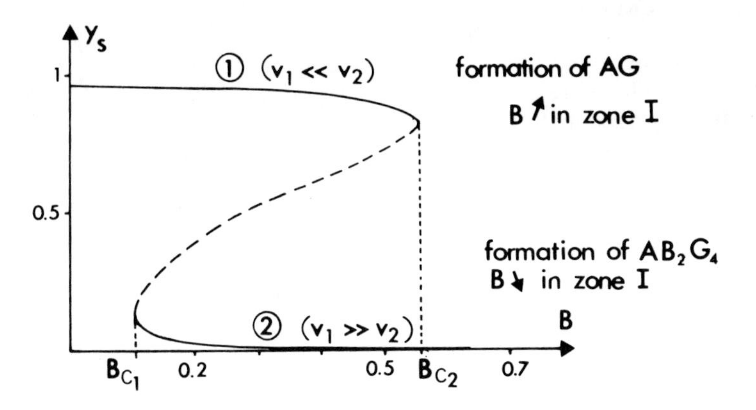

Figure 7. The bistable steady-state of the interface reaction.

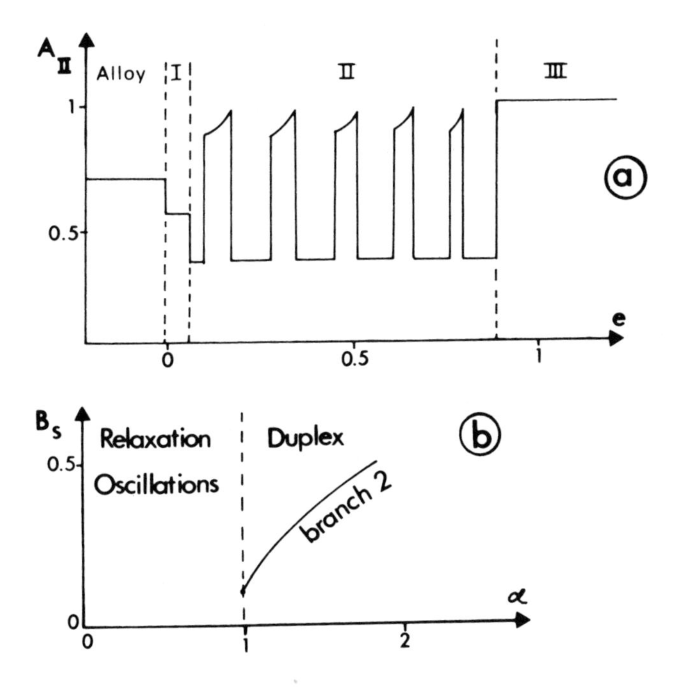

Figure 8. Numerical simulations of the scale growth
(a) Concentration profile in the oxide scale.
(b) The scaled diffusion coefficient of the specie A, α as a bifurcation parameter (transition multilayer →duplex).

mical instability.

The first is developped from the energetic approach of Timoshenko [11, 12, 13]. This establishes two conditions, namely for opening and propagation of cracks ; these conditions fix the critical thickness above which an opened crack propagates, i.e. at this thickness the oxide separates from the metal (see Figure 9).

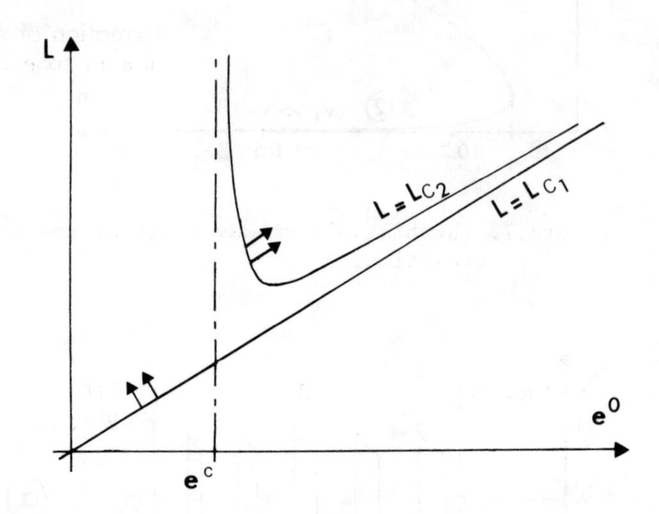

Figure 9. Conditions for opening and propagation of cracks plotted in a (L, e^o) diagram.

- Condition for opening of cracks

$$L^2 \geqslant L^2_{c_1} = 1.03 \frac{e^{o2}}{\varepsilon^o}$$

- Condition for propagation of cracks

$$L^4 \geqslant L^4_{c_2} = \frac{1.01\ e^{o4}}{0.96\ \varepsilon^{o2} - \frac{\gamma}{Ee^o}}$$

e^o : adherent oxide thickness
L : crack radius
E :Young modulus of the oxide
γ : adhesion energy per surface unit
ε^o : deformation in the interface.

The second is developped from a chemical approach of the rupture according to the Tobolsky-Eyring theory [14, 15]. This model considers that the activation energy for the interfacial bond breaking reaction depends on the chemical and mechanical state of the interface. When this dependancy is strong enough, two solutions are possible, one at low fraction of unsticked interface, the other corresponding to an almost total unsticking. The transition from one to the other is expressed by the brutal separation of the oxide layer and defines its critical thickness (see Figure 10).

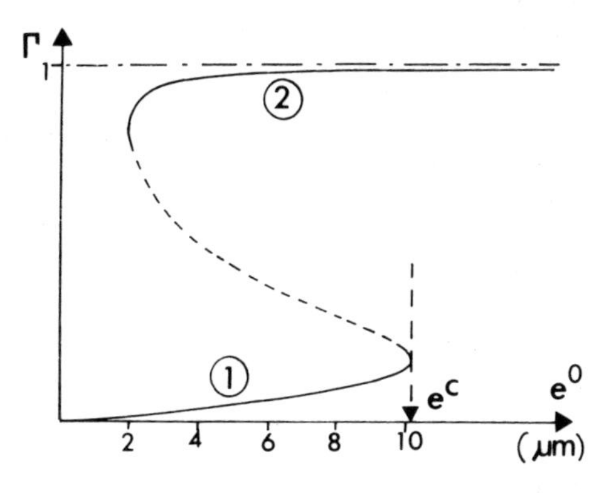

Figure 10. Variation of the unsticked surface fraction Γ versus oxide thickness e^o in steady conditions :

$$\frac{\Gamma e^{-\nu\Gamma}}{1-\Gamma} = e^{-\frac{\Delta G'}{RT}}$$

ν unsticking cooperativity coefficient

ΔG' free enthalpy of the interfacial bond breaking reaction

For the figure, $\nu = 7$, $\Delta G' = 18$ kcal.mol^{-1}.

CONCLUSION

All the experimental and theoretical examples treated in this paper demonstrate very well the fact that, beyond the specific study of multilayered scales, the contribution of solid dynamics represents an enrichment of the dissipative structures theory, which is more than a simple additional illustration.

Either the existence of reaction interface, involving a delocalized reaction-diffusion coupling or the continuous elaboration of a new solid material or the rheology and mechanochemistry of the solids and their interfaces, ... introduce new subjects, new types of self-organization... but really the correlation of the microscopic order of the crystal and of the macroscopic pattern of the dissipative structure remains the most fascinating study to realize.

NOMENCLATURE

C_p	specific heat
D	effective diffusion coefficient of the cation
D_T	thermal diffusivity
E	activation energy
e	scale thickness
e_R	dimensionless scale thickness
h	heat transfer coefficient at the solid-gas wall
k	rate constant of surface reaction
Q	heat of reaction
R	gas constant
T	local temperature
t	time
v	growth rate of the oxide scale
x	space variable
β	equilibrium constant volume cation $\leftrightarrows$ interface cation
Θ	dimensionless temperature
λ	thermal conductivity
ρ	mass density of the diffusing specie
ρ_{ox}	oxide density
τ	dimensionless time
$\hat{}$	indicates a surface quantity.

REFERENCES

1. Bénard, J. 1962, L'oxydation des métaux, Gauthier-Villars.
2. Barret, P. 1973, Cinétique hétérogène, Gauthier-Villars.
3. Fromhold, A.T. Jr. 1976 Theory of Metal Oxidation, North Holland.
4. Bertrand, G., Chaix, J.M., Larpin J.P. 1982, Mat. Res. Bull. 17, pp. 69-76.
5. Lions, J., Trottier, J.P., Foucault, M. 1980, Mémoires et Etudes Scientifiques, Revue de Métallurgie, pp. 743-755.
6. Jarraya, K., Chaix, J.M., Larpin, J.P., Bertrand, G. 1982, C.R. Acad. Sci. Paris 294, Série II, pp. 1365-1368.
7. Chaix, J.M., Hennenberg, M., Bertrand, G. 1983, J. Chim. Phys. 79, pp. 781-797.
8. Garcia, E.A., Lucas, X., Béranger, G., Lacombe, P. 1974, C.R. Acad. Sc. Paris 278C, pp. 837-840.
9. Sarrazin, P., Coddet, C. 1974, Corrosion Sci. 14, pp. 83-89.
10. Boissonade, J., De Kepper, P., 1980, J. Phys. Chem. 84, pp. 501-506.
11. Timoshenko, S. 1940, Theory of plates and shells, Mc Graw-Hill.
12. Wells, C.H., Follansbee, P.S., Dils, R.R. 1975, Stress Effects and the Oxidation of Metals, Cathcart J.V. Ed., the Metallurgical Society of AIME, pp. 220-244.
13. Bui, H.D. 1977, Mécanique de la Rupture, Masson.

14. Tobolsky, A., Eyring, H., 1943, J. Chem. Phys. 11, pp. 125-134.
15. Henderson, C.B.,Graham, P.H., Robinson, C.N. 1970, Int. Journal of Fracture Mech., pp. 33-40.

INDEX